Genomic and Precision Medicine

Oncology

Genomic and Precision Medicine

Oncology

Third Edition

Edited by

Geoffrey S. Ginsburg
All of Us Research Program National Institutes of Health, Bethesda, MD, United States

Huntington F. Willard
Genome Medical, Inc., South San Francisco, CA, United States

John H. Strickler
Department of Medicine, Division of Medical Oncology, Duke University Medical Center, Durham, NC, United States

Matthew S. McKinney
Department of Medicine, Hematologic Malignancies and Cellular Therapy, Duke University Medical Center, Durham, NC, United States

ELSEVIER

ACADEMIC PRESS
An imprint of Elsevier

ISBN: 978-0-12-800684-9

For Information on all Academic Press publications
visit our website at https://www.elsevier.com/books-and-journals

Publisher: Stacy Masucci
Acquisitions Editor: Megan Ashdown
Editorial Project Manager: Sam W. Young
Production Project Manager: Sreejith Viswanathan
Cover Designer: Christian J. Bilbow

Typeset by MPS Limited, Chennai, India

Working together
to grow libraries in
developing countries

www.elsevier.com • www.bookaid.org

Contents

List of contributors

Matthew L. Anderson Department of Obstetrics and Gynecology, University of South Florida Morsani College of Medicine and H. Lee Moffitt Cancer Center and Research Institute, Tampa, FL, United States

Georgia M. Beasley Department of Surgery, Duke University, Durham, NC, United States

Andrew Berchuck Department of Obstetrics and Gynecology, Duke University, Durham, NC, United States; Duke Cancer Institute, Durham, NC, United States

Jonathan R. Brody Departments of Surgery and Cell, Developmental & Cancer Biology, Brenden-Colson Center for Pancreatic Care Knight Cancer Institute, Oregon Health and Science University, Portland, OR, United States

Kristen K. Ciombor Division of Hematology/Oncology, Department of Internal Medicine, Vanderbilt University Medical Center, Nashville, TN, United States

Jeffrey Clarke Division of Medical Oncology, Department of Medical Oncology, Duke University, Durham, NC, United States

Sandeep Dave Division of Hematologic Malignancies and Cellular Therapy, Department of Medicine, Duke University Medical Center, Durham, NC, United States

Matthew Emmet Department of Medicine, Massachusetts General Hospital, Boston, MA, United States

Norma E. Farrow Department of Surgery, Duke University, Durham, NC, United States

Ahmed Galal Division of Hematologic Malignancies and Cellular Therapy, Department of Medicine, Duke University Medical Center, Durham, NC, United States

Michelle F. Green Department of Pathology, Duke University Medical Center, Durham, NC, United States

Jennifer H. Gross Department of Otolaryngology-Head & Neck Surgery, Emory University School of Medicine, Atlanta, GA, United States

Sendhilnathan Hari Ramalingam Duke University School of Medicine, Durham, North Carolina, United States

Kathleen Harnden Department of Medical Oncology, Inova Schar Cancer Institute, Fairfax, VA, United States; Department of Medicine, University of Virginia, Charlottesville, VA, United States

James Isaacs Department of Medicine, Duke University Medical Center, Durham, NC, United States

Jingquan Jia Division of Medical Oncology, Department of Medicine, Duke University Medical Center, Durham, NC, United States

Samuel Klempner Harvard Medical School, Massachusetts General Hospital, Boston, MA, United States

David M. Nanus Division of Hematology and Medical Oncology, Weill Cornell Medical College, New York, NY, United States

Ankit Madan SOVAH Cancer Center, Danville, VA, United States

Lauren Mauro Department of Medical Oncology, Inova Schar Cancer Institute, Fairfax, VA, United States; Department of Medicine, University of Virginia, Charlottesville, VA, United States

Matthew S. McKinney National Institutes of Health, Bethesda, MD, United States

Arnav Mehta Department of Medicine, Massachusetts General Hospital, Boston, MA, United States

Mary Katherine Montes de Oca Department of Obstetrics and Gynecology, Duke University, Durham, NC, United States

Kunle Odunsi Department of Gynecologic Oncology, Roswell Park Cancer Center, Buffalo, NY, United States

Lori A. Orlando Duke Center for Applied Genomics & Precision Medicine, Duke Department of Medicine, Duke University, Durham, NC, United States

Philip R.O. Payne Institute for Informatics, School of Medicine, Washington University in St. Louis, St. Louis, MO, United States

Tanja Pejovic Department of Obstetrics and Gynecology, Oregon Health Sciences University, Portland, OR, United States

Angela Pennisi Department of Medical Oncology, Inova Schar Cancer Institute, Fairfax, VA, United States; Department of Medicine, University of Virginia, Charlottesville, VA, United States

Michael J. Pishvaian Department of Oncology, Johns Hopkins University School of Medicine, SKCC, Washington, DC, United States

Rebecca Ann Previs Department of Obstetrics and Gynecology, Duke University, Durham, NC, United States; Duke Cancer Institute, Durham, NC, United States

Timothy E. Reddy Center for Genomic and Computational Biology, Duke University, Durham, NC, United States

Marc T. Roth Medical Oncology, Sarah Cannon Cancer Institute, Kansas City, MO, United States

Rachna Shroff Division of Hematology and Oncology, Department of Medicine, University of Arizona College of Medicine, Tucson, AZ, United States

Daniel Spinosa Department of Obstetrics and Gynecology, Duke University, Durham, NC, United States

John H. Strickler National Institutes of Health, Bethesda, MD, United States

Vanessa C. Stubbs Department of Otolaryngology Head & Neck Surgery, Rutgers-Robert Wood Johnson School of Medicine, New Brunswick, NJ, United States

Aaron Therien Department of Surgery, Duke University, Durham, NC, United States

Giovana R. Thomas Division of Head & Neck Oncologic and Robotic Surgery, Department of Otolaryngology-Head & Neck Surgery, University of Miami Miller School of Medicine, Miami, FL, United States

Douglas S. Tyler Department of Surgery, University of Texas Medical Branch, Galveston, TX, United States

Catherine H. Watson Vanderbilt University Medical Center, Nashville, TN, United States

R. Ryanne Wu Duke Center for Applied Genomics & Precision Medicine, Duke Department of Medicine, Duke University, Durham, NC, United States

Jason Zhu Levine Cancer Institute, Charlotte, NC, United States

Preface

In the past decade, the oncology community has witnessed a proliferation of actionable molecular targets and novel targeted therapies. This rapidly expanding treatment landscape has coincided with the broad adoption of comprehensive genomic profiling as a routine clinically indicated test. In some cases these actionable molecular targets predict resistance to therapies and help avoid medically futile treatments. In other cases, these targets predict exquisite sensitivity to novel therapies. These changes have undoubtedly improved survival and quality of life for many patients, but many challenges remain.

With the proliferation of molecular targets and therapies, it is increasingly difficult for clinicians to maintain state-of-the art practice. For patients with non-small-cell lung cancer alone, current National Comprehensive Cancer Network (NCCN) guidelines list no fewer than seven actionable genomic targets, in addition to PD-L1 expression testing. With each target comes a multitude of associated therapies, each with its own clinical efficacy and toxicity data. Additionally, some targets (e.g., *NTRK*, tumor mutational burden, microsatellite instability) are targetable by therapies via tumor-agnostic approvals. However, in other cases, tumor site of origin in the context of a specific mutation (e.g., $BRAF^{V600E}$) has a dramatic impact on therapeutic efficacy. To master the molecular landscape of cancer, rapid dissemination of knowledge and continuous lifelong learning is required. Improvements in information technology, artificial intelligence, and treatment pathways can help automate and support clinical decision-making, but clinical judgment remains paramount. The importance of clinical judgment is unlikely to change in the near future.

Despite rapid advancements in precision cancer medicine, many patients with cancer are still left behind. Too many molecular targets still remain "undruggable," despite major advances in drug screening and medicinal chemistry. Even when breakthroughs are made, clinical benefit in many cases is transient, and outgrowth of resistance rapidly emerges. This problem is particularly acute for solid tumors—such as colorectal cancer—which are characterized by inter- and intratumoral heterogeneity. The extent of this heterogeneity was underappreciated only a decade ago. To capture this heterogeneity, access to tumor tissue is critical. However, the cost and risks associated with serial tumor tissue biopsies limit routine clinical use.

To better understand the genomic drivers of treatment resistance and heterogeneity, new diagnostic technologies have emerged. One of these diagnostic technologies—analysis of cell-free DNA—is feasible from routine blood collection. In some malignancies, analysis of cell-free DNA may better capture heterogeneity and longitudinal changes in a tumor's mutational profile. Although these diagnostic technologies are informative and may alter disease management, they are also resource intensive. The cost of comprehensive molecular profiling has fallen in recent years, but third-party payers and other critical stakeholders are slow to support technologies that add cost without a clear economic rationale. To achieve the potential of precision cancer medicine, comprehensive molecular profiling assays must deliver value. Moreover, the biopharmaceutical industry must be willing to invest in the development of active therapies for rare targets.

Finally, there is a risk that rapid advancements in precision cancer medicine will leave uninsured and underinsured patients behind. In the United States, these disparities disproportionately impact rural and urban populations living in poverty, as well as underrepresented racial and ethnic groups. Outside the United States, these disparities are particularly acute in societies with high-income inequality and limited resources. A concerted effort is needed to ensure that precision cancer medicine diagnostic and treatment strategies are available to all people facing cancer.

Despite these myriad challenges, there is reason for optimism. Genomic alterations that were thought to be "undruggable" are now being effectively targeted with novel therapies. In 2021 the US Food and Drug Administration approved sotorasib for patients with $KRAS^{G12C}$-mutated non-small-cell lung cancer. This achievement brings hope that other $KRAS$ mutations and other "undruggable targets" will one day have effective therapies. Moreover, the emergence of immunotherapies and their associated biomarkers has revolutionized cancer medicine. In this third edition, much of the new content addresses the impact of immunotherapy on precision cancer medicine strategies. Progress in precision cancer medicine can feel painfully slow—particularly for patients and their families facing cancer—but the advancements ushered in by novel immunotherapies and targeted therapies have in some cases vastly improved treatment outcomes.

Finally, in the past decade, access to comprehensive molecular profiling has also improved. As the cost of genomic profiling has fallen, the use of targeted next-generation sequencing, whole exome sequencing, whole transcriptome sequencing, germline testing, and cell-free DNA assays has rapidly expanded. With enhanced testing comes a vast amount of data to collect and understand. Multiple academic, biopharma, and commercial entities are attempting to link clinical outcomes with molecular data. Examples of large efforts that have been pursued in the past decade include The Cancer Genome Atlas (TCGA), the American Association for Cancer Research (AACR) Project GENIE, which is a public registry of real-world clinical data and others. With enhanced interoperability, data sharing, and artificial intelligence, it is hoped that insights from clinically annotated molecular data will drive the next generation of cancer breakthroughs.

It is with this background that we present the third edition of *Genomic and Precision Medicine: Oncology*. This volume provides a comprehensive overview of precision cancer medicine across a broad range of solid tumor and hematologic malignancies. Each chapter is organized to cover the application of genomics and personalized medicine tools and technologies including the following topics:

1. Risk Assessment and Susceptibility
2. Diagnosis and Prognosis
3. Pharmacogenomics and Precision Therapeutics
4. Emerging and Future Opportunities in the Field

Additionally, this edition includes chapters dedicated to important topics impacting precision cancer medicine, including molecular tumor boards, clinical decision-making support, bioinformatics, information technology, and epigenetics/epigenomics.

It is hoped that this edition will provide vital information for oncologists, scientists, geneticists, medical providers, and other key stakeholders who dedicate their lives to improving cancer outcomes. Precision cancer medicine is dynamic, but the core principles presented in this edition are timeless. We hope that this content will provide an invaluable resource as you endeavor to study and implement precision cancer medicine treatment strategies. It is with a desire for progress and optimism for the future that we present this third edition.

John H. Strickler
Matthew S. McKinney

Chapter 1

Introduction and overview of cancer precision medicine

Matthew S. McKinney and John H. Strickler
National Institutes of Health, Bethesda, MD, United States

Background

Precision medicine offers significant promise in improving and lengthening the lives of those afflicted with cancer. As a whole, cancer is the second most common cause of death in the United States with 1.7M new cases and ~600,000 cancer deaths annually (Henley, Ward, & Scott, 2020). Palliation or cure of malignancy is often associated with significant morbidity and cost, particularly compared to other conditions such as cardiovascular disease or infectious diseases. In contrast to these ailments, there has been relatively slow progress in improving cancer-related death rates and other outcomes. Survival in patients with malignancy is often limited by significant inter- and intratumoral molecular heterogeneity (Dagogo-Jack & Shaw, 2018; Vogelstein et al., 2013) that drives resistance to therapy and dictates the use of cytotoxic agents not targeted to the tumor's underlying molecular drivers. The use of conventional chemotherapy often results in excess toxicity, and rarely results in cure.

Given the shortcomings of many currently available strategies for treating cancer, there has been a significant effort to understand the molecular and genomic underpinnings across the hundreds of existing cancer subtypes so that more effective and safer treatment programs can be designed. The use of "-omics" approaches (genomics, transcriptomics, proteomics, metabolomics, and epigenomics) is transforming our understanding of the biological basis of cancer on an ever-accelerating basis.

The effort to incorporate "-omics" approaches into cancer treatment has been fueled in many cases by exponential gains in computational power for deconvoluting the results of massively parallel genomic sequencing as well as more effective high throughput drug and immunotherapy design and screening technologies (Freedman, Klabunde, & Wiant, 2018; Grewal & Stephan, 2013; Pettersson, Lundeberg, & Ahmadian, 2009). The nexus of

Genomic and Precision Medicine. DOI: https://doi.org/10.1016/B978-0-12-800684-9.00008-3

these efforts has led to acceleration of discoveries capable of leveraging our understanding of the biological basis of cancer to produce life-saving therapeutics with reduced side effects. Indeed, the use of multiplexed genomic assays and precision medicine treatment approaches has become standard of care in many clinical scenarios across almost all cancer histologic subtypes. From the period of 2017−20, more than a dozen new FDA-approved therapies have the requirement for an associated biomarker (Abida, Patnaik, & Campbell, 2020; Abou-Alfa, Sahai, & Hollebecque, 2020; André, Ciruelos, & Rubovszky, 2019; de Bono, Mateo, & Fizazi, 2020; Doebele, Drilon, & Paz-Ares, 2020; Drilon, Oxnard, & Tan, 2020; Golan, Hammel, & Reni, 2019; Kopetz, Grothey, & Yaeger, 2019; Loriot, Necchi, & Park, 2019; Marcus, Lemery, Keegan, & Pazdur, 2019; Wolf, Seto, & Han, 2020). Additionally, ∼40% of oncology clinical trials ongoing in 2019 include a molecular biomarker as inclusion criteria (IQVIA, 2019). The plethora of −omics-based discoveries have been translated into new clinical standards of care

One way that precision cancer medicine has been implemented at cancer centers and treatment networks including our own has been the establishment of database technology to store complex genomic assay information as well as the establishment of "molecular tumor boards" that provide expertise in matching genomic alterations to therapies based on the rapidly evolving landscape of clinical genomics assays and our armamentarium of targeted cancer therapies (Brown & Elenitoba-Johnson, 2020; Dalton, Forde, & Kang, 2017; Johnson, Khotskaya, & Brusco, 2017; Massard, Michiels, & Ferté, 2017; Pishvaian, Blais, & Bender, 2019).

This overview seeks to introduce concepts related to the genomic/molecular basis of cancer, the technology used to assay genomic/molecular alterations in the clinical settings and outline frameworks for implementing precision cancer therapy. The ultimate goal is to improve patient outcomes while better understanding the biological underpinning of disease. The field of precision cancer medicine is rapidly evolving and is now only beginning to fulfill the promise of matching tumor sequencing results to highly active molecularly targeted therapies.

Overview of hereditary and somatic alterations as the basis of cancer and genomic heterogeneity

One of the most important concepts in cancer genomics and precision medicine is the appreciation that many important drivers of malignancy exist as somatic alterations in cancer genomes. Human genomes themselves are incredibly complex with almost 50,000 annotated genes encoded in 6200 Mbp (the size of a diploid human genome) (Venter, Adams, & Myers, 2001) and this produces the vast array of phenotypic differences in human beings, including variations in the risk for and phenotype of human diseases. Germline genetic variation is important to determining an Fraumeni syndrome (Li & Fraumeni, 1969; Strickler, Loree, & Ahronian, 2018; Varley, 2003) as well as dictating the response to targeted agents (such as with synthetic lethal PARP inhibition strategies) (Abida et al., 2020; de Bono et al., 2020; Golan et al., 2019). Finally, the risk of treatment-related toxicity can be related to parmacogenomic variation in genes important in the metabolism of cytotoxic and targeted agents. However, cancer genomes contain additional somatic alterations that additionally work to drive tumorigenesis in tandem with the phenotype of their underlying host genome. In contrast to their host's genome, cancers acquire complex somatic alterations including single nucleotide variations (SNVs or gene "mutations"), gene copy number alterations (CNVs including gene amplifications and deletions often manifest as aneuploidy), gene fusions, and altered gene expression profiles that often stem from acquired deregulation of epigenetic control of gene transcription.

Additionally, there is significant genomic heterogeneity that exists across cancer histologic subtypes (intertumoral heterogeneity) as well across the individual cells comprising a patient's tumor and/or metastases (intratumoral heterogeneity) (Strickler et al., 2018). Metastatic tumors also exhibit clonal selection for new driver alterations or tumor suppressor loss in the setting of selection pressure induced by therapeutics and this process is an important feature in treatment resistance. Presumably, this process is fostered by the profound genomic instability found in cancers and additional genomic alterations may accumulate over time and provide a mechanism by which tumors can evade precision cancer medicine approaches. Therapy-driven clonal evolution and development of treatment resistance with the appearance of novel genomic drivers is not unlike the theory of evolution of species described by Charles Darwin (Fig. 1.1). Darwin's framework seems appropriate to understand the clonal evolution displayed by cancers both during their formation and metastasis as well as in response to selection pressure induced by targeted therapies. Interestingly much of our knowledge of the clonal architecture of cancer's behavior in this regard has been illuminated by the increasing availability of multiplexed genomic assays that can be repeated in longitudinal manner, as individual patients undergo therapeutic trials and response assessments. The concept of tumor evolution has important implications for how we assay the molecular/genomic characteristics of cancers and how we track and respond to the development of resistance. Thus it is important to understand how the complexity of cancer genomes differs from that of nontransformed tissue states in terms of fully appreciating the challenges and opportunity of precision cancer medicine.

The molecular heterogeneity of cancer has produced both opportunities and frustration in the endeavor of precision cancer medicine. With the advent and evolution of massively parallel sequencing technologies, it is now

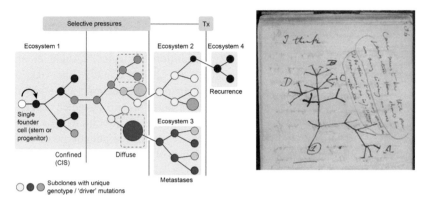

FIGURE 1.1 Cancer subclones form and evolve with therapy (left panel) in a manner consistent with Darwinian evolution (right panel depicting Darwin's field notes from Origin of Species). *Source: Nature volume 481, pages 306–313 (2012).*

possible to perform assays such as whole genome or whole transcriptome sequencing in a relatively cost-efficient manner allowing both discovery at the level of basic and translational research as well as clinical implementation of validated clinical assays. In fact, the advances in our ability to perform multiplexed genomic assays with technologies such as whole genome, targeted exome, or RNA sequencing (RNA-seq) panels have outpaced gains expected with geometric expansion of our capabilities to perform such assays as suggested by Moore's law. These advances have been made possible by massive increases in computational power on microchips over time (Pettersson et al., 2009). Because of this, the use of next generation sequencing (NGS) technology is now pervasive across basic, translational, and clinical research efforts in cancer and has led to an explosion in our knowledge of the underlying genomic and epigenomic alterations driving cancer subtypes. The clinical implementation of these assays has important implications for cancer care as well as how clinical trials of investigational agents are designed and implemented.

Clinical implementation of multiplexed molecular/genomic panels and therapeutic choice

The recent development of next generation genomic sequencing (NGS) (Shendure, Porreca, & Reppas, 2005) has both revolutionized our view of somatic genetic alterations in human cancer and created myriad possibilities for targeted precision medicine approaches (de Bono & Ashworth, 2010). In this respect, recent tumor agnostic FDA approvals of molecularly targeted therapies (Drilon, Laetsch, & Kummar, 2018; Le, Durham, & Smith, 2017) based on specific genomic alterations emphasizes the fact that novel approaches are needed to interpret and leverage multiplexed genomic data.

Clinical grade NGS data are fundamentally different from conventional diagnostic testing in terms of scope (these tests assay genomic alterations over many thousands of DNA base pairs) and a high degree of knowledge is needed for successful interpretation of the data generated by such assays. Given these developments, there has been a wide effort to direct clinicians to utilize precision medicine approaches. Per Dr. Richard Schilsky (previous American Society of Clinical Oncology President): "Knowledge of the molecular profile of the tumor is necessary to guide selection of therapy for patient"; similar statements have been made by guidelines forming organizations (Razelle, Colevas, & Anthony, 2015). The most basic concept in precision medicine is the ability to select patients for which a drug/intervention is the most beneficial with least toxicity. An example of this approach utilizing precision cancer medicine would be selection of treatments in a histology agnostic manner based on molecular alterations detected by NGS profiling.

A basic precision cancer medicine approach would be the development of a small molecular inhibitor to block oncogene signaling driven by a specific molecular alteration (most often defined genetically by DNA sequencing) within cancer cells and the use of directed pharmacotherapy in patients with malignancies bearing that alteration (Fig. 1.2). The earliest example of this would be the use of imatinib mesylate in BCR-ABL1 rearranged chronic myelogenous leukemia (CML), whereby in the IRIS study comparing imatinib mesylate to interferon-based therapy there was a significant survival benefit with less toxicity via targeting BCR-ABL1. BCR-ABL1 inhibitor

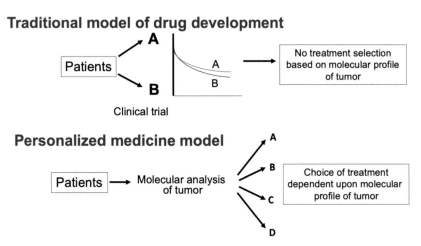

FIGURE 1.2 Traditional model of cancer therapy based on histology versus a "basket" histology agnostic approach. In the traditional approach, therapies are applied without selection in regard to underlying molecular or genomic markers. In "personalized" or "precision" cancer medicine, therapies are selected based on knowledge of the underlying molecular profile of the patient's tumor with the goal of enhancing therapeutic effectiveness and decreasing toxicities.

development has transformed the treatment landscape in CML and saved thousands of lives through the use of targeted inhibitors. Such strategies often avoid significant toxicity because the target of the small molecular inhibitor is a specific fusion protein product. These molecular targets can thus be effectively targeted with agents that have minimal if any toxicity related to manipulation of normal cellular pathways. Other more recent examples where targeting fusion products have improved outcomes over standard chemotherapy include ALK inhibitors in ALK-rearranged lung carcinoma (Solomon, Mok, & Kim, 2014) and anaplastic large cell lymphoma (Mossé, Voss, & Lim, 2017) as well as inhibitors of FGFR in various histologic types fusions(Doebele et al., 2020). There are now countless examples of therapeutic strategies targeted to the products of other genetic alterations, such as SNVs, indels, or CNVs.

Consideration for monitoring and managing toxicity is important in precision cancer medicine, as many molecular targets in cancer cells also have important biological roles in nonmalignant tissue. A salient example of this is the FDA-approval of alpelisib for estrogen receptor positive breast cancer with activating *PIK3CA* kinase mutations. In this patient population, alpelisib showed a significant improvement in progression-free survival when added to fulvestrant (André et al., 2019). However, a significant incidence of hyperglycemia due to on target inhibition of phosphoinositide 3-kinase (PI3K) signaling important to the regulation of cellular metabolism and insulin resistance was noted. Investigation of strategies to mitigate hyperglycemia is underway (Glucagon Receptor Inhibition to Enable Breast Cancer Patients to Benefit From PI3K Inhibitor Therapy).

Molecular profiling of tumor DNA may also identify genomic alterations associated with treatment resistance. When these alterations are identified, patients can be spared ineffective and potentially toxic therapies. Treatment resistance markers include examples such as *KRAS* and *NRAS* mutations in colon cancer, which predict resistance to antiepidermal growth factor (EGFR) antibodies (Lièvre, Bachet, & Boige, 2008) or clonal evolution with the development of MET amplification to drive resistance to EGFR tyrosine kinase inhibitors in non-small cell lung cancer (Gao, Li, Jin, Jiang, & Ding, 2019; Schmid, Früh, Peters, & Targeting, 2020). Similarly, genomic drivers of resistance to inhibitors of BCR-ABL have been defined and guide treatment changes.

In both of these cases, clinical NGS panels and guidelines for treatment decisions are now standard of care.

Opportunities and challenges in precision oncology

The past decade has seen a tremendous number of new discoveries in cancer biology first by the use of transcriptome/gene expression technologies and later by the development of NGS technologies paired with advanced computational data. Next generation multiplexed sequencing technology has reached the clinical space and novel precision medicine agents have been FDA approved in a tissue histology agnostic manner. These recent approvals are dependent on the availability and the use of precision medicine assays. Clinical guidelines for treatment of various cancer subtypes have rapidly evolved to incorporate the use of precision medicine to improve therapeutic options. In many respects, the accumulation of data in the field has outpaced the ability of clinicians to understand how best to utilize NGS results. Unfortunately, uptake of NGS testing for actionable cancer alterations consistent with national guidelines appears to be incomplete with a significant fraction of patients not receiving testing for targetable alterations (Schink, Trosman, & Weldon, 2014). This phenomenon may be a result of regional variations in practice and variations in payor coverage for molecular testing. There are also differences in molecular testing based on demographic factors, such as age, gender, and race. Additionally, only a small (often less than 5%) percentage of eligible patients who received genomic testing are enrolled in targeted "basket" clinical trials (Meric-Bernstam, Brusco, & Shaw, 2015). Thus there are significant improvements to be made to our implementation and utilization of precision cancer medicine.

The rapid expansion of precision cancer medicine approaches also presents challenges for clinicians and institutions. A recent survey of US oncologists found that uptake of molecular testing was significantly aided by the formation of institutional "molecular tumor boards." Tumor boards incorporating multidisciplinary discussion including disease-based specialists, molecular pathologists, medical geneticists, pharmacy and clinical trial staff have been employed at many institutions and our experience is that these groups are integral to providing "just in time" treatment recommendations (de Moor, Gray, Mitchell, Klabunde, & Freedman, 2020). Significant infrastructure and availability of expertise is needed to support precision multidisciplinary discussion as well as manage the many data points each NGS-based assay generates. Ideally, NGS data are kept in a registry that can interface with molecular tumor boards as well as clinical decision support software and clinical trial matching tools. Our experience is that without such tools, clinicians may be left unaware of off-label treatments and clinical trial options (Green, Bell, & Hubbard, 2021).

Economic and payor issues are also rising in importance in the field of cancer precision medicine. Healthcare spending in the United States reached $3.8 trillion (or $11,500 per person) in 2019, accounting for 17.7% of the national GDP (National Health Expenditures, 2019). Healthcare costs and the sustainability of the US healthcare system are a significant concern. Related to precision medicine, NGS assays can be expensive; similarly, the cost to develop new drugs has been estimated to be billions of dollars, and these expenses are often passed to patients or insurers and other stakeholders. There are significant concerns that these costs may drive disparities in access to care, and that advanced, expensive new technologies may drive "financial toxicity" for patients (Carrera, Kantarjian, & Blinder, 2018; Chino & Zafar, 2019; Tran & Zafar, 2018). These issues will continue to be challenging in the field of precision cancer medicine as we expand the breath of clinical genomics testing, and the field moves to utilization of an armamentarium of precision medicine therapeutics.

Summary

Precision medicine promises to transform our approach to the diagnosis, prognostication, and treatment of patients with cancer. Cancer continues to be a significant cause of mortality and morbidity but incremental gains are being made through laboratory discovery and the implementation of molecular precision medicine approaches.

References

Abida, W., Patnaik, A., Campbell, D., et al. (2020). Rucaparib in men with metastatic castration-resistant prostate cancer harboring a BRCA1 or BRCA2 gene alteration. *Journal of Clinical Oncology*, *38*(32), 3763−3772. Available from https://doi.org/10.1200/jco.20.01035.

Abou-Alfa, G. K., Sahai, V., Hollebecque, A., et al. (2020). Pemigatinib for previously treated, locally advanced or metastatic cholangiocarcinoma: A multicentre, open-label, phase 2 study. *The Lancet Oncology*, *21*(5), 671−684. Available from https://doi.org/10.1016/s1470-2045(20)30109-1.

André, F., Ciruelos, E., Rubovszky, G., et al. (2019). Alpelisib for PIK3CA-mutated, hormone receptor-positive advanced breast cancer. *The New England Journal of Medicine*, *380*(20), 1929−1940. Available from https://doi.org/10.1056/NEJMoa1813904.

Brown, N. A., & Elenitoba-Johnson, K. S. J. (2020). Enabling precision oncology through precision diagnostics. *Annual Review of Pathology: Mechanisms of Disease*, *15*(1), 97−121. Available from https://doi.org/10.1146/annurev-pathmechdis-012418-012735.

Carrera, P. M., Kantarjian, H. M., & Blinder, V. S. (2018). The financial burden and distress of patients with cancer: Understanding and stepping-up action on the financial toxicity of cancer treatment. *CA: A Cancer Journal for Clinicians*, *68*(2), 153−165. Available from https://doi.org/10.3322/caac.21443.

Chino, F., & Zafar, S. Y. (2019). Financial toxicity and equitable access to clinical trials. *American Society of Clinical Oncology Educational Book, American Society of Clinical Oncology Annual Meeting*, *39*, 11−18. Available from https://doi.org/10.1200/edbk_100019.

Dagogo-Jack, I., & Shaw, A. T. (2018). Tumour heterogeneity and resistance to cancer therapies. *Nature Reviews Clinical Oncology*, *15*(2), 81−94. Available from https://doi.org/10.1038/nrclinonc.2017.166.

Dalton, W. B., Forde, P. M., Kang, H., et al. (2017). Personalized medicine in the oncology clinic: Implementation and outcomes of the Johns Hopkins Molecular Tumor Board. *JCO Precision Oncology*, 2017. Available from https://doi.org/10.1200/po.16.00046.

de Bono, J., Mateo, J., Fizazi, K., et al. (2020). Olaparib for metastatic castration-resistant prostate cancer. *New England Journal of Medicine*, *382*(22), 2091−2102. Available from https://doi.org/10.1056/NEJMoa1911440.

de Bono, J. S., & Ashworth, A. (2010). Translating cancer research into targeted therapeutics. *Nature, 467*(7315), 543−549. Available from https://doi.org/10.1038/nature09339.

de Moor, J. S., Gray, S. W., Mitchell, S. A., Klabunde, C. N., & Freedman, A. N. (2020). Oncologist confidence in genomic testing and implications for using multimarker tumor panel tests in practice. *JCO Precision Oncology, 4.* Available from https://doi.org/10.1200/po.19.00338.

Doebele, R. C., Drilon, A., Paz-Ares, L., et al. (2020). Entrectinib in patients with advanced or metastatic NTRK fusion-positive solid tumours: Integrated analysis of three phase 1−2 trials. *The Lancet Oncology, 21*(2), 271−282. Available from https://doi.org/10.1016/S1470-2045(19)30691-6.

Drilon, A., Laetsch, T. W., Kummar, S., et al. (2018). Efficacy of larotrectinib in TRK fusion-positive cancers in adults and children. *The New England Journal of Medicine, 378*(8), 731−739. Available from https://doi.org/10.1056/NEJMoa1714448.

Drilon, A., Oxnard, G. R., Tan, D. S. W., et al. (2020). Efficacy of selpercatinib in RET fusion-positive non-small-cell lung cancer. *The New England Journal of Medicine, 383*(9), 813−824. Available from https://doi.org/10.1056/NEJMoa2005653.

Freedman, A. N., Klabunde, C. N., Wiant, K., et al. (2018). Use of next-generation sequencing tests to guide cancer treatment: Results from a nationally representative survey of oncologists in the United States. *JCO Precision Oncology* (2), 1−13. Available from https://doi.org/10.1200/po.18.00169.

Gao, J., Li, H. R., Jin, C., Jiang, J. H., & Ding, J. Y. (2019). Strategies to overcome acquired resistance to EGFR TKI in the treatment of non-small cell lung cancer. *Clinical & Translational Oncology, 21*(10), 1287−1301. Available from https://doi.org/10.1007/s12094-019-02075-1.

Glucagon receptor inhibition to enable breast cancer patients to benefit from PI3K inhibitor therapy (REMD-477). <https://ClinicalTrials.gov/show/NCT04330625>.

Golan, T., Hammel, P., Reni, M., et al. (2019). Maintenance olaparib for germline BRCA-mutated metastatic pancreatic cancer. *The New England Journal of Medicine, 381*(4), 317−327. Available from https://doi.org/10.1056/NEJMoa1903387.

Green, M., Bell, J., Hubbard, C., et al. (2021). Implementation of a Molecular Tumor Registry to Support the Adoption of Precision Oncology Within an Academic Medical Center: The Duke University Experience. *JCO Precision Oncology, 5*(Sep 16), 1493−1506. Available from https://doi.org/10.1200/PO.21.00030.

Grewal, A., & Stephan, D. A. (2013). Diagnostics for personalized medicine: What will change in the era of large-scale genomics studies? *Personalized Medicine, 10*(8), 835−848. Available from https://doi.org/10.2217/pme.13.82.

Henley, S. J., Ward, E. M., Scott, S., et al. (2020). Annual report to the nation on the status of cancer, part I: National cancer statistics. *Cancer, 126*(10), 2225−2249. Available from https://doi.org/10.1002/cncr.32802.

IQVIA. (2019). *Global oncology trends 2019.*

Johnson, A., Khotskaya, Y. B., Brusco, L., et al. (2017). Clinical use of precision oncology decision support. *JCO Precision Oncology, 2017.* Available from https://doi.org/10.1200/po.17.00036.

Kopetz, S., Grothey, A., Yaeger, R., et al. (2019). Encorafenib, binimetinib, and cetuximab in BRAF V600E−mutated colorectal cancer. *New England Journal of Medicine, 381*(17), 1632−1643. Available from https://doi.org/10.1056/NEJMoa1908075.

Le, D. T., Durham, J. N., Smith, K. N., et al. (2017). Mismatch repair deficiency predicts response of solid tumors to PD-1 blockade. *Science (New York, NY), 357*(6349), 409−413. Available from https://doi.org/10.1126/science.aan6733.

Li, F. P., & Fraumeni, J. F., Jr. (1969). Soft-tissue sarcomas, breast cancer, and other neoplasms. A familial syndrome? *Annals of Internal Medicine*, *71*(4), 747−752. Available from https://doi.org/10.7326/0003-4819-71-4-747.

Lièvre, A., Bachet, J. B., Boige, V., et al. (2008). KRAS mutations as an independent prognostic factor in patients with advanced colorectal cancer treated with cetuximab. *Journal of Clinical Oncology*, *26*(3), 374−379. Available from https://doi.org/10.1200/jco.2007.12.5906.

Loriot, Y., Necchi, A., Park, S. H., et al. (2019). Erdafitinib in locally advanced or metastatic urothelial carcinoma. *The New England Journal of Medicine*, *381*(4), 338−348. Available from https://doi.org/10.1056/NEJMoa1817323.

Marcus, L., Lemery, S. J., Keegan, P., & Pazdur, R. (2019). FDA approval summary: Pembrolizumab for the treatment of microsatellite instability-high solid tumors. *Clinical Cancer Research*, *25*(13), 3753−3758. Available from https://doi.org/10.1158/1078-0432.Ccr-18-4070.

Massard, C., Michiels, S., Ferté, C., et al. (2017). High-throughput genomics and clinical outcome in hard-to-treat advanced cancers: Results of the MOSCATO 01 trial. *Cancer Discovery*, *7*(6), 586. Available from https://doi.org/10.1158/2159-8290.CD-16-1396.

Meric-Bernstam, F., Brusco, L., Shaw, K., et al. (2015). Feasibility of large-scale genomic testing to facilitate enrollment onto genomically matched clinical trials. *Journal of Clinical Oncology*, *33*(25), 2753−2762. Available from https://doi.org/10.1200/jco.2014.60.4165.

Mossé, Y. P., Voss, S. D., Lim, M. S., et al. (2017). Targeting ALK with crizotinib in pediatric anaplastic large cell lymphoma and inflammatory myofibroblastic tumor: A Children's Oncology Group Study. *Journal of Clinical Oncology*, *35*(28), 3215−3221. Available from https://doi.org/10.1200/jco.2017.73.4830.

National Health Expenditures. (2019). *Highlights*. <https://www.cms.gov/files/document/highlights.pdf>.

Pettersson, E., Lundeberg, J., & Ahmadian, A. (2009). Generations of sequencing technologies. *Genomics*, *93*(2), 105−111. Available from https://doi.org/10.1016/j.ygeno.2008.10.003.

Pishvaian, M. J., Blais, E. M., Bender, R. J., et al. (2019). A virtual molecular tumor board to improve efficiency and scalability of delivering precision oncology to physicians and their patients. *JAMIA Open*, *2*(4), 505−515. Available from https://doi.org/10.1093/jamiaopen/ooz045.

Razelle, K., Colevas, A. D., Anthony, O., et al. (2015). NCCN oncology research program's investigator steering committee and NCCN best practices committee molecular profiling surveys. *Journal of the National Comprehensive Cancer Network*, *13*(11), 1337−1346. Available from https://doi.org/10.6004/jnccn.2015.0163.

Schink, J. C., Trosman, J. R., Weldon, C. B., et al. (2014). Biomarker testing for breast, lung, and gastroesophageal cancers at NCI designated cancer centers. *Journal of the National Cancer Institute*, *106*(10). Available from https://doi.org/10.1093/jnci/dju256.

Schmid, S., Früh, M., Peters, S., & Targeting, M. E. T. (2020). In EGFR resistance in non-small-cell lung cancer-ready for daily practice? *The Lancet Oncology*, *21*(3), 320−322. Available from https://doi.org/10.1016/s1470-2045(19)30859-9.

Shendure, J., Porreca, G. J., Reppas, N. B., et al. (2005). Accurate multiplex polony sequencing of an evolved bacterial genome. *Science (New York, NY)*, *309*(5741), 1728−1732. Available from https://doi.org/10.1126/science.1117389.

Solomon, B. J., Mok, T., Kim, D. W., et al. (2014). First-line crizotinib vs chemotherapy in ALK-positive lung cancer. *The New England Journal of Medicine*, *371*(23), 2167−2177. Available from https://doi.org/10.1056/NEJMoa1408440.

Strickler, J. H., Loree, J. M., Ahronian, L. G., et al. (2018). Genomic landscape of cell-free DNA in patients with colorectal cancer. *Cancer Discovery, 8*(2), 164−173. Available from https://doi.org/10.1158/2159-8290.Cd-17-1009.

Tran, G., & Zafar, S. Y. (2018). Financial toxicity and implications for cancer care in the era of molecular and immune therapies. *Annals of Translational Medicine, 6*(9), 166. Available from https://doi.org/10.21037/atm.2018.03.28.

Varley, J. M. (2003). Germline TP53 mutations and Li-Fraumeni syndrome. *Human Mutation, 21*(3), 313−320. Available from https://doi.org/10.1002/humu.10185.

Venter, J. C., Adams, M. D., Myers, E. W., et al. (2001). The sequence of the human genome. *Science (New York, NY), 291*(5507), 1304−1351. Available from https://doi.org/10.1126/science.1058040.

Vogelstein, B., Papadopoulos, N., Velculescu, V. E., Zhou, S., Diaz, L. A., Jr., & Kinzler, K. W. (2013). Cancer genome landscapes. *Science (New York, NY), 339*(6127), 1546−1558. Available from https://doi.org/10.1126/science.1235122.

Wolf, J., Seto, T., Han, J.-Y., et al. (2020). Capmatinib in MET exon 14−mutated or MET-amplified non−small-cell lung cancer. *New England Journal of Medicine, 383*(10), 944−957. Available from https://doi.org/10.1056/NEJMoa2002787.

Chapter 2

From data to knowledge: an introduction to biomedical informatics

Philip R.O. Payne

Institute for Informatics, School of Medicine, Washington University in St. Louis, St. Louis, MO, United States

Introduction

The field of Biomedical Informatics (BMI) has emerged over the last several decades as a driving force behind the ability of the biological and healthcare research and delivery communities relative to the ability to harness and make sense of every increasing volume and complexities of incumbent data. For the purposes of clarity in the remainder of this chapter, we define BMI as follows (per the conventions put forth by the American Medical Informatics Association or AMIA):

> *Biomedical informatics (BMI) is the interdisciplinary field that studies and pursues the effective uses of biomedical data, information, and knowledge for scientific inquiry, problem solving, and decision making, driven by efforts to improve human health. (Kulikowski et al., 2012)*

It is important to note that the scientific field of BMI as defined above is both different from and complementary to the areas of Computer and Quantitative Science. Concretely, the aforementioned fields emphasize theories and methods focused upon mechanisms to collect, manage, analyze, and report upon data, while BMI emphasized the broader issues of how we generate and use such data-driven insight given any number of driving problems, yielding information (e.g., contextualized data), and actionable knowledge (e.g., information that is delivered in the right format, place, and time to enable decision-making or analogous processes). The complementarity of these constituent areas is often encountered when BMI practioners use methods and technology solutions derived from the Computer and Quantitative Science domains in order to collect, store, assess, and present large-scale

Genomic and Precision Medicine. DOI: https://doi.org/10.1016/B978-0-12-800684-9.00010-1

and/or heterogeneous data, as are frequently encountered throughout the bio-medical and healthcare settings. Building upon these foundations, BMI prac-tioners are able to use domain-specific theories and methods in order to interpret and reason upon those data and generate and deliver actionable knowledge at multiple end-points, such as the laboratory, point-of-care, or population health settings. In this chapter, we present a broad model for the critical evaluation and understanding of how such methods are selected and applied. In doing so, we hope to equip readers with the ability to navigate an ever-evolving armamentarium of such methods and approaches in a thought-ful and systematic manner.

A primer on the role of biomedical informatics in the era of precision approaches to research, healthcare delivery, and population health

Across a spectrum from basic science investigation, to clinical research, healthcare delivery, and ultimately population health, it is becoming increasingly important to apply data-driven and precision approaches that leverage the best available scientific knowledge (Auffray, Chen, & Hood, 2009; Hood & Friend, 2010; Hood & Perlmutter, 2004). This evolving model is predicated on the establishment of close and well-instrumented interconnections between research, healthcare delivery, and population health investigators, practioners, and their respective methodologies. Unfortunately and at the present time, prevailing approaches to the afore-mentioned domains are usually not well integrated, resulting in data analytic and decision-making processes that neither take advantage of up-to-date scientific knowledge nor well-validated techniques or approaches to reasoning, sense making, and the delivery of information or knowledge in a manner consistent with predisposing or enabling sociocultural frameworks (Ahn, Tewari, Poon, & Phillips, 2006; Auffray et al., 2009; Hood & Friend, 2010). Fortunately, there exists a constantly growing body of multi-method and knowledge synthesis techniques incumbent to the field of BMI that can overcome the preceding challenges—for example, exploring poten-tial linkages been bio-molecular and clinical phenotypes as well as drug-targeting information so as to design and delivery highly tailored and precision therapeutics for a number of important disease states in a partially or fully in silico manner (Ahn et al., 2006; Hood & Perlmutter, 2004; Pattin et al., 2015; Payne, Embi, & Sen, 2009; Payne, Johnson, Starren, Tilson, & Dowdy, 2005; Regan et al., 2016; Regan & Payne, 2015; Schadt & Bjorkegren, 2012). Examples of the type of problems and corresponding BMI theories and methods that can be used to address such information and knowledge needs are provided in Table 2.1. It is of note that evolving and increasingly commonly utilized BMI methods and techniques seek to bridge the gap between humans, available domain knowledge, and large-scale

TABLE 2.1 Examples of application domains and problem areas addressed by BMI theories and methods.

Application domain	Related problem areas addressed by BMI theories and methods
Biomedical Data Science	• Data collection, storage, management, and dissemination • Data integration, harmonization, and sharing • Syntactic and semantic standards development and application (basic theories and methods) • Knowledge-based systems design and application such as those related to the use of artificial intelligence and/or cognitive computing in biomedical settings (basic theories and methods)
Bioinformatics and Computational Biology	• Analysis and interpretation of the output of bio-molecular phenotyping instruments (e.g., genomics, proteomics, metabolomics, etc.) • Derivation and evaluation of biological networks • Annotation and enrichment of biological data sets using public data resources • Bio-molecular data visualization, exploration, and hypothesis generation/testing in the preceding problem areas • Syntactic and semantic standards development and application (as applied to biological data types) • Knowledge-based systems design and application such as those related to the use of artificial intelligence and/or cognitive computing in biomedical settings (as applied to biological data types)
Clinical and Translational Informatics	• Design, implementation and management of data, information, and knowledge management systems for use in both patient-focused research and care settings (e.g., Clinical Research Management Systems, Electronic Health Records, Personal Health Records, Data Warehouses, etc.) • Clinical decision support systems design and evaluations • Syntactic and semantic standards development and application (as applied to clinical data and research data types) • Knowledge-based systems design and application such as those related to the use of artificial intelligence and/or cognitive computing in biomedical settings (as applied to clinical data and research data types)

(Continued)

TABLE 2.1 (Continued)

Application domain	Related problem areas addressed by BMI theories and methods
Imaging Informatics	• Design, implementation and management of data, information, and knowledge management systems for use the collection, storage, transaction, and delivery of data generated via imaging instruments • Computer-aided interpretation and feature extraction from biomedical images (for hypothesis discovery and clinical decision-making purposes) • Syntactic and semantic standards development and application (as applied to imaging data types)
Public Health and/or Population Health Informatics	• Epidemiological surveillance using patient level and higher-order data sets combined across care and population settings • Design, delivery, and evaluation of tailored health communication and intervention measures at a population-level • Geographic and/or temporal reasoning across population-level data types for hypothesis generation and/or testing purposes • Syntactic and semantic standards development and application (as applied to population-level data types) • Knowledge-based systems design and application such as those related to the use of artificial intelligence and/or cognitive computing in biomedical settings (as applied to population-level data types)

or heterogeneous data sets, so as to enable the efficient, timely, and empirically defensible conduct of research, healthcare delivery, and/or population-health interventions (Embi & Payne, 2014; Han et al., 2015; Pattin et al., 2015; Payne et al., 2005, 2009; Payne & Embi, 2015a). Such approaches and methods are regularly being recognized as serving a central role in the conduct of studies or clinical care delivery paradigms predicated on a genomic medicine approach to health and wellness (Ahn et al., 2006; Barabasi & Oltvai, 2004; Butcher, Berg, & Kunkel, 2004; Hood & Perlmutter, 2004; Hripcsak et al., 2015; Payne et al., 2009; Payne & Embi, 2015b; Regan et al., 2016; Regan & Payne 2015). These gaps in knowledge and practice, as well as the promise afforded by BMI theories and methods in such contexts, serve as the motivation for the review and recommendations made in the remainder of this discussion.

A framework for selecting, understanding, and assessing biomedical informatics methods

As was introduced in the preceding section, BMI theories and methods are broadly aligned with a central dogma for the field in which data are contextualized, so as to produce information, and that information is in-turn delivered in an actionable manner as knowledge. In this dogma, the process of moving from data to information involves the *augmentation* of data with contextualizing information—for example, in the case of clinical data corresponding to a given patient, such augmentation could involve the coupling of individual data points with metadata that identifies the patient, the measurement methods used to generate said data, and pertinent temporal characteristics of that measure. Similarly, the process of rendering such information as actionable knowledge involves the *delivery* of information to the right end-users in the right format and time, so as to support the intended decision-making or otherwise data-driven process. For each of the phases of the preceding central dogma, a range of methods can be employed. This conceptual model is illustrated in Fig. 2.1 and explained below. At a high level, these methods can be aligned with the end-points of: (1) data generation; (2) information generation; or (3) knowledge generation. In addition, a broad

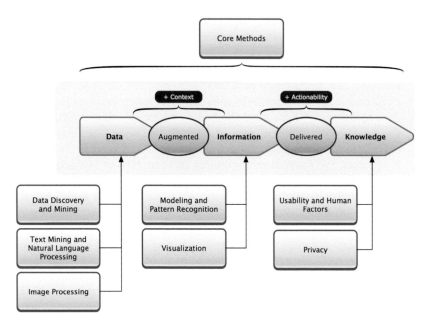

FIGURE 2.1 Overview of the central dogma of the field of Biomedical Informatics (BMI), concerned with the generation of information and knowledge from foundational data, and the alignment of methodologies to support or enable the constituent activities comprising that dogma.

class of crosscutting (or core) methods can be used across this entire spectrum of activities.

Methods aligned with data generation

- *Data discovery and mining*: One of the initial steps in almost any biomedical or health-focused data analysis project is the identification, extraction, and normalization or quality assurance of source data sets that exhibit some type of definable structure. Such structured source data can be distributed across any number of resources such as (but not limited to): (1) structured or semistructured databases (e.g., relational, hierarchical, or object-oriented data structures); (2) linked open data; (3) metadata repositories; and/or (4) transactional or otherwise operational data repositories. A number of critical methods can be employed to address the identification and downstream processing of such data-centric targets so as to support follow-on analyses, such as (but not limited to): (1) logical and platform-independent data modeling and the linkage of such models to physical repositories via approaches such as object-relational mapping; (2) the use of physical and metadata mining algorithms to derive the structure and content of heretofore informally or undefined data repositories; (3) the application of semantic reasoning algorithms to link data resources to formal syntactic and/or semantic definitions of common data elements, controlled terminologies, or more complex and semantically rich ontologies; and (4) the use of data mining techniques such as column-level indexing and graph-based data linkage invocation, so as to support the structural and/or content-level integration of heterogeneous but related data resources. All of these methods, and many others that apply to the fundamental problem area of data discovery and mining, involve the platform-independent modeling, linkage, and integration of data based upon the generation or curation of some type of syntactic and/or semantic model that serves to define the fundamental characteristics of said data sets (Bellazzi, Gabetta, & Leonardi, 2013; Embi, Hebert, Gordillo, Kelleher, & Payne, 2013; Holzinger & Jurisica, 2014; Payne & Embi, 2015b; Regan et al., 2016; Regan & Payne, 2015).
- *Text mining and natural language processing (NLP)*: In a similar manner to the methods and approaches used to identify and extract structured data resources, there exist a body of techniques that can be used to generate and codify structured data based upon the context of unstructured narrative text, such as that found in the biomedical literature or health records (to name a few of many such sources). Such methods usually rely upon statistical and/or rule-based artificial intelligence algorithms to identify and assess the lexical, semantic, and other higher-order meanings (e.g., sentiment, negation of concepts,

certainty, etc.) of such narrative text, and then codify its meaning using highly expressive data structures and semantic annotations (the later often involving the annotation of text using formal terminology or ontology-defined concepts). Ultimately, these classes of methods allow for the creation of structured data where it is absent, thus enabling the cross-linkage of those data with others derived from structured resources, so as to facilitate integrative follow-on analysis. That being said, such text mining or NLP techniques are infrequently 100% accurate, given their artificial intelligence and probabilistic "roots," and therefore can introduce potential biases, "noise," and/or errors into such downstream analyses. As such, they should be used carefully and with full awareness of such potential impacts on the data analytics process they are incumbent to (Bellazzi et al., 2013; Ohno-Machado, Nadkarni, & Johnson, 2013).

- *Image processing:* Again, and as was the case in the preceding discussion of Text Mining and/or NLP approaches, there are additional sources of data to be found in the unstructured data that comprises a variety of imaging modalities. Such imaging resources can span multiple granularities, from bio-molecules to human anatomy and physiology, with each such granularity and level presenting its own unique challenges. However, regardless of such scales, the basic principles that define image processing involve: (1) the conversion of what is often an analog or otherwise continuous scale measurement at a discrete juncture (such as a pixel or otherwise spatially oriented data-point) into a structured and actionable data element; (2) the application of search-space constraints to render further analysis of image contents computationally tractable or otherwise understandable from a spatial or structural standpoint; and (3) the identification of patterns in such data using statistical or machine learning methods; ultimately resulting in the generation of discrete measures of features of interest that are definitional to the initial image. For example, such a process could be used to evaluate a clinical image, via which an initial analog signal derived from an imaging instrument is converted into a spatially aligned matrix of pixel-level measures, that matrix is then segmented to isolate an anatomical area of interest, and then pattern recognition techniques are used to identify aberrations in the area of interest as compared to some predefined reference standard image, all for the purposes of diagnostic decision-making. Again and mirroring the prior discussion of text mining or NLP techniques, image processing algorithms are also infrequently 100% accurate, and therefore can introduce potential biases, "noise," and/or errors into downstream analyses. As such, they too should be used carefully and with full awareness of such potential impacts on the data analytics process involved (Bellazzi et al., 2013; Najarian & Splinter, 2012).

Methods aligned with information generation

- *Modeling and pattern recognition*: As was noted in a number of the preceding methods aligned with data generation, there is a frequent need in the broad biomedical and healthcare data analytics domain to extract structured data from otherwise unstructured sources so as to enable integrative analyses (Hripcsak et al., 2015; Payne & Embi, 2015a, 2015b). These same types of methods can also be employed to make sense of and contextualize resulting data by cross-linking it with complementary resources (e.g., metadata, an understanding of the biological or clinical basis for a given set of measurements, etc.) through a process of modeling and higher-order pattern recognition. In a broad-stroke, such methods can be thought of as focusing on multimodeling. These multimodeling and pattern recognition approaches are designed to overcome the limitations of reductionist approaches to scientific discovery, replacing decomposition focused problem solving with integrative network-based modeling and analysis techniques (Ahn et al., 2006; Barabasi & Oltvai, 2004). Systems-level analysis of complex problem domains ultimately enables the study of critical interactions that influence health and wellness across a scale from molecules to populations, and that are not observable when such systems are broken down into constituent components. The use of systems-level analysis methodologies is well supported by the foundational theory of vertical reasoning first proposed by Blois (1988). This theory holds that effective decision-making in the biomedical domain is predicated on the vertical integration of multiple scales and levels of reasoning. This fundamental premise is the basis for the correlative framework put forth by Tsafnat and colleagues that the ability to replicate expert reasoning relative to complex biomedical problems using computational agents requires the replication of such multiscale and integrative decision-making (Tsafnat & Coiera, 2009). In order to achieve such an outcome, Tsafnat posits that multiscale decision-making in an in silico context requires both: (1) the generation of component decision-making models at multiple scales; and (2) the similar generation of interchange layers that define important pair-wise connections between entities situated in two or more component models, often referred to as vertical linkages (Tsafnat & Coiera, 2009). Of note, this type of approach is extremely reliant upon graph-theoretic reasoning and representational models, using a network paradigm that allows for the application of logical reasoning operations spanning the entities and relationships that make up a multimodel (Barabasi & Oltvai, 2004).
- *Visualization*: While all of the preceding methods, spanning a spectrum from data discovery and integration to multimodeling and pattern recognition, have great promise in terms of enabling the analysis and understanding of complex and high-throughput biomedical and healthcare data,

they cannot address every use case or need. In fact, humans possesses certain and currently noncomputationally reproducible cognitive strengths in the areas of pattern recognition and multiscale reasoning across and between data sets. As such, there remains a strong need for the use of visualization methods to extract, present, and deliver data in a manner that leverages such unique cognitive strengths so as to support hybrid human−computer analytical processes. These visualization methods can involve simple approaches such as the generation of conventional data graphics, and increase in complexity to include multiscale and resolution delivery of complex data in immersive and potentially 3-dimensional environments (employing advanced computational and user-interaction technologies) (Boland et al., 2014; Embi et al., 2013; Holzinger & Jurisica, 2014; Sarkar, 2013; Turkay, Jeanquartier, Holzinger, & Hauser, 2014).

Methods aligned with knowledge generation

- *Usability and human factors*: Once data have been identified and contextualized as information, the next step in the BMI dogma is to deliver it to the right stakeholder in the right format and temporality so as to yield actionable knowledge. Doing so requires that we understand the core usability and human-centric dimensions that serve to define the information needs and environmental factors that will influence or predispose the actionability of said information. The domains of human−computer interaction, user experience design, workflow assessment and modeling, and cognitive science (to name a few of many) provide a variety of methods to be employed in this capacity. Such approaches can include the formal modeling and instrumentation of "real world" workflows so as to understand how technologies are to be used and what environmental factors influence such utilization, to the detailed assessment of the usability of technology-based interventions intended to deliver necessary information to targeted recipients, to the qualitative and quantitative evaluation of internalized decision-making models employed by human beings when acting upon said information (and how that may impact their ability to do so, given the preceding workflow and technical issues that may be elucidated during the course of a study or implementation effort). Many of these areas represent the "fuzzy" science that exists at the implementation-level of BMI, wherein a combination of quantitative and qualitative methods must be used to triangulate an otherwise difficult to measure "ground truth" and understand how to achieve actionability relative to a given information resource (Boland et al., 2013; Sarkar, 2013).
- *Privacy*: Finally, and importantly given the increasing integration of biomolecular and clinical phenotyping in translational science or medicine contexts, it is imperative that the delivery of actionable knowledge

account for or otherwise incorporate methods to ensure appropriate patient confidentiality or privacy. These types of methods can involve the use of de-identification algorithms that can be applied to patient-derived data sets to reduce and/or eliminate re-identification potential and therefore enable research using that data in those cases where direct patient consent is infeasible, or to support the application of rule-based or knowledge-based constraints at the point-of-care so as to ensure that actionable knowledge is only delivered to individuals who have a justifiable and ethically defensible reason for accessing that information. These types of methods remain one of the most emergent in the field of BMI, and simultaneously, are perhaps some of the most important given the need to bridge basic science and clinical practice in the era of genomic medicine (Gkoulalas-Divanis, Loukides, Xiong, & Sun, 2014; Hersh et al., 2013).

Crosscutting (core) methods

While the methods introduced in the preceding discussion are organized to roughly aligned with the major steps in the BMI dogma introduced at the outset of this section, there also exist a number of crosscutting (or core) methods that can support or enable a full spectrum of data analytics, presentation, and delivery needs. These methods can include but are not limited to:

- data structure and algorithm design,
- multilevel conceptual data modeling,
- knowledge engineering and management,
- probabilistic modeling and analyses, and
- implementation and application of leaning systems such as classifiers.

A comprehensive and pragmatic treatment of these methodologies can be found in the excellent text provided by Sarkar and colleagues (Sarkar, 2013).

Additional readings

A further exploration of the methods and approaches enumerated in the preceding discussion can be found in the selected articles indicated in Table 2.2.

The relationship between problem solving and methods selection

When selecting appropriate (and often times multimethod) approaches to biomedical problem solving, it is helpful to position such decision-making in the broader context of biomedical or healthcare problem solving. Of note, such a general problem-solving model applies equally to research and

TABLE 2.2 Selected readings focusing upon common or emergent BMI methodologies, organized by thematic areas and focus.

Authors	Thematic focus	Application focus	Title	Publication	Year
Sarkar I.N.	Core Methods	Board	*Methods in biomedical informatics: a pragmatic approach*	Methods in Biomedical Informatics: A Pragmatic Approach	2013
Holzinger A., Jurisica I.	Data Discovery and Mining	Board	*Knowledge discovery and data mining in biomedical informatics: The future is in integrative machine learning solutions*	Interactive Knowledge Discovery and Data Mining in Biomedical Informatics	2014
Hersh W.R., Cimino J., Payne P.R.O., Embi P., Logan J., Weiner M., et al	Data Discovery and Mining	Healthcare Delivery	*Recommendations for the use of operational electronic health record data in comparative effectiveness research: a case-driven exploration*	eGEMs (Generating Evidence & Methods to improve patient outcomes)	2013
Embi P.J., Hebert C., Gordilli G., Kelleher K., Payne P.R.O.	Knowledge Engineering	Clinical Research	*Knowledge management and informatics considerations for comparative effectiveness research: a case-driven exploration*	Medical Care	2013
Holmes J.H.	Modeling and Pattern Recognition	Basic Science	*Methods and applications of evolutionary computation in biomedicine*	Journal of Biomedical Informatics	2014

(Continued)

TABLE 2.2 (Continued)

Authors	Thematic focus	Application focus	Title	Publication	Year
Jiang X., Cai B., Xue D., Lu X., Cooper G.F., Neapolitan R.E.	Modeling and Pattern Recognition	Clinical Research	*A comparative analysis of methods for predicting clinical outcomes using high-dimensional genomic data sets*	Journal of the American Medical Informatics Association	2014
Liu M., Hinz E.R.M., Matheny M.E., Denny J.C., Schildcrout J.S., Miller R. A., et al	Modeling and Pattern Recognition	Healthcare Delivery	*Comparative analysis of pharmacovigilance methods in the detection of adverse drug reactions using electronic medical records*	Journal of the American Medical Informatics Association	2013
Gkoulalas-Divanis A., Loukides G., Xiong L., Sun J.	Privacy	Healthcare Delivery	*Informatics methods in medical privacy*	Journal of Biomedical Informatics	2014
Ohno-Machado, Nadkarni P., Johnson K.	Text Mining and Natural Language Processing	Board	*Natural language processing: algorithms and tools to extract computable information from EHRs and from the biomedical literature*	Journal of the American Medical Informatics Association	2013
Boland M.R., Rusanov A., So Y., Lopez-Jimenez C., Busacca L., Steinman R. C., et al	Usability and Human Factors	Board	*From expert-derived user needs to user-perceived ease of use and usefulness: A two-phase mixed-methods evaluation framework*	Journal of Biomedical Informatics	2013
Turkay C., Jeanquartier F., Holzinger H.	Visualization	Board	*On computationally-enhanced visual analysis of heterogeneous data and its application in biomedical informatics*	Interactive Knowledge Discovery and Data Mining in Biomedical Informatics	2014

operational scenarios. This approach is shown in Fig. 2.2 and briefly explained below:

Problem-solving process

The biomedical- and healthcare-related problem-solving process initiates with the definition of a problem and associated use case (e.g., the stakeholders involve, the end-points those stakeholders wish to achieve, and the metrics or other measurements of success therein). Such use cases easily apply across a variety of problem spaces from research (e.g., hypothesis generation and testing) to clinical care (e.g., diagnostic decision-making and therapy planning) to population health (e.g., intervention planning and evaluation of outcomes). Subsequently the problem-solving process involves the identification and engagement of necessary data sets, the concomitant identification and engagement contextual resources such as pertinent domain knowledge or prior work, and the derivation of appropriate delivery mechanisms for ensuing actionable knowledge generated via the problem-solving

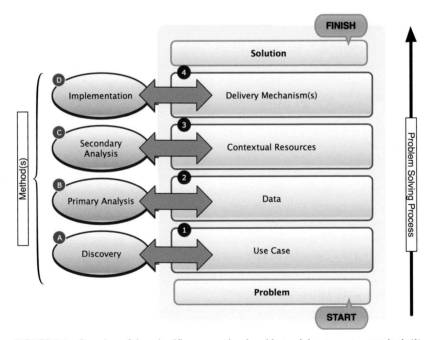

FIGURE 2.2 Overview of the scientific or operational problem-solving process, comprised: (1) use case definition; (2) data discovery; (3) contextualization of said data; and (4) identification and use of appropriate delivery mechanisms so as to arrive at a solution to the underlying problem. In this same model, core BMI theories and methods can be aligned with: (A) discovery activities, supporting step 1; (B) primary analyses, supporting step 2; (C) secondary analyses, supporting step 3; and (D) implementation approaches, supporting step 4.

process (combining the prior components) so as to achieve success given the underlying use case (e.g., a solution to the initial problem).

Alignment with biomedical informatics methods

Given the preceding problem-solving process model, the methods introduced previously can be assigned to one or more categories, corresponding with: (1) the initial discovery phase for the identification of the definitional characteristics underlying the use case; (2) the primary analysis phase for the discovery, integration, and assessment of source data; (3) the secondary analysis phase for the contextualization of initial data and analytics results so as to generate information; and (4) the implementation phase, via which appropriate delivery mechanisms, aligned with the underlying use case and success criteria can be selected and evaluated in a meaningful manner. For each of the methods we have introduced, alignment can be induced with none or more than one of these phases, depending on problem and use-case specific needs and requirements.

Conclusion

As was introduced at the outset of this chapter, the field BMI focuses first and foremost on multimethod approaches to generating contextualized information and actionable knowledge from a variety of biological and healthcare-relevant data types. In order to achieve such outcomes, BMI practitioners adopt and adapt methods drawn from the computational, quantitative, and qualitative sciences. In this chapter, we have reviewed a selection of such methods and the rationale for how they can be applied to address driving biological and clinical problems drawn from the "real world." These types of use cases span a range from the bio-molecular characterization of disease states to the comprehensive phenotyping of patients to the promotion of population health. Ultimately, this review provides the basis for critically understanding and evaluating the ways in which such multimethod approaches can be optimally utilized so as to advance biomedical research, clinical practice, and population-level health

Glossary and abbreviations

- Biomedical Informatics (BMI): "Biomedical informatics (BMI) is the interdisciplinary field that studies and pursues the effective uses of biomedical data, information, and knowledge for scientific inquiry, problem solving, and decision-making, driven by efforts to improve human health." (Kulikowski et al., 2012)
- Data: facts or information used usually to calculate, analyze, or plan something (source: Merriam Webster Dictionary)

- Information: knowledge that you get about someone or something; facts or details about a subject; knowledge obtained from investigation, study, or instruction (source: Merriam Webster Dictionary)
- Knowledge: the fact or condition of knowing something with familiarity gained through experience or association; acquaintance with or understanding of a science, art, or technique; the sum of what is known, the body of truth, information, and principles acquired by humankind (source: Merriam Webster Dictionary)
- Methodology: a set of methods, rules, or ideas that are important in a science or art, a particular procedure or set of procedures (source: Merriam Webster Dictionary)

References

Ahn, A. C., Tewari, M., Poon, C. S., & Phillips, R. S. (2006). The limits of reductionism in medicine: Could systems biology offer an alternative? *PLoS Medicine, 3*(6), 709−713.

Auffray, C., Chen, Z., & Hood, L. (2009). Systems medicine: The future of medical genomics and healthcare. *Genome Medicine, 1*(1), 2.1−2.11.

Barabasi, A. L., & Oltvai, Z. N. (2004). Network biology: Understanding the cell's functional organization. *Nature Reviews. Genetics, 5*, 101−113, February.

Bellazzi, R., Gabetta, M., & Leonardi, G. (2013). Engineering principles in biomedical informatics. *Methods in Biomedical Informatics: A Pragmatic Approach, 313.*

Blois, M. (1988). Medicine and the nature of vertical reasoning. *The New England Journal of Medicine, 381*(13), 847−851.

Boland, M. R., Rusanov, A., So, Y., Lopez-Jimenez, C., Busacca, L., Steinman, R. C., Bakken, S., Bigger, J. T., & Weng, C. (2014). From expert-derived user needs to user-perceived ease of use and usefulness: A two-phase mixed-methods evaluation framework. *Journal of Biomedical Informatics, 52*, 141−150.

Butcher, E. C., Berg, E. L., & Kunkel, E. J. (2004). Systems biology in drug discovery. *Nature Biotechnology, 22*(10), 1253−1259.

Embi, P. J., Hebert, C., Gordillo, G., Kelleher, K., & Payne, P. R. O. (2013). Knowledge management and informatics considerations for comparative effectiveness research: A case-driven exploration. *Medical Care, 51*, S38−S44.

Embi, P. J., & Payne, P. R. O. (2014). Advancing methodologies in Clinical Research Informatics (CRI): Foundational work for a maturing field. *Journal of Biomedical Informatics, 52*, 1−3.

Gkoulalas-Divanis, A., Loukides, G., Xiong, L., & Sun, J. (2014). Informatics methods in medical privacy. *Journal of Biomedical Informatics, 50*, 1−3.

Han, D., Wang, S., Jiang, C., Jiang, X., Kim, H.-E., Sun, J., et al. (2015). Trends in biomedical informatics: Automated topic analysis of JAMIA articles. *Journal of the American Medical Informatics Association, 22*(6), 1153−1163.

Hersh, W. R., Cimino, J., Payne, P. R. O., Embi, P., Logan, J., Weiner, M., et al. (2013). Recommendations for the use of operational electronic health record data in comparative effectiveness research. *eGEMs (Generating Evidence & Methods to improve patient outcomes), 1*(1), 14.

Holzinger, A., & Jurisica, I. (2014). *Knowledge discovery and data mining in biomedical informatics: The future is in integrative, interactive machine learning solutions. Interactive knowledge discovery and data mining in biomedical informatics* (pp. 1−18). Springer.

Hood, L., & Friend, S. H. (2010). Predictive, personalized, preventive, participatory (P4) cancer medicine. *Nature Reviews Clinical Oncology, 8*, 184−187.

Hood, L., & Perlmutter, R. M. (2004). The impact of systems approaches on biological problems in drug discovery. *Nature Biotechnology, 22*(10), 1215−1217.

Hripcsak G., Duke J.D., Shah N.H., Reich C.G., Huser V., Schuemie M.J., et al. Observational Health Data Sciences and Informatics (OHDSI): Opportunities for observational researchers. *Studies in Health Technology and Informatics*, 216:574-578.

Kulikowski, C. A., Shortliffe, E. H., Currie, L. M., Elkin, P. L., Hunter, L. E., Johnson, T. R., et al. (2012). AMIA Board white paper: Definition of biomedical informatics and specification of core competencies for graduate education in the discipline. *Journal of the American Medical Informatics Association, 19*, 931−938.

Najarian, K., & Splinter, R. (2012). *Biomedical signal and image processing*. CRC Press.

Ohno-Machado, L., Nadkarni, P., & Johnson, K. (2013). Natural language processing: Algorithms and tools to extract computable information from EHRs and from the biomedical literature. *Journal of the American Medical Informatics Association, 20*(5), 805.

Pattin, K. A., Greene, A. C., Altman, R. B., Cohen, K. B., Wethington, E., & GÖRg, C. (Eds.), (2015). Training the next generation of quantitative biologists in the era of big data. *Pacific Symposium on Biocomputing, 20*, 488−492.

Payne, P. R., Embi, P. J., & Sen, C. K. (2009). Translational informatics: Enabling high throughput research paradigms. *Physiological Genomics*.

Payne, P. R., Johnson, S. B., Starren, J. B., Tilson, H. H., & Dowdy, D. (2005). Breaking the translational barriers: The value of integrating biomedical informatics and translational research. *Journal of Investigative Medicine, 53*(4), 192−200.

Payne, P. R. O., & Embi, P. J. (2015a). *An introduction to translational informatics and the future of knowledge-driven healthcare. Translational informatics* (pp. 3−19). Springer.

Payne, P. R. O., & Embi, P. J. (2015b). *Driving clinical and translational research using biomedical informatics. Translational informatics* (pp. 99−117). Springer.

Regan, K., Abrams, Z., Sharpnack, M., Srivastava, A., Huang, K. U. N., Shah, N., et al. (Eds.), (2016). *Discovery of molecularly targeted therapies*.

Regan, K., & Payne, P. R. O. (2015). From molecules to patients: The clinical applications of translational bioinformatics. *Yearbook of Medical Informatics, 10*(1), 164.

Sarkar, I. N. (2013). *Methods in biomedical informatics: A pragmatic approach*. Academic Press.

Schadt, E. E., & Bjorkegren, J. L. (2012). Network-enabled wisdom in biology, edicine, and health care. *Science Translational Medicine, 4*(115), 115rv1.

Tsafnat, G., & Coiera, E. W. (2009). Computational reasoning across multiple models. *Journal of the American Medical Informatics Association, 16*(6), 768−774.

Turkay, C., Jeanquartier, F., Holzinger, A., & Hauser, H. (2014). *On computationally-enhanced visual analysis of heterogeneous data and its application in biomedical informatics. Interactive knowledge discovery and data mining in biomedical informatics* (pp. 117−140). Springer.

Chapter 3

The functional genome: epigenetics and epigenomics

Timothy E. Reddy

Center for Genomic and Computational Biology, Duke University, Durham, NC, United States

Introduction

The primary sequence of the human genome is the order in which the nucleotides adenine (A), cytosine (C), guanine (G), and thymine (T) occur along individual's chromosomes. The genome sequence of an individual is almost entirely the same in every cell. How a common genome sequence generates the panoply of distinct cell types and environmental response is therefore a major area of research. The answer, in part, relies on the addition of chemical modifications to the genome sequence and to the proteins that package the genome inside of cells. Collectively, those modifications are known as the epigenome.

While the genome is largely constant between different cells of a person's body, the epigenome can vary drastically between cell types and can change over time and in response to the environment. There is also evidence that changes in the epigenome can be transmitted from parent to offspring, creating the potential for heritable transmission without modifying the genome sequence itself. This chapter will describe the basic components of the epigenome; how the epigenome is established and maintained; the role of the epigenome in health and disease; and current and future directions in epigenetic and epigenomic research.[1]

The composition of the epigenome

The epigenome has two primary components: chemical modifications of the individual nucleotides that make up the genome and chemical

---

1. The terms "epigenetics" and "epigenomics" are often used interchangeably. In this chapter, we will use "epigenetics" to refer to the study of chemical modifications at individual loci in the genome, and "epigenomics" to refer to the study of how those modifications are established and altered across the entire human genome. The scale at which epigenetics becomes epigenomics is not well defined, hence the inconsistent and often interchangeable use of the terms.

Genomic and Precision Medicine. DOI: https://doi.org/10.1016/B978-0-12-800684-9.00017-4
29

modifications to the histone proteins that package the genome in cells. Collectively, those modifications are referred as "epigenetic marks." Nucleotide modifications in the human genome are primarily limited to the methylation and hydroxymethylation of cytosine, the latter of which was first demonstrated to occur in humans in 2009 (Kriaucionis & Heintz, 2009). In sharp contrast, the histone modifications present in human cell are far more diverse. While many epigenetic marks have been associated with changes in how the genome is used in a cell, it is important to note that direct causality between epigenetic marks and genome function is typically not well established. Overcoming technical hurdles to interrogate potential causal relationships remains an active area of investigation.

Chemically modified nucleotides

Various covalently modified nucleotides occur throughout the tree of life. In humans, those modifications are thought to be limited to the addition of a methyl group to the $5'$ carbon in cytosine (5-methylcytosine, 5mC) and the subsequent addition of a hydroxy group to that methyl group to form 5-hydroxymethylcytosine (5hmC). The process by which these modifications occur is summarized in Fig. 3.1.

5-Methylcytosine

Human cytosine methylation occurs almost exclusively when a cytosine is followed by a guanine in the genome. The pair is commonly referred to as a CpG dinucleotide, where the "p" represents the phosphodiester bond between adjacent nucleotides. 5mC was first discovered as an indicator of tightly compacted regions of the genome known as heterochromatin (Miller, Schnedl, Allen, & Erlanger, 1974). Genes in heterochromatin are generally not expressed, and 5mC was therefore considered an epigenetic mark that silences gene expression. More recent studies have expanded on those initial findings to reveal that the 5mC may have different effects in different regions of the genome. Outside of genes, 5mC is typically associated with the exclusion of regulatory proteins known as transcription factors that control the expression of nearby genes. That finding is consistent with the role of 5mC in silencing gene expression. Conversely, within genes, there is growing evidence that 5mC is positively correlated with expression levels (Ball et al., 2009). While it is not yet known if gene body methylation is a cause or an effect of increased gene expression, it is clear that the relationship between 5mC and gene expression is complex and dependent on genomic location.

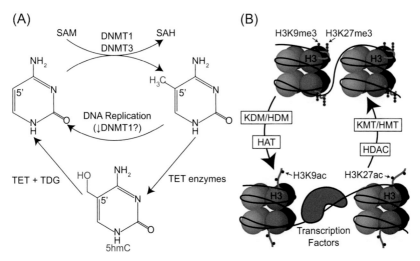

FIGURE 3.1 Overview of the enzymatic steps and enzymes involved in establishing and modifying the epigenome. Each *arrow* indicates an enzymatic step and is labeled with enzyme(s) known or thought to catalyze that step. (A) Schematic of the methylation and demethylation of cytosine in the human genome. (B) Schematic of the modification of histone methylation and acetylation. The *DNMT1* and *DNMT3* enzymes catalyze the transfer of a methyl group from SAM to the 5′ carbon of cysteine, resulting in 5mC and *S*-adenosyl-l-homocysteine (SAH). The mechanisms by which 5mC is converted back to cytosine are less clear. One mechanism involves rounds of DNA replication that, perhaps with reduced *DNMT1* levels, fail to reproduce the methylation state of the parent cell. A second possible mechanism may involve the conversion of 5mC to 5hmC and other modified forms of cysteine by the ten-eleven translocation methylcytosine dioxygenase 1 (*TET1*). 5hmC may then be directly converted to cysteine via subsequent unknown enzymatic steps. Alternatively, 5hmC may evade recognition by the pathway that recapitulates 5mC after mitosis, leading to a DNA-synthesis−dependent loss of 5mC.

The establishment and maintenance of 5mC

Cytosines are methylated by the DNA methyltransferase (DNMT) family of enzymes (Bestor, 2000). The DNMTs transfer a methyl group from *S*-adeno-syl-l-methionine (SAM) to the 5′ carbon of cytosine residues in the genome (Adams, McKay, Craig, & Burdon, 1979). There are two known subfamilies of DNMTs in the human genome with different properties. The DNMT3 sub-family consists of *DNMT3A* and *DNMT3B* and is responsible for methylating cytosine de novo (Okano, Bell, Haber, & Li, 1999). Once methylation is established on one strand of the double helix, the DNMT1 subfamily, made up solely of *DNMT1*, is thought to methylate the corresponding cytosine on the opposite strand (Glickman, Flynn, & Reich, 1997; Song, Rechkoblit, Bestor, & Patel, 2011). In that capacity, *DNMT1* is sometimes considered to be a postre-plication maintenance DNMT. *DNMT1* also has the ability to methylate DNA de novo in certain regions of the genome (Yoder, Soman, Verdine, & Bestor, 1997), suggesting that the division of labor between *DNMT1*, *DNMT3A*, and

DNMT3B is neither strict nor simple (Bestor, 2000). The formerly named DNMT2 enzyme was found to methylate transfer RNA (tRNA) and not DNA (Durdevic, Mobin, Hanna, Lyko, & Schaefer, 2013; Goll et al., 2006). For that reason, DNMT2 has since been renamed the tRNA aspartic acid methyltransferase 1 (TRDMT1). RNA methylation by *TRDMT1* may be related to a more extensive RNA-based epigenome that will not be discussed in this chapter.

The mechanism by which cytosine methylation is erased remains a highly active area of research. Two predominant mechanisms have emerged (Chen & Riggs, 2011; Franchini, Schmitz, & Petersen-Mahrt, 2012). The passive model of cytosine demethylation relies on a failure to methylate newly synthesized DNA during mitosis, perhaps involving the inhibition of *DNMT1*. If the cytosine methylation is systematically not established at a given genomic location after DNA replication, then the number of chromosomes with cytosine methylation at that location in the two daughter cells will be reduced by half. With subsequent rounds of mitosis and new DNA synthesis, the vast majority of daughter cells will eventually have nonmethylated cytosines at that location (Franchini et al., 2012). The appeal of the passive model in humans is that it does not depend on the presence of an active cytosine demethylation enzyme or pathway that has long eluded discovery. The requirement for mitosis, however, restricts passive demethylation to dividing cells and does not explain, for example, the rapid loss of methylation that occurs on the paternal genome after a sperm cell fertilizes an egg but before the first rounds of mitotic cell division (Mayer, Niveleau, Walter, Fundele, & Haaf, 2000; Morgan, Santos, Green, Dean, & Reik, 2005).

More recently, increasing evidence also supports a second, active mechanism of demethylation that does not depend on cell division. The base-excision-repair hypothesis states that methylated cytosines are excised from the genome and replaced by nonmodified cytosines. Meanwhile, the enzymatic demethylation hypothesis states that the methyl group on 5mC is directly removed from the cytosine by a demethylase enzyme. Demethylation by base-excision repair has been demonstrated in plants, but has not been shown to occur in the human genome (Zhu, 2009). Enzymatic removal of the methyl group from 5mC by a single enzyme is thought to be energetically unlikely. Instead, the possibility of a demethylation pathway that relies on successive modifications of the 5mC by several enzymes has recently gained attention. The ten-eleven translocation (TET) enzymes convert 5mC into 5hmC, which may be the first step in a chain of further modifications that ultimately result in unmodified cytosine (He et al., 2011; Ito et al., 2011; Zhu, 2009). Human enzymes that convert 5hmC or 5hmC derivatives to cytosine, however, have not yet been found.

Monoallelic cytosine methylation

Over the majority of the genome, DNA methylation occurs on both copies of the same chromosome or not at all. If a cytosine is methylated on the maternally inherited chromosome, then the homologous cytosine on the paternally inherited allele is also methylated. There are two major exceptions to that rule, both of which are essential for human health: X inactivation and imprinting.

The defining genetic difference between typical females and males is the presence of two X chromosomes in the female genome, and one X and one Y in the male genome. The number of functional copies of the X is held more or less equivalent between the sexes, however, through a process known as X inactivation (Barr & Bertram 1949; Lyon, 1961). In female cells, a combination of epigenetic mechanisms silences most (but not all) of the gene expression from one of the two X chromosomes. That process begins with the expression of the long noncoding RNA gene *XIST* from one of the two X chromosomes. *XIST* expression triggers extensive cytosine methylation, heterochromatin formation, and gene silencing on the same X chromosome (Brown et al., 1991; Clemson, McNeil, Willard, & Lawrence, 1996; Penny, Kay, Sheardown, Rastan, & Brockdorff, 1996). The initial choice of which X to inactivate is typically but not always random and is made independently by a small number of cells early in embryogenesis (Amos-Landgraf et al., 2006; Augui, Nora, & Heard, 2011; Lyon, 1961). After cells make the initial choice of which X to inactivate, that choice is transmitted to daughter cells, leading to mosaic patterns of which X is inactivated in the adult. A classic visual example is the patches of color in the coats of calico cats. Those patches are created by mosaic inactivation of coat color genes located on the feline X chromosome.

Monoallelic DNA methylation and gene silencing also occur via a process known as imprinting (Ferguson-Smith, 2011). Unlike for X inactivation, imprinting occurs in a parent-of-origin dependent manner. In some loci, the maternal chromosome is inactivated and the paternal chromosome is active; in other loci, the reverse is true. A classic example is the H19 locus which, as for X inactivation, requires the expression of a long noncoding RNA from the imprinted allele. In the H19 locus, the long noncoding RNA H19 is expressed from the maternally inherited chromosome, leading to silencing of gene expression and extensive cytosine methylation in the surrounding region on the same chromosome (Bartolomei, Zemel, & Tilghman, 1991; Davis, Yang, McCarrey, & Bartolomei, 2000; Forne et al., 1997). How exactly the H19 RNA directs imprinting is not fully understood and may involve mechanisms distinct from those that cause X inactivation (Ferguson-Smith, 2011). There are currently over 100 loci in the human genome with evidence for imprinting, and the rate at which imprinted loci were discovered rapidly increased with the ability to detect monoallelic gene expression using genome-wide approaches (Morison, Ramsay, & Spencer, 2005; Pollard et al., 2008). The rate of discovery has recently slowed, however, and there is growing evidence that nearly all imprinted genes in humans have been identified (DeVeale, van der Kooy, & Babak, 2012; Wang, Soloway, & Clark, 2011).

5-Hydroxymethylcytosine

The discovery of 5hmC in the human brain was the first evidence that a second nucleotide modification was present in the human epigenome. In contrast to 5mC that is found in every cell type, it is currently thought that 5hmC is predominantly limited to the brain and to early stages of development (Kriaucionis & Heintz, 2009; Tahiliani et al., 2009; Wen & Tang, 2014). Those findings are very recent, however, and there is much to be learned about the distribution of 5hmC in the rest of the human body. As described above, one potential biological function of 5hmC is as part of a demethylation pathway. Another is that binding of regulatory factors to 5hmC may influence gene regulation (Jin, Kadam, & Pfeifer, 2010). Both of these hypotheses are currently areas of active investigation.

Histone modifications

The human genome is wrapped around complexes of histone proteins in a structure known as the nucleosome. Covalent modifications to the histone proteins—the second part of the epigenome—have been associated with how tightly the nucleosomes are packed together and whether the proteins that control gene expression can bind the genome (Bannister & Kouzarides, 2011; Kouzarides, 2007). Regions in which the nucleosomes are tightly packed are known as heterochromatin. As discussed earlier, heterochromatin is also highly enriched for 5mC and depleted for expressed genes (Grewal & Moazed, 2003; Straub, 2003). Genomic regions that are not heterochromatic are called euchromatin. In the euchromatic regions of the genome, nucleosomes are less densely packed, allowing the binding of RNA polymerase and regulatory proteins to control gene expression. Differences between heterochromatin or euchromatin may be an important component of cell and tissue differentiation (Gaspar-Maia, Alajem, Meshorer, & Ramalho-Santos, 2011).

The histone code

The histones that form the core of the nucleosome are an octameric complex that consists of two copies each of histone H2A, H2B, H3, and H4. The histones are encoded in numerous clusters throughout the genome, and there are several slightly variant versions of each histone protein in the genome (Andrews & Luger, 2011). Within each histone protein, there are numerous sites where the amino acids can be modified by the covalent addition of small molecule groups. The most well-understood histone modifications involve the addition of between one and three methyl groups or an acetyl group to specific lysine residues on the N-terminal tail of histone H3. Those modifications will be discussed in some depth and are diagramed in Fig. 3.2. However, many other modifications have been discovered across all four core histone proteins.

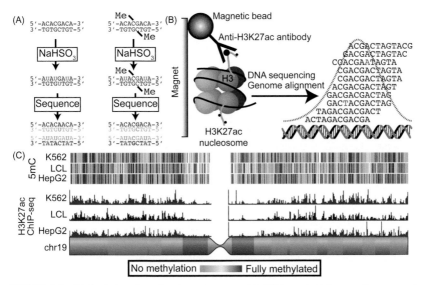

FIGURE 3.2 Methods for observing the epigenome. (A) DNA methylation can be measured with bisulfite sequencing, in which unmethylated cytosine is converted to uracil by sodium bisulfite ($NaHSO_3$), whereas methylated cytosine is protected from the reaction. Sequencing of the converted DNA can then reveal the $NaHSO_3$-induced mutations that mark the unmethylated cytosines. (B) Chromatin Immunoprecipitation (ChIP) can be used to isolate histones with specific epigenetic modifications such as H3K27ac. In ChIP, an antibody specific to the modification is conjugated to a magnetic bead, enabling immunoprecipitation of the histones with that modification. The DNA bound to that histone can then be observed with high-throughput DNA sequencing. Mapping the resulting sequences to the genome can then be used to determine the genomic location of the modified histone. That combination of ChIP with high-throughput sequencing is known as ChIP-seq. (C) Both bisulfite sequencing and ChIP-seq can be used to measure the epigenome genome-wide. As an example, the distribution of cytosine methylation (top) and H3K27ac (bottom) across human chromosome 19 as determined by the ENCODE project is shown. In heterochromatic regions of the genome, depicted at top, nucleosomes are tightly packed and typically marked with H3K9me3 and H3K27me3. During the transition to euchromatin, depicted at the bottom, a combination of histone demethylases and histone acetyltransferases act to replace the methyl modification with acetyl modifications. One hypothesis is that the acetyl groups neutralize the positive charge on histone tails, leading to less tightly bound histones and a chromatin state that is more permissive to the binding of transcription factors. In the transition from euchromatin to heterochromatin, the histone modifications are reversed by histone deacetylases and histone methyltransferases.

Those modifications include arginine methylation, serine phosphorylation, sumolyation, ubiquitination, etc. (Bannister & Kouzarides, 2011). The modifications are typically written using a shorthand notation that specifies (1) the histone, (2) the amino acid, and (3) the covalent modification. For example, the acetylation of histone H3 lysine 27 is expressed as H3K27ac. Similarly, H3K4me3 means that lysine 4 on histone H3 has three methyl groups.

A growing body of evidence suggests that different histone modifications each associate with different states of the genome, a hypothesis known as the histone code (Berger, 2002; Strahl & Allis, 2000). For example, heterochromatin is not only enriched for 5mC but also enriched for H3K27me3 and H3K9me3 (Hansen et al., 2008; Trojer & Reinberg, 2007). Meanwhile, in euchromatin, different sets of histone marks have been associated with different genomic activities. For example, the start sites of actively expressed genes are enriched in H3K4me3 and H3K9ac (Bernstein et al., 2005; Hon, Hawkins, & Ren, 2009; Liang et al., 2004). Different modifications of the same amino acid are sometimes associated with opposing activities. For example, the mutually exclusive modifications H3K27ac and H3K27me3 are enriched at active and poised distal regulatory elements, respectively (Creyghton et al., 2010; Rada-Iglesias et al., 2011). Finally, marks on different amino acids that signal contradictory activities may be indicative of regions for which the terminal state is yet to be determined. The genomic regions enriched for both the activation-associated mark H3K4me3 and the repression-associated mark H3K27me3 are a well-studied example. Such bivalent regions often identify developmentally important genes that, concurrent with loss of one of those marks, will later become actively expressed in one cell type and repressed in another (Azuara et al., 2006; Bernstein et al., 2006; Mikkelsen et al., 2007; Mohn et al., 2008; Rada-Iglesias et al., 2011).

Establishing and removing histone modifications

As in the case of DNA methylation, multiple enzymes contribute to establishing and modifying histone modifications (Bannister & Kouzarides, 2011). Those enzymes typically do not bind DNA by themselves, but instead are recruited to specific sites in the genome as part of larger regulatory complexes. The E1A-binding protein p300, for example, plays a major role in regulating genes from distal enhancers. The *EP300* gene includes a histone acetyltransferase domain that establishes H3K27ac when recruited to enhancers (Vo & Goodman, 2001). Meanwhile, the polycomb repressive complexes establish and maintain H3K27me3 via their histone methyltransferase activity (Schwartz & Pirrotta, 2013). It is possible that the histone modifications established by these enzymes are not necessary for altering genome activity and are instead simply a consequence of their recruitment to the genome for other purposes. However, growing evidence suggests that some histone modifications do indeed mechanistically contribute to the regulation of nearby gene expression (Mendenhall et al., 2013; Witzgall, O'Leary, Leaf, Onaldi, & Bonventre, 1994).

Epigenetic mechanisms of disease

The epigenome has a major role in gene regulation during differentiation and development (Agger et al., 2007; Cantone & Fisher, 2013; Reik, Dean, & Walter, 2001). It is therefore not surprising that numerous diseases also involve disruption of the enzymes that establish and maintain the epigenome (Brookes & Shi, 2014; Hendrich & Bickmore, 2001; Portela & Esteller, 2010). Examples of some of the most well-studied epigenetic diseases are described below.

Diseases involving DNA methylation

Global changes in DNA methylation are a hallmark of cancer, especially at advanced stages (Baylin & Herman, 2000; Baylin & Jones, 2011; Jones & Baylin, 2007; Laird & Jaenisch, 1996). The extent to which that methylation contributes causally to cancer remains to be fully understood. Growing evidence indicates that, in some instances, changes in cytosine methylation established prior to tumor development can serve as early markers of tumor formation (Anjum et al., 2014; Palmisano et al., 2000). Whether causative or not, the association between changes in DNA methylation and cancer has prognostic value in some cases (Brock et al., 2008; Wei et al., 2006) and has been used for early detection (Lange et al., 2012; Lasseigne et al., 2014; Oh et al., 2013). Chemotherapeutic drugs have also been developed that target DNA methylation. The most well known of those drugs are based on the cytosine analog 5-azacytidine that globally reduces genomic 5mC (Christman, 2002). While the primary mechanism by which 5-azacytidine blocks cancer cell replication is through inhibiting DNA synthesis, losses in cytosine methylation may contribute to additional beneficial effects (Daskalakis et al., 2002; Plumb, Strathdee, Sludden, Kaye, & Brown, 2000). The inhibition of DNA methylation by 5-azacytidine has also been harnessed by researchers to reactivate the expression of genes that have been silenced during differentiation or other processes (Chiu & Blau, 1985; Ginder, Whitters, & Pohlman, 1984).

Mutations in the enzymes that establish and maintain 5mC and in the proteins that bind 5mC have also been implicated in severe but rare diseases (Table 3.1). Mutations in each of the DNMT1 and DNMT3 genes have been connected to developmental neurological, sensory, and immune defects (Jin et al., 2008; Klein et al., 2011, 2013; Tatton-Brown et al., 2014). Meanwhile, one of the most well-studied epigenetic diseases, Rett syndrome, results not from mutations in the DNMTs but instead in a protein that binds 5mC in the genome. Children born with Rett syndrome have severely affected neurological development that manifests as impaired communication and motor skills with onset within the first year of life (Dunn & MacLeod, 2001). In nearly all cases, Rett syndrome results from the spontaneous mutation of the methyl-CpG binding protein *MECP2* (Amir et al., 1999). While much of the underlying mechanism has yet to be resolved, a leading hypothesis is that binding of *MECP2* to methylated regions of the genome contributes to the regulation of nearby genes in a way that is essential for neuronal development. MECP2 is located on the X chromosome. Likely due to prenatal

TABLE 3.1 Diseases caused by mutations in enzymes that alter 5mC and 5hmC.

Gene	Enzymatic activity	Disease associated with mutation	Reference (PubMed ID)
Genes involved or potentially involved in DNA methylation			
DNMT1	DNA methyltransferase	Cerebellar ataxia, deafness, and narcolepsy, autosomal dominant	22328086
		Neuropathy, hereditary sensory, type IE	21532572, 23365052
DNMT3A	DNA methyltransferase	Tatton−Brown−Rahman syndrome	24614070
DNMT3B	DNA methyltransferase	Immunodeficiency-centromeric instability-facial anomalies syndrome 1	10647011 and others
TET2	Methylcytosine dioxygenase	Myelodysplastic syndrome, somatic	21057493, 19474426, and others

lethality in males who only have one X, nearly all Rett syndrome occurs in females.

Defects in imprinting can also contribute to disease either through a loss of imprinting leading to aberrantly increased gene expression or through silencing of both alleles leading to a loss of gene expression. Beckwith−Wiedemann syndrome (BWS), a rare developmental disorder, is an example of imprinting loss leading to increased gene expression and developmental disorder (Weksberg, Shuman, & Beckwith, 2010; Wiedemann, 1969). Children with BWS have abnormally large birth size and a predisposition to cancer. The primary cause of BWS involves loss of imprinting in the maternally imprinted H19 locus described above (Brown et al., 1996; Reik et al., 1994). In some cases, loss of imprinting results from the inheritance of two active copies of H19 from the father. In other cases, genetic mutations in the region lead to a loss of imprinting (Weksberg et al., 2010). Regardless of the mechanism, with imprinting lost, genes in the region become expressed from both chromosomes as opposed to from a single paternal copy. The result is elevated expression of a growth factor and a cell-cycle gene in the region, which likely explains the physical manifestation of BWS (Lee et al., 1999; Ogawa et al., 1993; Weksberg, Shen, Fei, Song, & Squire, 1993).

Examples of imprinting-related loss of gene expression leading to developmental abnormalities include Prader−Willi syndrome (PWS) and

Angelman syndrome (AS) (Horsthemke & Wagstaff, 2008). In both cases, the underlying cause involves a combination of genetic mutation on one copy of a gene and imprinting of the other copy. While both PWS and AS are caused by mutations in the same locus on chromosome 15, the genes involved and the phenotypes are distinct. In PWS, loss-of-function mutations on the paternal chromosome combined with imprinting on the maternal chromosome results in no functional copies of several genes in the child. Conversely, in AS, genetic loss-of-function mutations occur on the maternally inherited copy of UBE3A while the paternal copy is epigenetically silenced. Together, the two examples show that, because of a lack of redundancy in the genes expressed from imprinted loci, mutations in the single expressed copy of imprinted genes can have severe effects on phenotype.

Diseases involving histone modifications

There are dozens of histone-modifying enzymes known to exist in the human genome, but relatively few human phenotypes have been linked to mutations in those enzymes (Table 3.2). Speculatively, that could be due to either redundancy in the histone-modifying proteins or, conversely, due to the severity of consequences associated with those mutations. However, as genome sequencing has become more common, an increasing number of rare congenital disorders have been mapped to specific histone-modifying enzymes. Three examples are provided here. Weaver syndrome, characterized by a combination of skeletal and cognitive abnormalities, is an example in which the causal mutation is in an enzyme *EZH2* that is responsible for the methylation of H3K27 (Gibson et al., 2012). Mutations that abrogate the function or expression of the histone deacetylase *HDAC4* have been linked to brachydactyly-mental retardation syndrome, a syndrome that includes developmental delays and skeletal and craniofacial abnormalities (Williams et al., 2010). In cases of X-linked Claes−Jensen type mental retardation, mutations in the lysine demethylase KDM5C are thought to be causative (Kerr, Gedeon, Mulley, & Turner, 1992). Together, these results show that defects in histone-modifying enzymes can and do cause disease, but those diseases appear to have extremely low prevalence in the human population.

Several other disorders are associated not with the histone-modifying enzymes directly, but instead with mutations in other genes that lead to altered histone modifications. A classic example is the neurodegenerative and autosomal dominant Huntington's disease (HD) (Walker, 2007). The genetic cause of HD is the expansion of trinucleotide CAG repeats in the DNA sequence of the huntingtin (*HTT*) gene. Those repeats lead to polyglutamines in the HTT protein, ultimately leading to progressive neurodegenerative consequences. Patients with HD are now well documented to have epigenetic abnormalities, including a loss of histone acetylation (McFarland et al., 2012; Sadri-Vakili & Cha, 2006). The mechanism by which mutant

TABLE 3.2 Diseases caused by mutations in enzymes that alter histone acetylation or methylation.

Gene	Enzymatic activity	Disease associated with mutation	Reference (PubMed ID)
Genes involved in histone acetylation			
CBP	Histone acetyltransferase	Rubinstein–Taybi syndrome	7630403
P300	Histone acetyltransferase	Rubinstein–Taybi syndrome 2	17299436, 7630403, 19353645
		Colorectal cancer, somatic	10700188, 21390126, and others
KAT6B	Histone acetyltransferase	Genitopatellar syndrome	22077973, 22265017
		SBBYSS syndrome	23436491
HDAC4	Histone deacetylase	Brachydactyly-mental retardation syndrome	
HDAC6	Histone deacetylase	Chondrodysplasia	20181727
HDAC8	Histone deacetylase	Cornelia de Lange syndrome 5	22885700
		Wilson–Turner syndrome	22889856
Genes involved in histone methylation			
EZH2	Histone methyltransferase	Weaver syndrome	22177091
KMT2A	Histone methyltransferase	Wiedemann–Steiner syndrome	22795537
KMT2D	Histone methyltransferase	Kabuki syndrome 1	20711175
EHMT1	Histone methyltransferase	Kleefstra syndrome	15805155, 19264732, 16826528
KDM5C	Histone demethylase	Mental retardation, X-linked, syndromic, Claes–Jensen type	15586325, 21575681
KDM6A	Histone demethylase	Kabuki syndrome 2	22197486

HTT contributes to that hypoacetylation is still being determined. Current evidence supports a model in which mutant HTT alters genomic recruitment of histone-modifying genes (Nucifora et al., 2001). That mechanism may generalize to other neurodegenerative diseases (Gusella & MacDonald, 2000; Nucifora et al., 2001).

Small molecules that inhibit histone deacetylases, known as HDAC inhibitors, have been developed under the hypothesis that correcting the epigenome will improve outcomes in so-affected patients. In early successes, HDAC inhibitors have been approved as a second-line treatment for certain types of lymphomas, and several additional cancer trials are currently underway (Minucci & Pelicci, 2006; West & Johnstone, 2014). There is also evidence that HDAC inhibitors improve HD-associated symptoms in animal models, but human trials have not been performed (Ferrante et al., 2003; Hockly et al., 2003; Jia, Morris, Williams, Loring, & Thomas, 2015; Steffan et al., 2001). Basic research studies have also used HDAC inhibitors to investigating the cause and effect relationships between the epigenome and gene regulation (Glaser et al., 2003; Lopez-Atalaya, Ito, Valor, Benito, & Barco, 2013). Those studies have met challenges due to the broad genome-wide effects of those compounds combined with potential effects resulting from inhibition of the deacetylation of nonhistone proteins. It is likely that more targeted approaches currently under development will overcome some of those challenges, leading to a more nuanced understanding of the cause and effect of epigenetic histone modifications.

Epigenetic responses to the environment

Even though the epigenome is largely established during cell differentiation early in development, interactions between the environment and the epigenome are important at all stages in life (Jirtle & Skinner, 2007). The epigenome both contributes to determining how cells in the body respond to the environment and, vice-versa, the environment can lead to changes in the epigenome. Environmental effects on the epigenome can accumulate over a lifetime and potentially contribute to changes in disease risk later in life. To provide deeper insight into the interactions between the epigenome and the environment, several examples will be discussed in detail.

Epigenetic effects on environmental responses

It is well established that changes in histone modifications are associated with changes in gene regulation. While establishing cause and effect relationships remains challenging, recent studies of steroid hormone responses show that preestablished epigenetic marks contribute to determining the gene expression responses to those hormones. One of the classic mechanisms by which some steroid hormones act involves binding to receptors inside of cells and causing those receptors to bind the genome and regulate gene expression (Evans, 1988; Ribeiro, Kushner, & Baxter, 1995; Tsai & O'Malley, 1994). The specific places that those receptors bind and the genes that they regulate differ between cell types. Several hypotheses about the determinants of such tissue-specific responses have been posited (Gross & Cidlowski, 2008). Among the possible explanations, recent studies have shown that epigenetic states associated with gene activation are strongly predictive of where in the genome hormone receptors bind in different types of cells (Burd & Archer, 2013; Gertz et al., 2013; He et al., 2012; John et al., 2008, 2011; Magnani & Lupien, 2014). Meanwhile, in the relatively small number of genomic regions where hormone receptors bind despite an absence of those epigenetic marks, histone acetylation is typically gained after hormone treatment (Chen, Lin, Xie, Wilpitz, & Evans, 1999; Gadaleta & Magnani, 2014; Ito, Barnes, & Adcock, 2000; Sharma & Fondell, 2002). Together, these results demonstrate the likely possibility that the epigenome both controls and is controlled by the binding of regulatory proteins that govern responses to the environment.

Epigenetic changes over a lifetime

Studies of human populations and mouse models clearly show that maternal environment during gestation can have epigenetic impacts on the developing fetus. One such example is the Dutch famine during the winter of 1944–45, also known as the Dutch hunger winter. German blockades of Dutch ports combined with frozen canals due to extremely low temperatures led to a severe shortage of food in German-held areas of the Netherlands. Millions of Dutch were exposed to severe famine lasting from November 1944 until liberation in May 1944.

Studies of the effects that the famine on survivors revealed striking health consequences on babies who were in utero at the time. For example, those babies had substantially elevated risk for metabolic and cardiovascular disease later in life (Roseboom, de Rooij, & Painter, 2006; Schulz, 2010). Because of the famine-affected people of all ages and social classes, attribution of the effects can be made to the famine independent of socioeconomic explanations. The Dutch hunger winter studies therefore provide some of the strongest evidence that the maternal environment during pregnancy can have substantial and long-lasting effects on the health of the offspring. Several studies have expanded upon those initial findings, and it is now known that a wide variety of exposures both during gestation and immediately after birth can have life-long impacts on the child's long-term health (Dolinoy, Weidman, & Jirtle, 2007; Feil & Fraga, 2011; McGowan et al., 2009).

A key next challenge, then, is to determine the mechanisms by which fetal environment impacts life-long health. Strong evidence supports the hypothesis that the epigenome is a major contributor to those effects. Follow-up studies of the Dutch hunger winter found changes in DNA methylation at both imprinted and nonimprinted loci 60 years after the famine (Heijmans et al., 2008; Tobi et al., 2009). A similar effect has been demonstrated in mice, in which maternal grooming during the first weeks after birth contributes to epigenetic reprogramming of the stress response pathways and worse health later in life. The effects of maternal grooming can be partially reversed by histone deacetylase inhibitors, a finding that strengthens the argument that the fetal environment acting on the epigenome contributes to the long-term effects (Weaver et al., 2004).

Exposure to environmental agents later in life can also lead to epigenetic changes that may contribute to health (Dolinoy, Huang, & Jirtle, 2007; Jaenisch & Bird 2003; Reamon-Buettner, Mutschler, & Borlak, 2008; Toledo-Rodriguez et al., 2010). Because the epigenome is responsive to a wide variety of environmental exposures, it is unsurprising that studies have found substantial age-related epigenetic changes (Fraga, Agrelo, & Esteller, 2007; Fraga & Esteller, 2007; Horvath, 2013). Taken together, it is now clear that epigenetic reprogramming occurs in early development, continues throughout life, and is likely involved in a wide variety of traits and diseases.

Heritability of the epigenome

The epigenome can be transmitted both via genetic and nongenetic mechanisms. Genetic transmission can trivially involve a mutation that disrupts a CpG dinucleotide, thus preventing methylation of that cytosine. Additional studies have shown that non-CpG mutations can also alter methylation in a region (Gertz et al., 2011). In those cases, the change in DNA sequence may alter the recruitment of regulatory factors that, in turn, may contribute to altered epigenetic state. Finally, it is important to note that cytosine methylation increases the rate of cytosine to thymidine mutations and, via that mechanism, the epigenome can influence the genome sequence over long timescales (Holliday & Grigg, 1993).

Examples of epigenetic inheritance that do *not* have a genetic basis are more elusive. One clear example is imprinting, where the epigenetic state is transmitted to child in a parent-dependent manner (Morgan, Sutherland, Martin, & Whitelaw, 1999). There are also growing examples in which environmental exposures in one generation lead to changes in disease risk in subsequent generations (Skinner, 2011). One example already discussed is the Dutch hunger winter study, in which the second generation was also shown to have increased adverse health outcomes (Painter et al., 2008). Several studies have elaborated on that finding to show that epigenetic state established via other environmental signals can be passed on to offspring both via the maternal and the paternal lineage (Anway, Cupp, Uzumcu, & Skinner, 2005; Dolinoy, Das, Weidman, & Jirtle, 2007; Morgan et al., 1999; Pembrey et al., 2006; Waterland et al., 2006). An intuitive potential mechanism is that incomplete erasure of DNA methylation during early development allows some methylation to be directly passed to the offspring (Li, 2002; Morgan et al., 2005). It is also possible that parental DNA methylation state is transmitted via more complex pathways involving protein intermediates. With rapidly increasing ability to measure genetic epigenetic changes genome-wide with few cells, the required studies to dissect those mechanisms may soon be possible.

Genome-wide observation of the epigenome

The ability to measure the epigenome has been revolutionized over the past decade. Whereas it has long been possible to measure the epigenome at specific targeted locations in the human genome, technological advances in DNA sequencing (Mardis, 2008; Shendure & Ji, 2008) have made it possible to measure the epigenome at every location in the genome in a single experiment (Lister et al., 2009; Schones & Zhao, 2008). Examples of two of the most commonly used strategies and the data that they produce are shown in Fig. 3.2. That increase in measurement ability has been a major driver for new understanding into the basic principles governing the epigenome. The result is that we now have highly detailed maps of the epigenetic differences between individuals (Banovich et al., 2014; Bell et al., 2011; Fraser, Lam, Neumann, & Kobor, 2012; Kundaje et al., 2015); between different cells and tissues in the human body (Heintzman et al., 2009; Leung et al., 2015); of the epigenetic responses to the environment (Feil & Fraga, 2011); and of epigenetic alterations that may be contributing to a variety of diseases (Michels et al., 2013; Rakyan, Down, Balding, & Beck, 2011; Robertson, 2005). The ability to harness such genome-wide approaches to study the epigenome is quickly becoming an essential skill for epigenetic and epigenomic research.

Integrative epigenomics and other future areas of study

The ability to comprehensively measure the state of the epigenome has led to a new era of discovery. Major efforts such as the Roadmap Epigenome Project (Kundaje et al., 2015) and the ENCODE Project (Consortium, 2012) have coordinated multinational teams of researchers to map a wide diversity of epigenetic marks across multiple cell types. As a result of those and related efforts, vast public repositories of epigenetic information are now freely available. Researchers have begun to rely on those data in various ways such as predicting how specific genes are regulated and investigating how gene regulation contributes to disease (Farh et al., 2015; Karlic, Chung, Lasserre, Vlahovicek, & Vingron, 2010). Initial studies to integrate across different epigenetic assays have confirmed that certain combinations of epi-genetic marks are often acquired at the same genomic locations, and that many of those marks are associated with the regulation of nearby genes (Zhou, Goren, & Bernstein, 2011). Those projects have revealed numerous new associations between the genome, the epigenome, gene regulation, and human health. Ongoing studies are actively investigating possible mechanisms explaining those newly identified associations. Meanwhile, as sequencing of whole human genomes is becoming routine, an increasing number of rare diseases may be mapped to mutations in epigenome-modifying enzymes that are currently not implicated in disease. Combined with expanding application of high-throughput epigenomic assays to various health and basic research questions, there is likely to soon be major advances in understanding how the epigenome contributes to health and diseases.

References

Adams, R. L., McKay, E. L., Craig, L. M., & Burdon, R. H. (1979). Mouse DNA methylase: Methylation of native DNA. *Biochimica et Biophysica Acta, 561*(2), 345−357.

Agger, K., Cloos, P. A., Christensen, J., Pasini, D., Rose, S., & Rappsilber, J. (2007). UTX and JMJD3 are histone H3K27 demethylases involved in HOX gene regulation and development. *Nature, 449*(7163), 731−734.

Amir, R. E., Van den Veyver, I. B., Wan, M., Tran, C. Q., Francke, U., & Zoghbi, H. Y. (1999). Rett syndrome is caused by mutations in X-linked MECP2, encoding methyl-CpG-binding protein 2. *Nature Genetics, 23*(2), 185−188.

Amos-Landgraf, J. M., Cottle, A., Plenge, R. M., Friez, M., Schwartz, C. E., & Longshore, J. (2006). X chromosome-inactivation patterns of 1,005 phenotypically unaffected females. *American Journal of Human Genetics, 79*(3), 493−499.

Andrews, A. J., & Luger, K. (2011). Nucleosome structure(s) and stability: Variations on a theme. *Annual Review Biophysics, 40*, 99−117.

Anjum, S., Fourkala, E. O., Zikan, M., Wong, A., Gentry-Maharaj, A., & Jones, A. (2014). A BRCA1-mutation associated DNA methylation signature in blood cells predicts sporadic breast cancer incidence and survival. *Genome Medicine, 6*(6), 47.

Anway, M. D., Cupp, A. S., Uzumcu, M., & Skinner, M. K. (2005). Epigenetic transgenerational actions of endocrine disruptors and male fertility. *Science (New York, N.Y.), 308*(5727), 1466−1469.

Augui, S., Nora, E. P., & Heard, E. (2011). Regulation of X-chromosome inactivation by the X-inactivation centre. *Nature Reviews Genetics, 12*(6), 429−442.

Azuara, V., Perry, P., Sauer, S., Spivakov, M., Jorgensen, H. F., & John, R. M. (2006). Chromatin signatures of pluripotent cell lines. *Nature Cell Biology, 8*(5), 532−538.

Ball, M. P., Li, J. B., Gao, Y., Lee, J. H., LeProust, E. M., & Park, I. H. (2009). Targeted and genome-scale strategies reveal gene-body methylation signatures in human cells. *Nature Biotechnology, 27*(4), 361−368.

Bannister, A. J., & Kouzarides, T. (2011). Regulation of chromatin by histone modifications. *Cell Research, 21*(3), 381−395.

Banovich, N. E., Lan, X., McVicker, G., van de Geijn, B., Degner, J. F., & Blischak, J. D. (2014). Methylation QTLs are associated with coordinated changes in transcription factor binding, histone modifications, and gene expression levels. *PLoS Genetics, 10*(9), e1004663.

Barr, M. L., & Bertram, E. G. (1949). A morphological distinction between neurones of the male and female, and the behaviour of the nucleolar satellite during accelerated nucleoprotein synthesis. *Nature, 163*(4148), 676.

Bartolomei, M. S., Zemel, S., & Tilghman, S. M. (1991). Parental imprinting of the mouse H19 gene. *Nature, 351*(6322), 153–155.

Baylin, S. B., & Herman, J. G. (2000). DNA hypermethylation in tumorigenesis: epigenetics joins genetics. *Trends in Genetics, 16*(4), 168–174.

Baylin, S. B., & Jones, P. A. (2011). A decade of exploring the cancer epigenome-biological and translational implications. *Nature Reviews Cancer, 11*(10), 726–734.

Bell, J. T., Pai, A. A., Pickrell, J. K., Gaffney, D. J., Pique-Regi, R., & Degner, J. F. (2011). DNA methylation patterns associate with genetic and gene expression variation in HapMap cell lines. *Genome Biology, 12*(1), R10.

Berger, S. L. (2002). Histone modifications in transcriptional regulation. *Current Opinion in Genetics & Development, 12*(2), 142–148.

Bernstein, B. E., Kamal, M., Lindblad-Toh, K., Bekiranov, S., Bailey, D. K., & Huebert, D. J. (2005). Genomic maps and comparative analysis of histone modifications in human and mouse. *Cell, 120*(2), 169–181.

Bernstein, B. E., Mikkelsen, T. S., Xie, X., Kamal, M., Huebert, D. J., & Cuff, J. (2006). A bivalent chromatin structure marks key developmental genes in embryonic stem cells. *Cell, 125*(2), 315–326.

Bestor, T. H. (2000). The DNA methyltransferases of mammals. *Human Molecular Genetics, 9*(16), 2395–2402.

Brock, M. V., Hooker, C. M., Ota-Machida, E., Han, Y., Guo, M., & Ames, S. (2008). DNA methylation markers and early recurrence in stage I lung cancer. *The New England Journal of Medicine, 358*(11), 1118–1128.

Brookes, E., & Shi, Y. (2014). Diverse epigenetic mechanisms of human disease. *Annual Review of Genetics, 48*, 237–268.

Brown, C. J., Ballabio, A., Rupert, J. L., Lafreniere, R. G., Grompe, M., & Tonlorenzi, R. (1991). A gene from the region of the human X inactivation centre is expressed exclusively from the inactive X chromosome. *Nature, 349*(6304), 38–44.

Brown, K. W., Villar, A. J., Bickmore, W., Clayton-Smith, J., Catchpoole, D., & Maher, E. R. (1996). Imprinting mutation in the Beckwith–Wiedemann syndrome leads to biallelic IGF2 expression through an H19-independent pathway. *Human Molecular Genetics, 5*(12), 2027–2032.

Burd, C. J., & Archer, T. K. (2013). Chromatin architecture defines the glucocorticoid response. *Molecular and Cellular Endocrinology, 380*(1–2), 25–31.

Cantone, I., & Fisher, A. G. (2013). Epigenetic programming and reprogramming during development. *Nature Structural & Molecular Biology, 20*(3), 282–289.

Chen, H., Lin, R. J., Xie, W., Wilpitz, D., & Evans, R. M. (1999). Regulation of hormone-induced histone hyperacetylation and gene activation via acetylation of an acetylase. *Cell, 98*(5), 675–686.

Chen, Z. X., & Riggs, A. D. (2011). DNA methylation and demethylation in mammals. *The Journal of Biological Chemistry, 286*(21), 18347–18353.

Chiu, C. P., & Blau, H. M. (1985). 5-Azacytidine permits gene activation in a previously noninducible cell type. *Cell, 40*(2), 417–424.

Christman, J. K. (2002). 5-Azacytidine and 5-aza-2'-deoxycytidine as inhibitors of DNA methylation: Mechanistic studies and their implications for cancer therapy. *Oncogene, 21*(35), 5483–5495.

Clemson, C. M., McNeil, J. A., Willard, H. F., & Lawrence, J. B. (1996). XIST RNA paints the inactive X chromosome at interphase: Evidence for a novel RNA involved in nuclear/chromosome structure. *The Journal of Cell Biology, 132*(3), 259–275.

Consortium, T. E. P. (2012). An integrated encyclopedia of DNA elements in the human genome. *Nature, 489*(7414), 57–74.

Creyghton, M. P., Cheng, A. W., Welstead, G. G., Kooistra, T., Carey, B. W., & Steine, E. J. (2010). Histone H3K27ac separates active from poised enhancers and predicts developmental state. *Proceedings of the National Academy of Sciences of the United States of America, 107*(50), 21931–21936.

Daskalakis, M., Nguyen, T. T., Nguyen, C., Guldberg, P., Kohler, G., & Wijermans, P. (2002). Demethylation of a hypermethylated P15/INK4B gene in patients with myelodysplastic syndrome by 5-Aza-2'-deoxycytidine (decitabine) treatment. *Blood, 100*(8), 2957–2964.

Davis, T. L., Yang, G. J., McCarrey, J. R., & Bartolomei, M. S. (2000). The H19 methylation imprint is erased and re-established differentially on the parental alleles during male germ cell development. *Human Molecular Genetics, 9*(19), 2885–2894.

DeVeale, B., van der Kooy, D., & Babak, T. (2012). Critical evaluation of imprinted gene expression by RNA-Seq: A new perspective. *PLoS Genetics, 8*(3), e1002600.

Dolinoy, D. C., Das, R., Weidman, J. R., & Jirtle, R. L. (2007). Metastable epialleles, imprinting, and the fetal origins of adult diseases. *Pediatric Research, 61*(5), 30r–37r, Pt 2.

Dolinoy, D. C., Huang, D., & Jirtle, R. L. (2007). Maternal nutrient supplementation counteracts bisphenol A-induced DNA hypomethylation in early development. *Proceedings of the National Academy of Sciences of the United States of America, 104*(32), 13056–13061.

Dolinoy, D. C., Weidman, J. R., & Jirtle, R. L. (2007). Epigenetic gene regulation: Linking early developmental environment to adult disease. *Reproductive Toxicology (Elmsford, N.Y.), 23*(3), 297–307.

Dunn, H. G., & MacLeod, P. M. (2001). Rett syndrome: Review of biological abnormalities. *The Canadian Journal of Neurological Sciences. Le Journal Canadien des Sciences Neurologiques, 28*(1), 16–29.

Durdevic, Z., Mobin, M. B., Hanna, K., Lyko, F., & Schaefer, M. (2013). The RNA methyltransferase Dnmt2 is required for efficient Dicer-2-dependent siRNA pathway activity in *Drosophila. Cell Reports, 4*(5), 931–937.

Evans, R. M. (1988). The steroid and thyroid hormone receptor superfamily. *Science (New York, N.Y.), 240*(4854), 889–895.

Farh, K. K., Marson, A., Zhu, J., Kleinewietfeld, M., Housley, W. J., & Beik, S. (2015). Genetic and epigenetic fine mapping of causal autoimmune disease variants. *Nature, 518*(7539), 337–343.

Feil, R., & Fraga, M. F. (2011). Epigenetics and the environment: Emerging patterns and implications. *Nature Reviews Genetics, 13*(2), 97–109.

Ferguson-Smith, A. C. (2011). Genomic imprinting: The emergence of an epigenetic paradigm. *Nature Reviews Genetics, 12*(8), 565–575.

Ferrante, R. J., Kubilus, J. K., Lee, J., Ryu, H., Beesen, A., & Zucker, B. (2003). Histone deacetylase inhibition by sodium butyrate chemotherapy ameliorates the neurodegenerative phenotype in Huntington's disease mice. *The Journal of Neuroscience, 23*(28), 9418–9427.

Forne, T., Oswald, J., Dean, W., Saam, J. R., Bailleul, B., & Dandolo, L. (1997). Loss of the maternal H19 gene induces changes in Igf2 methylation in both cis and trans. *Proceedings of the National Academy of Sciences of the United States of America, 94*(19), 10243–10248.

Fraga, M. F., Agrelo, R., & Esteller, M. (2007). Cross-talk between aging and cancer: The epigenetic language. *Annals of the New York Academy of Sciences, 1100,* 60−74.

Fraga, M. F., & Esteller, M. (2007). Epigenetics and aging: The targets and the marks. *Trends in Genetics, 23*(8), 413−418.

Franchini, D. M., Schmitz, K. M., & Petersen-Mahrt, S. K. (2012). 5-Methylcytosine DNA demethylation: More than losing a methyl group. *Annual Review of Genetics, 46,* 419−441.

Fraser, H. B., Lam, L. L., Neumann, S. M., & Kobor, M. S. (2012). Population-specificity of human DNA methylation. *Genome Biology, 13*(2), R8.

Gadaleta, R. M., & Magnani, L. (2014). Nuclear receptors and chromatin: An inducible couple. *Journal of Molecular Endocrinology, 52*(2), R137−R149.

Gaspar-Maia, A., Alajem, A., Meshorer, E., & Ramalho-Santos, M. (2011). Open chromatin in pluripotency and reprogramming. *Nature Reviews Molecular Cell Biology, 12*(1), 36−47.

Gertz, J., Savic, D., Varley, K. E., Partridge, E. C., Safi, A., & Jain, P. (2013). Distinct properties of cell-type-specific and shared transcription factor binding sites. *Molecular Cell, 52*(1), 25−36.

Gertz, J., Varley, K. E., Reddy, T. E., Bowling, K. M., Pauli, F., & Parker, S. L. (2011). Analysis of DNA methylation in a three-generation family reveals widespread genetic influence on epigenetic regulation. *PLoS Genetics, 7*(8), e1002228.

Gibson, W. T., Hood, R. L., Zhan, S. H., Bulman, D. E., Fejes, A. P., & Moore, R. (2012). Mutations in EZH2 cause Weaver syndrome. *American Journal of Human Genetics, 90*(1), 110−118.

Ginder, G. D., Whitters, M. J., & Pohlman, J. K. (1984). Activation of a chicken embryonic globin gene in adult erythroid cells by 5-azacytidine and sodium butyrate. *Proceedings of the National Academy of Sciences of the United States of America, 81*(13), 3954−3958.

Glaser, K. B., Staver, M. J., Waring, J. F., Stender, J., Ulrich, R. G., & Davidsen, S. K. (2003). Gene expression profiling of multiple histone deacetylase (HDAC) inhibitors: Defining a common gene set produced by HDAC inhibition in T24 and MDA carcinoma cell lines. *Molecular Cancer Therapeutics, 2*(2), 151−163.

Glickman, J. F., Flynn, J., & Reich, N. O. (1997). Purification and characterization of recombinant baculovirus-expressed mouse DNA methyltransferase. *Biochemical and Biophysical Research Communications, 230*(2), 280−284.

Goll, M. G., Kirpekar, F., Maggert, K. A., Yoder, J. A., Hsieh, C. L., & Zhang, X. (2006). Methylation of tRNAAsp by the DNA methyltransferase homolog Dnmt2. *Science (New York, N.Y.), 311*(5759), 395−398.

Grewal, S. I., & Moazed, D. (2003). Heterochromatin and epigenetic control of gene expression. *Science (New York, N.Y.), 301*(5634), 798−802.

Gross, K. L., & Cidlowski, J. A. (2008). Tissue-specific glucocorticoid action: A family affair. *Trends in Endocrinology and Metabolism, 19*(9), 331−339.

Gusella, J. F., & MacDonald, M. E. (2000). Molecular genetics: Unmasking polyglutamine triggers in neurodegenerative disease. *Nature Reviews Neuroscience, 1*(2), 109−115.

Hansen, K. H., Bracken, A. P., Pasini, D., Dietrich, N., Gehani, S. S., & Monrad, A. (2008). A model for transmission of the H3K27me3 epigenetic mark. *Nature Cell Biology, 10*(11), 1291−1300.

He, H. H., Meyer, C. A., Chen, M. W., Jordan, V. C., Brown, M., & Liu, X. S. (2012). Differential DNase I hypersensitivity reveals factor-dependent chromatin dynamics. *Genome Research, 22*(6), 1015−1025.

He, Y. F., Li, B. Z., Li, Z., Liu, P., Wang, Y., & Tang, Q. (2011). Tet-mediated formation of 5-carboxylcytosine and its excision by TDG in mammalian DNA. *Science (New York, N.Y.), 333*(6047), 1303−1307.

Heijmans, B. T., Tobi, E. W., Stein, A. D., Putter, H., Blauw, G. J., & Susser, E. S. (2008). Persistent epigenetic differences associated with prenatal exposure to famine in humans. *Proceedings of the National Academy of Sciences of the United States of America, 105*(44), 17046−17049.

Heintzman, N. D., Hon, G. C., Hawkins, R. D., Kheradpour, P., Stark, A., & Harp, L. F. (2009). Histone modifications at human enhancers reflect global cell-type-specific gene expression. *Nature, 459*(7243), 108−112.

Hendrich, B., & Bickmore, W. (2001). Human diseases with underlying defects in chromatin structure and modification. *Human Molecular Genetics, 10*(20), 2233−2242.

Hockly, E., Richon, V. M., Woodman, B., Smith, D. L., Zhou, X., & Rosa, E. (2003). Suberoylanilide hydroxamic acid, a histone deacetylase inhibitor, ameliorates motor deficits in a mouse model of Huntington's disease. *Proceedings of the National Academy of Sciences of the United States of America, 100*(4), 2041−2046.

Holliday, R., & Grigg, G. W. (1993). DNA methylation and mutation. *Mutation Research, 285* (1), 61−67.

Hon, G. C., Hawkins, R. D., & Ren, B. (2009). Predictive chromatin signatures in the mammalian genome. *Human Molecular Genetics, 18*(R2), R195−R201.

Horsthemke, B., & Wagstaff, J. (2008). Mechanisms of imprinting of the Prader−Willi/Angelman region. *American Journal of Medical Genetics Part A, 146a*(16), 2041−2052.

Horvath, S. (2013). DNA methylation age of human tissues and cell types. *Genome Biology, 14* (10), R115.

Ito, K., Barnes, P. J., & Adcock, I. M. (2000). Glucocorticoid receptor recruitment of histone deacetylase 2 inhibits interleukin-1beta-induced histone H4 acetylation on lysines 8 and 12. *Molecular and Cellular Biology, 20*(18), 6891−6903.

Ito, S., Shen, L., Dai, Q., Wu, S. C., Collins, L. B., & Swenberg, J. A. (2011). Tet proteins can convert 5-methylcytosine to 5-formylcytosine and 5-carboxylcytosine. *Science (New York, N.Y.), 333*(6047), 1300−1303.

Jaenisch, R., & Bird, A. (2003). Epigenetic regulation of gene expression: How the genome integrates intrinsic and environmental signals. *Nature Genetics, 33*, 245−254, Suppl.

Jia, H., Morris, C. D., Williams, R. M., Loring, J. F., & Thomas, E. A. (2015). HDAC inhibition imparts beneficial transgenerational effects in Huntington's disease mice via altered DNA and histone methylation. *Proceedings of the National Academy of Sciences of the United States of America, 112*(1), E56−E64.

Jin, B., Tao, Q., Peng, J., Soo, H. M., Wu, W., & Ying, J. (2008). DNA methyltransferase 3B (DNMT3B) mutations in ICF syndrome lead to altered epigenetic modifications and aberrant expression of genes regulating development, neurogenesis and immune function. *Human Molecular Genetics, 17*(5), 690−709.

Jin, S. G., Kadam, S., & Pfeifer, G. P. (2010). Examination of the specificity of DNA methylation profiling techniques towards 5-methylcytosine and 5-hydroxymethylcytosine. *Nucleic Acids Research, 38*(11), e125.

Jirtle, R. L., & Skinner, M. K. (2007). Environmental epigenomics and disease susceptibility. *Nature Review Genetics, 8*(4), 253−262.

John, S., Sabo, P. J., Johnson, T. A., Sung, M. H., Biddie, S. C., & Lightman, S. L. (2008). Interaction of the glucocorticoid receptor with the chromatin landscape. *Molecular Cell, 29* (5), 611−624.

John, S., Sabo, P. J., Thurman, R. E., Sung, M. H., Biddie, S. C., & Johnson, T. A. (2011). Chromatin accessibility pre-determines glucocorticoid receptor binding patterns. *Nature Genetics, 43*(3), 264−268.

Jones, P. A., & Baylin, S. B. (2007). The epigenomics of cancer. *Cell, 128*(4), 683−692.

Karlic, R., Chung, H. R., Lasserre, J., Vlahovicek, K., & Vingron, M. (2010). Histone modification levels are predictive for gene expression. *Proceedings of the National Academy of Sciences of the United States of America, 107*(7), 2926−2931.

Kerr, B., Gedeon, A., Mulley, J., & Turner, G. (1992). Localization of non-specific X-linked mental retardation genes. *American Journal of Medical Genetics, 43*(1−2), 392−401.

Klein, C. J., Bird, T., Ertekin-Taner, N., Lincoln, S., Hjorth, R., & Wu, Y. (2013). DNMT1 mutation hot spot causes varied phenotypes of HSAN1 with dementia and hearing loss. *Neurology, 80*(9), 824−828.

Klein, C. J., Botuyan, M. V., Wu, Y., Ward, C. J., Nicholson, G. A., & Hammans, S. (2011). Mutations in DNMT1 cause hereditary sensory neuropathy with dementia and hearing loss. *Nature Genetics, 43*(6), 595−600.

Kouzarides, T. (2007). Chromatin modifications and their function. *Cell, 128*(4), 693−705.

Kriaucionis, S., & Heintz, N. (2009). The nuclear DNA base 5-hydroxymethylcytosine is present in Purkinje neurons and the brain. *Science (New York, N.Y.), 324*(5929), 929−930.

Kundaje, A., Meuleman, W., Ernst, J., Bilenky, M., Yen, A., & Heravi-Moussavi, A. (2015). Integrative analysis of 111 reference human epigenomes. *Nature, 518*(7539), 317−330.

Laird, P. W., & Jaenisch, R. (1996). The role of DNA methylation in cancer genetic and epigenetics. *Annual Review of Genetics, 30*, 441−464.

Lange, C. P., Campan, M., Hinoue, T., Schmitz, R. F., van der Meulen-de Jong, A. E., & Slingerland, H. (2012). Genome-scale discovery of DNA-methylation biomarkers for blood-based detection of colorectal cancer. *PLoS One, 7*(11), e50266.

Lasseigne, B. N., Burwell, T. C., Patil, M. A., Absher, D. M., Brooks, J. D., & Myers, R. M. (2014). DNA methylation profiling reveals novel diagnostic biomarkers in renal cell carcinoma. *BMC Medicine, 12*(1), 235.

Lee, M. P., DeBaun, M. R., Mitsuya, K., Galonek, H. L., Brandenburg, S., & Oshimura, M. (1999). Loss of imprinting of a paternally expressed transcript, with antisense orientation to KVLQT1, occurs frequently in Beckwith−Wiedemann syndrome and is independent of insulin-like growth factor II imprinting. *Proceedings of the National Academy of Sciences of the United States of America, 96*(9), 5203−5208.

Leung, D., Jung, I., Rajagopal, N., Schmitt, A., Selvaraj, S., & Lee, A. Y. (2015). Integrative analysis of haplotype-resolved epigenomes across human tissues. *Nature, 518*(7539), 350−354.

Li, E. (2002). Chromatin modification and epigenetic reprogramming in mammalian development. *Nature Reviews Genetics, 3*(9), 662−673.

Liang, G., Lin, J. C., Wei, V., Yoo, C., Cheng, J. C., & Nguyen, C. T. (2004). Distinct localization of histone H3 acetylation and H3-K4 methylation to the transcription start sites in the human genome. *Proceedings of the National Academy of Sciences of the United States of America, 101*(19), 7357−7362.

Lister, R., Pelizzola, M., Dowen, R. H., Hawkins, R. D., Hon, G., & Tonti-Filippini, J. (2009). Human DNA methylomes at base resolution show widespread epigenomic differences. *Nature, 462*(7271), 315−322.

Lopez-Atalaya, J. P., Ito, S., Valor, L. M., Benito, E., & Barco, A. (2013). Genomic targets, and histone acetylation and gene expression profiling of neural HDAC inhibition. *Nucleic Acids Research, 41*(17), 8072−8084.

Lyon, M. F. (1961). Gene action in the X-chromosome of the mouse (*Mus musculus* L.). *Nature, 190*, 372−373.

Magnani, L., & Lupien, M. (2014). Chromatin and epigenetic determinants of estrogen receptor alpha (ESR1) signaling. *Molecular and Cellular Endocrinology, 382*(1), 633−641.

Mardis, E. R. (2008). Next-generation DNA sequencing methods. *Annual Review of Genomics and Human Genetics, 9,* 387−402.

Mayer, W., Niveleau, A., Walter, J., Fundele, R., & Haaf, T. (2000). Demethylation of the zygotic paternal genome. *Nature, 403*(6769), 501−502.

McFarland, K. N., Das, S., Sun, T. T., Leyfer, D., Xia, E., & Sangrey, G. R. (2012). Genome-wide histone acetylation is altered in a transgenic mouse model of Huntington's disease. *PLoS One, 7*(7), e41423.

McGowan, P. O., Sasaki, A., D'Alessio, A. C., Dymov, S., Labonte, B., & Szyf, M. (2009). Epigenetic regulation of the glucocorticoid receptor in human brain associates with childhood abuse. *Nature Neuroscience, 12*(3), 342−348.

Mendenhall, E. M., Williamson, K. E., Reyon, D., Zou, J. Y., Ram, O., & Joung, J. K. (2013). Locus-specific editing of histone modifications at endogenous enhancers. *Nature Biotechnology, 31*(12), 1133−1136.

Michels, K. B., Binder, A. M., Dedeurwaerder, S., Epstein, C. B., Greally, J. M., & Gut, I. (2013). Recommendations for the design and analysis of epigenome-wide association studies. *Nature Methods, 10*(10), 949−955.

Mikkelsen, T. S., Ku, M., Jaffe, D. B., Issac, B., Lieberman, E., & Giannoukos, G. (2007). Genome-wide maps of chromatin state in pluripotent and lineage-committed cells. *Nature, 448*(7153), 553−560.

Miller, O. J., Schnedl, W., Allen, J., & Erlanger, B. F. (1974). 5-Methylcytosine localised in mammalian constitutive heterochromatin. *Nature, 251*(5476), 636−637.

Minucci, S., & Pelicci, P. G. (2006). Histone deacetylase inhibitors and the promise of epigenetic (and more) treatments for cancer. *Nature Reviews Cancer, 6*(1), 38−51.

Mohn, F., Weber, M., Rebhan, M., Roloff, T. C., Richter, J., & Stadler, M. B. (2008). Lineage-specific polycomb targets and de novo DNA methylation define restriction and potential of neuronal progenitors. *Molecular Cell, 30*(6), 755−766.

Morgan, H. D., Santos, F., Green, K., Dean, W., & Reik, W. (2005). Epigenetic reprogramming in mammals. *Human Molecular Genetics,* R47−R58.

Morgan, H. D., Sutherland, H. G., Martin, D. I., & Whitelaw, E. (1999). Epigenetic inheritance at the agouti locus in the mouse. *Nature Genetics, 23*(3), 314−318.

Morison, I. M., Ramsay, J. P., & Spencer, H. G. (2005). A census of mammalian imprinting. *Trends in Genetics, 21*(8), 457−465.

Nucifora, F. C., Sasaki, M., Peters, M. F., Huang, H., Cooper, J. K., & Yamada, M. (2001). Interference by huntingtin and atrophin-1 with cbp-mediated transcription leading to cellular toxicity. *Science (New York, N.Y.), 291*(5512), 2423−2428.

Ogawa, O., Eccles, M. R., Szeto, J., McNoe, L. A., Yun, K., & Maw, M. A. (1993). Relaxation of insulin-like growth factor II gene imprinting implicated in Wilms' tumour. *Nature, 362* (6422), 749−751.

Oh, T., Kim, N., Moon, Y., Kim, M. S., Hoehn, B. D., & Park, C. H. (2013). Genome-wide identification and validation of a novel methylation biomarker, SDC2, for blood-based detection of colorectal cancer. *The Journal of Molecular Diagnostics, 15*(4), 498−507.

Okano, M., Bell, D. W., Haber, D. A., & Li, E. (1999). DNA methyltransferases Dnmt3a and Dnmt3b are essential for de novo methylation and mammalian development. *Cell, 99*(3), 247−257.

Painter, R. C., Osmond, C., Gluckman, P., Hanson, M., Phillips, D. I., & Roseboom, T. J. (2008). Transgenerational effects of prenatal exposure to the Dutch famine on neonatal adiposity and health in later life. *BJOG: An International Journal of Obstetrics and Gynaecology, 115*(10), 1243−1249.

Palmisano, W. A., Divine, K. K., Saccomanno, G., Gilliland, F. D., Baylin, S. B., & Herman, J. G. (2000). Predicting lung cancer by detecting aberrant promoter methylation in sputum. *Cancer Research, 60*(21), 5954−5958.

Pembrey, M. E., Bygren, L. O., Kaati, G., Edvinsson, S., Northstone, K., & Sjostrom, M. (2006). Sex-specific, male-line transgenerational responses in humans. *European Journal of Human Genetics, 14*(2), 159−166.

Penny, G. D., Kay, G. F., Sheardown, S. A., Rastan, S., & Brockdorff, N. (1996). Requirement for Xist in X chromosome inactivation. *Nature, 379*(6561), 131−137.

Plumb, J. A., Strathdee, G., Sludden, J., Kaye, S. B., & Brown, R. (2000). Reversal of drug resistance in human tumor xenografts by 2'-deoxy-5-azacytidine-induced demethylation of the hMLH1 gene promoter. *Cancer Research, 60*(21), 6039−6044.

Pollard, K. S., Serre, D., Wang, X., Tao, H., Grundberg, E., & Hudson, T. J. (2008). A genome-wide approach to identifying novel-imprinted genes. *Human Genetics, 122*(6), 625−634.

Portela, A., & Esteller, M. (2010). Epigenetic modifications and human disease. *Nature Biotechnology, 28*(10), 1057−1068.

Rada-Iglesias, A., Bajpai, R., Swigut, T., Brugmann, S. A., Flynn, R. A., & Wysocka, J. (2011). A unique chromatin signature uncovers early developmental enhancers in humans. *Nature, 470*(7333), 279−283.

Rakyan, V. K., Down, T. A., Balding, D. J., & Beck, S. (2011). Epigenome-wide association studies for common human diseases. *Nature Reviews Genetics, 12*(8), 529−541.

Reamon-Buettner, S. M., Mutschler, V., & Borlak, J. (2008). The next innovation cycle in toxicogenomics: environmental epigenetics. *Mutation Research, 659*(1−2), 158−165.

Reik, W., Brown, K. W., Slatter, R. E., Sartori, P., Elliott, M., & Maher, E. R. (1994). Allelic methylation of H19 and IGF2 in the Beckwith−Wiedemann syndrome. *Human Molecular Genetics, 3*(8), 1297−1301.

Reik, W., Dean, W., & Walter, J. (2001). Epigenetic reprogramming in mammalian development. *Science (New York, N.Y.), 293*(5532), 1089−1093.

Ribeiro, R. C., Kushner, P. J., & Baxter, J. D. (1995). The nuclear hormone receptor gene super-family. *Annual Review of Medicine, 46*, 443−453.

Robertson, K. D. (2005). DNA methylation and human disease. *Nature Reviews Genetics, 6*(8), 597−610.

Roseboom, T., de Rooij, S., & Painter, R. (2006). The Dutch famine and its long-term consequences for adult health. *Early Human Development, 82*(8), 485−491.

Sadri-Vakili, G., & Cha, J. H. (2006). Mechanisms of disease: histone modifications in Huntington's disease. *Nature Clinical Practice Neurology, 2*(6), 330−338.

Schones, D. E., & Zhao, K. (2008). Genome-wide approaches to studying chromatin modifications. *Nature Reviews Genetics, 9*(3), 179−191.

Schulz, L. C. (2010). The Dutch Hunger Winter and the developmental origins of health and disease. *Proceedings of the National Academy of Sciences of the United States of America, 107*(39), 16757−16758.

Schwartz, Y. B., & Pirrotta, V. (2013). A new world of polycombs: Unexpected partnerships and emerging functions. *Nature Reviews Genetics, 14*(12), 853−864.

Sharma, D., & Fondell, J. D. (2002). Ordered recruitment of histone acetyltransferases and the TRAP/mediator complex to thyroid hormone-responsive promoters in vivo. *Proceedings of the National Academy of Sciences of the United States of America, 99*(12), 7934−7939.

Shendure, J., & Ji, H. (2008). Next-generation DNA sequencing. *Nature Biotechnology, 26*(10), 1135−1145.

Skinner, M. K. (2011). Environmental epigenomics and disease susceptibility. *EMBO Reports*, *12*(7), 620−622.

Song, J., Rechkoblit, O., Bestor, T. H., & Patel, D. J. (2011). Structure of DNMT1-DNA complex reveals a role for autoinhibition in maintenance DNA methylation. *Science (New York, N.Y.)*, *331*(6020), 1036−1040.

Steffan, J. S., Bodai, L., Pallos, J., Poelman, M., McCampbell, A., & Apostol, B. L. (2001). Histone deacetylase inhibitors arrest polyglutamine-dependent neurodegeneration in *Drosophila*. *Nature*, *413*(6857), 739−743.

Strahl, B. D., & Allis, C. D. (2000). The language of covalent histone modifications. *Nature*, *403*(6765), 41−45.

Straub, T. (2003). Heterochromatin dynamics. *PLoS Biology*, *1*(1), E14.

Tahiliani, M., Koh, K. P., Shen, Y., Pastor, W. A., Bandukwala, H., & Brudno, Y. (2009). Conversion of 5-methylcytosine to 5-hydroxymethylcytosine in mammalian DNA by MLL partner TET1. *Science (New York, N.Y.)*, *324*(5929), 930−935.

Tatton-Brown, K., Seal, S., Ruark, E., Harmer, J., Ramsay, E., & Del, S. (2014). Vecchio Duarte, Mutations in the DNA methyltransferase gene DNMT3A cause an overgrowth syndrome with intellectual disability. *Nature Genetics*, *46*(4), 385−388.

Tobi, E. W., Lumey, L. H., Talens, R. P., Kremer, D., Putter, H., & Stein, A. D. (2009). DNA methylation differences after exposure to prenatal famine are common and timing- and sex-specific. *Human Molecular Genetics*, *18*(21), 4046−4053.

Toledo-Rodriguez, M., Lotfipour, S., Leonard, G., Perron, M., Richer, L., & Veillette, S. (2010). Maternal smoking during pregnancy is associated with epigenetic modifications of the brain-derived neurotrophic factor-6 exon in adolescent offspring. *American Journal of Medical Genetics. Part B, Neuropsychiatric Genetics: The Official Publication of the International Society of Psychiatric Genetics*, *153b*(7), 1350−1354.

Trojer, P., & Reinberg, D. (2007). Facultative heterochromatin: Is there a distinctive molecular signature? *Molecular Cell*, *28*(1), 1−13.

Tsai, M. J., & O'Malley, B. W. (1994). Molecular mechanisms of action of steroid/thyroid receptor superfamily members. *Annual Review of Biochemistry*, *63*, 451−486.

Vo, N., & Goodman, R. H. (2001). CREB-binding protein and p300 in transcriptional regulation. *The Journal of Biological Chemistry*, *276*(17), 13505−13508.

Walker, F. O. (2007). Huntington's disease. *Lancet*, *369*(9557), 218−228.

Wang, X., Soloway, P. D., & Clark, A. G. (2011). A survey for novel imprinted genes in the mouse placenta by mRNA-seq. *Genetics*, *189*(1), 109−122.

Waterland, R. A., Dolinoy, D. C., Lin, J. R., Smith, C. A., Shi, X., & Tahiliani, K. G. (2006). Maternal methyl supplements increase offspring DNA methylation at Axin Fused. *Genesis (New York, N.Y.: 2000)*, *44*(9), 401−406.

Weaver, I. C., Cervoni, N., Champagne, F. A., D'Alessio, A. C., Sharma, S., & Seckl, J. R. (2004). Epigenetic programming by maternal behavior. *Nature Neuroscience*, *7*(8), 847−854.

Wei, S. H., Balch, C., Paik, H. H., Kim, Y. S., Baldwin, R. L., & Liyanarachchi, S. (2006). Prognostic DNA methylation biomarkers in ovarian cancer. *Clinical Cancer Research: An Official Journal of the American Association for Cancer Research*, *12*(9), 2788−2794.

Weksberg, R., Shen, D. R., Fei, Y. L., Song, Q. L., & Squire, J. (1993). Disruption of insulin-like growth factor 2 imprinting in Beckwith−Wiedemann syndrome. *Nature Genetics*, *5*(2), 143−150.

Weksberg, R., Shuman, C., & Beckwith, J. B. (2010). Beckwith−Wiedemann syndrome. *European Journal of Human Genetics*, *18*(1), 8−14.

Wen, L., & Tang, F. (2014). Genomic distribution and possible functions of DNA hydroxy-methylation in the brain. *Genomics, 104*(5), 341−346.

West, A. C., & Johnstone, R. W. (2014). New and emerging HDAC inhibitors for cancer treatment. *The Journal of Clinical Investigation, 124*(1), 30−39.

Wiedemann, H. R. (1969). [The EMG-syndrome: Exomphalos, macroglossia, gigantism and disturbed carbohydrate metabolism]. *Zeitschrift fur Kinderheilkunde, 106*(3), 171−185.

Williams, S. R., Aldred, M. A., Der Kaloustian, V. M., Halal, F., Gowans, G., & McLeod, D. R. (2010). Haploinsufficiency of HDAC4 causes brachydactyly mental retardation syndrome, with brachydactyly type E, developmental delays, and behavioral problems. *American Journal of Human Genetics, 87*(2), 219−228.

Witzgall, R., O'Leary, E., Leaf, A., Onaldi, D., & Bonventre, J. V. (1994). The Kruppel-associated box-A (KRAB-A) domain of zinc finger proteins mediates transcriptional repression. *Proceedings of the National Academy of Sciences of the United States of America, 91*(10), 4514−4518.

Yoder, J. A., Soman, N. S., Verdine, G. L., & Bestor, T. H. (1997). DNA (cytosine-5)-methyl-transferases in mouse cells and tissues. Studies with a mechanism-based probe. *Journal of Molecular Biology, 270*(3), 385−395.

Zhou, V. W., Goren, A., & Bernstein, B. E. (2011). Charting histone modifications and the functional organization of mammalian genomes. *Nature Reviews Genetics, 12*(1), 7−18.

Zhu, J. K. (2009). Active DNA demethylation mediated by DNA glycosylases. *Annual Review of Genetics, 43*, 143−166.

Chapter 4

Clinical decision support and molecular tumor boards

Michelle F. Green

Department of Pathology, Duke University Medical Center, Durham, NC, United States

Introduction

The field of precision oncology is based on the tenant that understanding a tumor's unique genetic profile can help establish diagnosis, inform prognosis, and guide treatment decisions. Due to a combination of rapid advances in next-generation sequencing (NGS) technology as well as decreasing costs and turnaround times, the promise of precision oncology has been increasingly realized in oncology patient care. This is exemplified by the rapid increase in FDA-approved targeted therapies associated with specific genomic alterations (Twomey, Brahme, & Zhang, 2017). In some cases, this has led to a paradigm shift away from treatment algorithms developed based on tissue of origin toward treatments based on the presence of specific molecular alterations (Hierro et al., 2019). While these new treatment algorithms may result in better outcomes for cancer patients, they present challenges to clinicians who are expected to process large amounts of emerging information related to genetic alterations, testing platforms, and tumor site agnostic therapy approvals. Molecular tumor boards (MTBs) can help bridge the gap between clinical practice and the rapid evolution of precision medicine, and have become an increasingly important physician support mechanism to guide the selection of appropriate patient populations and testing methods, interpretation of genetic data, and generation of clinically meaningful recommendations. In this chapter, we present an overview of the organization and functions of MTBs, as well as areas for continued growth and improvement (see Fig. 4.1).

Assembling the right team

Given the complexity of precision oncology, knowledge from several specialties is necessary to make informed decisions. Several institutions have

Genomic and Precision Medicine. DOI: https://doi.org/10.1016/B978-0-12-800684-9.00003-4

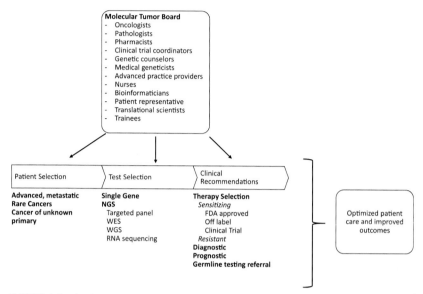

FIGURE 4.1 A schematic illustrating the composition of molecular tumor boards (MTBs) and their several roles, from patient and test selection to the generation of clinical recommendations.

reported the implementation of MTBs, and while composition varies, there is broad consensus that these groups should be multidisciplinary in nature (Basse et al., 2018; Dalton et al., 2017; Knepper et al., 2017; Lane et al., 2015; Luchini et al., 2020; Rolfo et al., 2018; Schwaederle et al., 2014; Tafe et al., 2015; van der Velden et al., 2017). At a minimum, medical oncologists from multiple disease groups and molecular pathologists should be present. Many MTBs also include genetic counselors, medical geneticists, clinical trial coordinators, advanced practice providers, nurses, bioinformaticians, patient representatives, radiation oncologists, surgical oncologists, translational scientists, and pharmacists. Given the increasing importance of precision medicine and the strong educational component of MTB, inclusion of trainees such as medical students, residents, and fellows is also advisable.

Patient and test selection

Outside of a few select cancer types, such as non-small cell lung carcinoma (NSCLC) (National Comprehensive Cancer Network, 2020), evidence-based guidelines informing the use of NGS testing are limited. While some MTBs actively participate in the decision to pursue molecular diagnostic testing (Harada et al., 2017), at most institutions this is left to the discretion of individual providers (Dalton et al., 2017; Knepper et al., 2017; Schwaederle et al., 2014; Tafe et al., 2015). Most NGS testing is performed in the advanced, metastatic disease state, while some institutions also focus on cancers that are rare, hard to treat, or have an unknown primary (Hirshfield et al., 2016; Khater et al., 2019). Timing also varies, with some providers ordering NGS testing at diagnosis of metastatic disease, while others may wait until after standard of care therapeutic options are exhausted (Rolfo et al., 2018). Thus one function of MTBs is to facilitate patient selection and timing of NGS testing, and this can occur though a formal review process, generation of institution-specific guidelines, or simply providing a forum for discussion.

Once a decision is made to pursue genomic testing, there are frequently multiple institution specific or commercial testing options available. In the early days of precision oncology, single gene assays were used to detect well-characterized genomic alterations associated with a high level of clinical significance. These tests are still used in several contexts, including *BRAF* for melanoma (National Comprehensive Cancer Network, 2020) and *KIT* and/or *PDGFRA* for gastrointestinal stromal tumors (National Comprehensive Cancer Network, 2020). However, as the number of gene mutations with associated clinical significance increases, it has become increasingly more practical and economical to adopt NGS platforms (Frampton et al., 2013; Pennell et al., 2019). Furthermore, NGS panels have the added ability to detect clinically meaningful genomic signatures including microsatellite instability, tumor mutation burden, and loss of heterozygosity (Chalmers et al., 2017; Swisher et al., 2017; Willis et al., 2019). NGS tests can include targeted gene panels, whole exome sequencing (WES), whole genome sequencing (WGS), and whole-transcriptome sequencing (RNA-Seq) (reviewed in Brown & Elenitoba-Johnson, 2020). Targeted gene panels, which exist in a range of sizes, are the most commonly used in oncology patient care. These panels generally have reduced costs and faster turnaround times as compared with WES and WGS, and cover genes with well-characterized or emerging clinical significance. Given their smaller size, targeted panels will detect fewer alterations than WES or WGS, but will require less effort to analyze and interpret. Smaller panel size also allows for greater depth of coverage and thus better detection of low-frequency variants.

When tumor tissue is limited or unavailable, so-called "liquid biopsies," where cell-free circulating tumor DNA (cfDNA) is extracted from peripheral blood and subjected to NGS analysis, may be of benefit. However, studies have reported varying levels of concordance between genomic profiles derived from paired tumor and cfDNA specimens or between results derived from different commercially available liquid biopsy tests (Bettegowda et al., 2014; Stetson et al., 2019; Torga & Pienta 2018; Zill et al., 2018). This could be the result of tumor heterogeneity, technical variations between platforms, or contamination from clonal hematopoiesis. Therefore results derived from liquid biopsy assays should be interpreted with caution, especially in cases with low cfDNA yield.

Recent studies using RNA sequencing to examine the landscape of gene fusions across cancer types have shown greater complexity than previously realized (Kumar-Sinha, Kalyana-Sundaram, & Chinnaiyan, 2015; Maher et al., 2009; Yoshihara et al., 2015). Several potent small molecule inhibitors, such as the NTRK inhibitors larotrectinib and entrectinib, have also been shown to result in high rates of durable responses in fusion positive cancers (Cocco, Scaltriti, & Drilon, 2018; Doebele et al., 2020; Drilon et al., 2018). Some gene fusions can be difficult to detect using DNA-based sequencing (Ozsolak & Milos, 2011), resulting in the increased popularity of RNA-based sequencing techniques. However, targetable gene fusions are relatively rare and can occur at low frequencies across several cancer types (Cocco et al., 2018). Therefore it may not be practical or cost effective to perform RNA sequencing for all cancer patients. Several groups have identified specific patient populations where fusion events are enriched, which may be helpful in the future to select for patients who can benefit from RNA sequencing (Benayed et al., 2019; Cocco et al., 2019; Ross et al., 2020).

In many cases, NGS testing is not routinely reimbursed by insurance companies. In addition, obtaining access to and covering the cost of targeted therapies, especially if used off-label, are a frequent barrier to precision oncology (Pagès et al., 2017; Schwaederle et al., 2014). Thus another role of MTBs may be to negotiate with outside testing companies and/or insurance companies to select the most appropriate testing platform and negotiate agreements for payment to insure patients are not subject to financial toxicity. MTBs can also be associated with clinical trials that enable NGS testing and treatment with targeted therapies, while others assist with obtaining drugs through expanded access or compassionate use programs (Beltran et al., 2015; Harada et al., 2017; Knepper et al., 2017; Meric-Bernstam et al., 2015; Réda et al., 2020; Rolfo et al., 2018).

Generating clinical recommendations

Therapy selection

A major bottleneck in precision oncology is the translation of genetic data into meaningful clinical recommendations (Danos et al., 2019; Good et al., 2014). One large retrospective study estimated that alterations associated with FDA-approved therapies are present in only 8.3% of cancer cases as of 2018. For biomarker-directed therapies that are already used as standard of care, such as EGFR- and ALK-targeted therapies in NSCLC (National Comprehensive Cancer Network, 2020), there is less of a need for MTB guidance. As the majority of alterations detected by NGS testing will not have matching FDA-approved therapies, MTBs can be especially useful to guide the use of off-label or investigational drugs. In these situations, careful review of the specific alteration present and disease context are critical for appropriate treatment selection. An example of this is the targeting of *BRAF* mutations in cancer. *BRAF* mutations at codon V600 occur in approximately

50% of malignant melanoma cases (Cancer Genome Atlas Network, 2015), and are well characterized to result in BRAF activation and promotion of downstream MAP kinase signaling. BRAF V600 specific inhibitors, either alone or in combination with MEK inhibitors, are now routinely used in the treatment of metastatic melanoma (Chapman et al., 2011; Dummer et al., 2018; Hauschild et al., 2012; Larkin et al., 2014; Long et al., 2014). Off-label use of BRAF/MEK inhibitors in other cancer types harboring BRAF V600 mutations has resulted in clinical benefit and additional FDA approvals in NSCLC (Odogwu et al., 2018), anaplastic thyroid cancer (Subbiah et al., 2017), and Erdheim Chester disease (Oneal et al., 2018). In contrast, early studies of BRAF inhibitors in *BRAF* V600E mutant colorectal cancer demonstrated a lack of clinical benefit (Corcoran et al., 2015; Kopetz et al., 2015). This was believed to be the result of innate resistance mediated through EGFR signaling. Clinical trials combining the BRAF V600 inhibitor encorafenib with the EGFR-directed monoclonal antibody cetuximab were able to overcome this innate resistance, with this combination receiving FDA approval in May 2020 (Kopetz et al., 2019). In addition to V600 mutations, there is a large variety of atypical *BRAF* mutations that can either activate or inactivate the BRAF protein, both resulting in activation of downstream MAP kinase signaling through different biological mechanisms (Heidorn et al., 2010). While BRAF V600 specific inhibitors are not efficacious for patients harboring atypical *BRAF* mutations, inhibitors targeting downstream MAP kinase targets such as MEK and ERK may be relevant (Yao et al., 2017). Therefore careful attention to the specific alteration detected and disease context are critical for appropriate therapy matching.

In addition to off-label drug use, MTBs are an important mechanism for matching patients with an increasing number of biomarker-specific clinical trials. Biomarker-selected clinical trials matching specific alterations with targeted therapies have reported actionable alterations in 27%−75% of cases (Le Tourneau et al., 2015; Massard et al., 2017; Meric-Bernstam et al., 2015; Réda et al., 2020; Trédan et al., 2019). However, the actual number of patients who received targeted therapies is considerably lower, ranging from 6% to 23%. There are several reasons for this difference including patient preference, poor performance status, failure to meet other clinical trial eligibility criteria, lack of open trial slots, and insurance denials.

Alterations that predict resistance to targeted therapies can be just as important as those that predict sensitivity, and are commonly detected through genetic profiling. These alterations can be present at baseline, acquired during the course of treatment, or both. There are several scenarios where detection of resistance mutations can guide treatment selections, including the lack of efficacy of EGFR−targeted therapies in colorectal cancer patients with *RAS* alterations (Dempke & Heinemann, 2010) and the use of osimertinib in NSCLC patients with secondary *EGFR* T790M alterations (Mok et al., 2016).

There are many resources available to aid MTBs with variant interpretation and therapeutic matching (see Table 4.1). NGS reports returned to the ordering provider, either from institutional or commercial laboratories, will likely contain information about available targeted therapies and clinical trials. In addition, several groups have developed somatic variant interpretation knowledgebases where evidence associating genomic alterations with clinical utility is curated from the scientific literature and presented in a concise, searchable format (Chakravarty et al., 2017; Griffith et al., 2017; Huang et al., 2017; Patterson et al., 2016; Tamborero et al., 2018). However, many of these resources rely on crowdsourcing and are highly discordant in the amount, quality, and structure of data they contain. The Variant Interpretation for Cancer Consortium (VICC) has attempted to address these challenges through the creation of a *meta*-knowledgebase containing harmonized variant interpretation from six prominent somatic alteration databases (Wagner et al., 2020). While these efforts are important and will ultimately lead to improvements in the adoption and implementation of precision oncology, the resultant *meta*-knowledgebase only contained approximately 13,000 aggregate interpretations, which is a small fraction of the somatic alterations known to occur in cancer (Tate et al., 2018). Furthermore, variant interpretation knowledgebases and commercial NGS reports do not take into consideration patient-specific information such as treatment history and relevant comorbidities. Currently, the most reliable and accurate method for variant interpretation and therapeutic matching is through expert analysis (Brown & Elenitoba-Johnson, 2020; Bungartz, Lalowski, & Elkin, 2018), with MTBs providing an important mechanism for holistic review of all relevant biomarker and patient data available to generate the most relevant clinical recommendations.

Another function of MTBs is to rank and prioritize all available therapeutic options, including both targeted therapies and standard chemotherapies. Multiple groups have suggested guidelines to rank and prioritize genomic alterations based on the strength of available clinical evidence (see Fig. 4.2). These include the joint consensus guidelines developed by the Association for Molecular Pathology (AMP), the American Society of Clinical Oncology (ASCO), the College of American Pathologists (CAP) (Li et al., 2017), the European Society for Medical Oncology (ESMO) Scale for Clinical Actionability of molecular targets (ESCAT) (Mateo et al., 2018), and the OncoKB knowledgebase levels of evidence scale (Chakravarty et al., 2017). All three guidelines rank clinical evidence for targeted therapies on a scale ranging from biomarker-specific FDA approvals down to compelling preclinical evidence. Variants predicting resistance to a target therapy are also included in the AMP/ASCO/CAP and OncoKB guidelines, while the ESMO ESCAT guidelines include additional categories for cotargeting approaches and negative evidence for actionability. The AMP/ASCO/CAP guidelines also extend beyond therapeutic implications, and include rankings for diagnostic and prognostic alterations. While these are all valuable systems, there

TABLE 4.1 A list of publically available resources for the analysis and interpretation of cancer variants.

Resource	Focus alterations	Aim	Host organization	URL
cBioPortal	Somatic cancer mutations	Repository for data generated from tumor sequencing studies, primarily the Cancer Genome Atlas (TCGA) program	Memorial Sloan Kettering Cancer Center	https://www.cbioportal.org/
GENIE	Somatic cancer mutations	Repository for clinical tumor sequencing data	American Association for Cancer Research	https://genie.cbioportal.org/login.jsp
COSMIC	Somatic cancer mutations	Repository for data generated from tumor sequencing studies	Wellcome Sanger Institute	https://cancer.sanger.ac.uk/cosmic
CiVIC	Somatic cancer mutations	Clinical interpretation of cancer variants	Washington University	https://civicdb.org/home
My Cancer Genome	Somatic cancer mutations	Clinical interpretation of cancer variants	Vanderbilt-Ingram Cancer Center	https://www.mycancergenome.org/
OncoKB	Somatic cancer mutations	Clinical interpretation of cancer variants	Memorial Sloan Kettering Cancer Center	https://www.oncokb.org/
Clinical Knowledgebase	Somatic cancer mutations	Clinical interpretation of cancer variants	The Jackson Laboratory	https://ckbhome.jax.org/
Cancer Genome Interpreter	Somatic cancer mutations	Clinical interpretation of cancer variants	Barcelona Biomedical Genomics Lab	https://www.cancergenomeinterpreter.org/home

(Continued)

TABLE 4.1 (Continued)

Resource	Focus alterations	Aim	Host organization	URL
Precision Medicine Knowledgebase	Somatic cancer mutations	Clinical interpretation of cancer variants	Englander Institute for Precision Medicine (EIPM)	https://pmkb.weill. cornell.edu/
gnomAD	Germline polymorphisms	Repository for harmonizing and aggregating germline sequencing data	Broad Institute	https://gnomad. broadinstitute.org/
ClinVar	Pathogenic germline alterations, some somatic cancer mutations	Repository for germline variants related to human disease	National Center for Biotechnology Information	https://www.ncbi.nlm. nih.gov/clinvar/

Note: Tumor sequencing repositories can be useful to determine the frequency of variants across cancer types, while germline repositories can help identify benign polymorphisms. Several knowledgebases associating cancer variants with targeted therapies are also available.

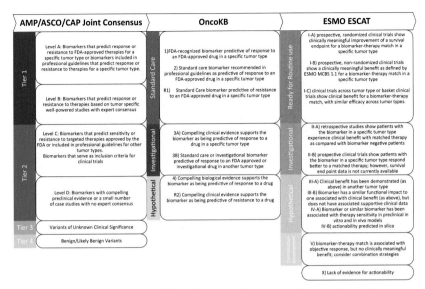

FIGURE 4.2 A comparison of guidelines for the ranking and prioritization of somatic sequence variants, including the AMP/ASCO/CAP joint consensus, ESMO ESCAT, and OncoKB recommendations.

is still a need for these recommendations to be widely adopted and incorporated into NGS clinical reporting. In the future, collective experiences gained through MTBs will help guide this process.

Genetic counseling and germline testing referrals

Emerging evidence suggests that unselected or minimally selected populations of cancer patients have increased incidences of germline alterations in genes associated with cancer predisposition, and that a significant portion of these patients would not otherwise meet criteria for germline testing (DeLeonardis et al., 2019; Norquist et al., 2016; Pearlman et al., 2017). Therefore an important outcome of MTB review is the recognition of potential germline variants in somatic tumor NGS sequencing results and coordination of referrals for genetic counseling and subsequent germline testing. This observation also underscores the importance of including genetic counselors and medical geneticists in MTBs, and has the potential to benefit both the patient and additional at-risk family members (Klek et al., 2020).

Diagnosis modification

In some situations, genomic profiling can be useful as an ancillary technique to establish or confirm diagnosis, especially in situations where tumor morphology and immunohistochemical profiling are inconclusive. Examples of this include soft tissue sarcoma where several subtypes harbor distinctive gene rearrangements (National Comprehensive Cancer Network, 2020). Genomic profiling is also becoming increasingly common for patients with myeloid malignancies where some genomic alterations are pathognomonic and required to establish diagnosis, while others are suggestive of a diagnosis or can be used to confirm an underlying clonal process (Arber, 2016). However, in most situations the presence of a specific genetic alteration is not sufficient to establish diagnosis on its own, and must be interpreted in the context of other diagnostic and clinical information.

Areas for improvement and concluding remarks

Molecular profiling of advanced solid tumors is becoming increasing common, with the portfolio of clinically meaningful alterations rapidly expanding. MTBs were developed as a physician support mechanism to guide the implementation of precision oncology. While structure varies between institutions, MTBs have important roles in guiding the use of NGS testing from patient and test selection to the interpretation of results and generation of clinical recommendations. In many situations, much of the work required to translate genetic finding into clinically meaningful action is dependent on manual expert review. As the volume of NGS testing continues to grow, this reliance on manual review will become increasing unsustainable. Thus there is an urgent need for the development of support systems to organize and harmonize clinical recommendations related to genetic profiling. Many academic institutions, as well as commercial vendors, are actively developing these systems, which should incorporate collective knowledge gained form MTBs (Good et al., 2014; Pishvaian et al., 2019; Singer et al., 2018; Tamborero et al., 2020; Tao et al., 2018, 2019).

Finally, while several large academic medical centers have established MTBs, smaller community-based practices may lack the expertise required for a successful program. Thus there is a need for expansion of MTBs into the community setting. Several large academic centers have started to fill this void by creating MTB networks that meet virtually to review cases and provide clinical recommendations (Levit et al., 2019; Rao et al., 2020). As the majority of cancer patients are treated in the community setting, this expansion of MTBs will allow for the promise of precision medicine to be realized in a much larger population.

References

Arber, D. A., Orazi, A., Hasserjian, R., Thiele, J., Borowitz, M. J., Le Beau, M. M., . . . Vardiman, J. W. (2016). The 2016 revision to the World Health Organization classification of myeloid neoplasms and acute leukemia. *Blood*, *127*(20), 2391–2405, 27069254.

Basse, C., Morel, C., Alt, M., Sablin, M. P., Franck, C., Pierron, G., ... Kamal, M. (2018). Relevance of a molecular tumour board (MTB) for patients' enrolment in clinical trials: Experience of the Institut Curie. *ESMO open*, *3*(3), e000339.

Beltran, H., Eng, K., Mosquera, J. M., Sigaras, A., Romanel, A., Rennert, H., ... Rubin, M. A. (2015). Whole-exome sequencing of metastatic cancer and biomarkers of treatment response. *JAMA Oncology*, *1*(4), 466−474.

Benayed, R., Offin, M., Mullaney, K., Sukhadia, P., Rios, K., Desmeules, P., ... Ladanyi, M. (2019). High yield of RNA sequencing for targetable kinase fusions in lung adenocarcinomas with no mitogenic driver alteration detected by DNA sequencing and low tumor mutation burden. *Clinical Cancer Research*, *25*(15), 4712.

Bettegowda, C., Sausen, M., Leary, R. J., Kinde, I., Wang, Y., Agrawal, N., ... Diaz, L. A., Jr. (2014). Detection of circulating tumor DNA in early- and late-stage human malignancies. *Science Translational Medicine*, *6*(224), 224ra24.

Brown, N. A., & Elenitoba-Johnson, K. S. J. (2020). Enabling precision oncology through precision diagnostics. *Annual Review of Pathology: Mechanisms of Disease*, *15*(1), 97−121.

Bungartz, K. D., Lalowski, K., & Elkin, S. K. (2018). Making the right calls in precision oncology. *Nature Biotechnology*, *36*(8), 692−696.

Cancer Genome Atlas N. (2015). Genomic classification of cutaneous melanoma. *Cell*, *161*(7), 1681−1696.

Chakravarty, D., Gao, J., Phillips, S., Kundra, R., Zhang, H., Wang, J., ... Schultz, N. (2017). OncoKB: A precision oncology knowledge base. *JCO Precision Oncology*, *1*, 1−16.

Chalmers, Z. R., Connelly, C. F., Fabrizio, D., Gay, L., Ali, S. M., Ennis, R., ... Frampton, G. M. (2017). Analysis of 100,000 human cancer genomes reveals the landscape of tumor mutational burden. *Genome Medicine*, *9*(1), 34.

Chapman, P. B., Hauschild, A., Robert, C., Haanen, J. B., Ascierto, P., Larkin, J., ... McArthur, G. A. (2011). Improved survival with vemurafenib in melanoma with BRAF V600E mutation. *The New England Journal of Medicine*, *364*(26), 2507−2516.

Cocco, E., Benhamida, J., Middha, S., Zehir, A., Mullaney, K., Shia, J., ... Hechtman, J. F. (2019). Colorectal carcinomas containing hypermethylated MLH1 promoter and wild-type BRAF/KRAS are enriched for targetable kinase fusions. *Cancer Research*, *79*(6), 1047−1053.

Cocco, E., Scaltriti, M., & Drilon, A. (2018). NTRK fusion-positive cancers and TRK inhibitor therapy. *Nature Reviews Clinical Oncology*, *15*(12), 731−747.

Corcoran, R. B., Atreya, C. E., Falchook, G. S., Kwak, E. L., Ryan, D. P., Bendell, J. C., ... Kopetz, S. (2015). Combined BRAF and MEK inhibition with dabrafenib and trametinib in BRAF V600-mutant colorectal cancer. *Journal of Clinical Oncology: Official Journal of the American Society of Clinical Oncology*, *33*(34), 4023−4031.

Dalton, W. B., Forde, P. M., Kang, H., Connolly, R. M., Stearns, V., Gocke, C. D., ... Lauring, J. (2017). Personalized medicine in the oncology clinic: Implementation and outcomes of the Johns Hopkins Molecular Tumor Board. *JCO Precision Oncology*, *2017*.

Danos, A. M., Krysiak, K., Barnell, E. K., Coffman, A. C., McMichael, J. F., Kiwala, S., ... Griffith, O. L. (2019). Standard operating procedure for curation and clinical interpretation of variants in cancer. *Genome Medicine*, *11*(1), 76.

DeLeonardis, K., Hogan, L., Cannistra, S. A., Rangachari, D., & Tung, N. (2019). When should tumor genomic profiling prompt consideration of germline testing? *Journal of Oncology Practice*, *15*(9), 465−473.

Dempke, W. C., & Heinemann, V. (2010). Ras mutational status is a biomarker for resistance to EGFR inhibitors in colorectal carcinoma. *Anticancer Research*, *30*(11), 4673−4677.

Doebele, R. C., Drilon, A., Paz-Ares, L., Siena, S., Shaw, A. T., Farago, A. F., . . . Demetri, G. D. (2020). Entrectinib in patients with advanced or metastatic NTRK fusion-positive solid tumours: Integrated analysis of three phase 1–2 trials. *The Lancet Oncology, 21*(2), 271–282.

Drilon, A., Laetsch, T. W., Kummar, S., DuBois, S. G., Lassen, U. N., Demetri, G. D., . . . Hyman, D. M. (2018). Efficacy of larotrectinib in TRK fusion-positive cancers in adults and children. *The New England Journal of Medicine, 378*(8), 731–739.

Dummer, R., Ascierto, P. A., Gogas, H. J., Arance, A., Mandala, M., Liszkay, G., . . . Flaherty, K. T. (2018). Encorafenib plus binimetinib vs vemurafenib or encorafenib in patients with BRAF-mutant melanoma (COLUMBUS): A multicentre, open-label, randomised phase 3 trial. *The Lancet Oncology, 19*(5), 603–615.

Frampton, G. M., Fichtenholtz, A., Otto, G. A., Wang, K., Downing, S. R., He, J., . . . Yelensky, R. (2013). Development and validation of a clinical cancer genomic profiling test based on massively parallel DNA sequencing. *Nature Biotechnology, 31*(11), 1023–1031.

Good, B. M., Ainscough, B. J., McMichael, J. F., Su, A. I., & Griffith, O. L. (2014). Organizing knowledge to enable personalization of medicine in cancer. *Genome Biology, 15*(8), 438.

Griffith, M., Spies, N. C., Krysiak, K., McMichael, J. F., Coffman, A. C., Danos, A. M., . . . Griffith, O. L. (2017). CIViC is a community knowledgebase for expert crowdsourcing the clinical interpretation of variants in cancer. *Nature Genetics, 49*(2), 170–174.

Harada, S., Arend, R., Dai, Q., Levesque, J. A., Winokur, T. S., Guo, R., . . . Yang, E. S. (2017). Implementation and utilization of the molecular tumor board to guide precision medicine. *Oncotarget, 8*(34), 57845–57854.

Hauschild, A., Grob, J. J., Demidov, L. V., Jouary, T., Gutzmer, R., Millward, M., . . . Chapman, P. B. (2012). Dabrafenib in BRAF-mutated metastatic melanoma: A multicentre, open-label, phase 3 randomised controlled trial. *Lancet, 380*(9839), 358–365.

Heidorn, S. J., Milagre, C., Whittaker, S., Nourry, A., Niculescu-Duvas, I., Dhomen, N., . . . Marais, R. (2010). Kinase-dead BRAF and oncogenic RAS cooperate to drive tumor progression through CRAF. *Cell, 140*(2), 209–221.

Hierro, C., Matos, I., Martin-Liberal, J., Ochoa de Olza, M., & Garralda, E. (2019). Agnostic-histology approval of new drugs in oncology: Are we already there? *Clinical Cancer Research, 25*(11), 3210–3219.

Hirshfield, K. M., Tolkunov, D., Zhong, H., Ali, S. M., Stein, M. N., Murphy, S., . . . Ganesan, S. (2016). Clinical actionability of comprehensive genomic profiling for management of rare or refractory cancers. *The Oncologist, 21*(11), 1315–1325.

Huang, L., Fernandes, H., Zia, H., Tavassoli, P., Rennert, H., Pisapia, D., . . . Elemento, O. (2017). The cancer precision medicine knowledge base for structured clinical-grade mutations and interpretations. *Journal of the American Medical Informatics Association, 24*(3), 513–519.

Khater, F., Vairy, S., Langlois, S., Dumoucel, S., Sontag, T., St-Onge, P., . . . Sinnett, D. (2019). Molecular profiling of hard-to-treat childhood and adolescent cancers. *JAMA Network Open, 2*(4), e192906.

Klek, S., Heald, B., Milinovich, A., Ni, Y., Abraham, J., Mahdi, H., . . . Funchain, P. (2020). Genetic counseling and germline testing in the era of tumor sequencing: A cohort study. *JNCI Cancer Spectr, 4*(3), pkaa018.

Knepper, T. C., Bell, G. C., Hicks, J. K., Padron, E., Teer, J. K., Vo, T. T., . . . Walko, C. M. (2017). Key lessons learned from Moffitt's molecular tumor board: The Clinical Genomics Action Committee experience. *The Oncologist, 22*(2), 144–151.

Kopetz, S., Grothey, A., Yaeger, R., Van Cutsem, E., Desai, J., Yoshino, T., . . . Tabernero, J. (2019). Encorafenib, binimetinib, and cetuximab in BRAF V600E–mutated colorectal cancer. *New England Journal of Medicine, 381*(17), 1632–1643.

Kopetz, S., Desai, J., Chan, E., Hecht, J. R., O'Dwyer, P. J., Maru, D., ... Saltz, L. (2015). Phase II pilot study of vemurafenib in patients with metastatic BRAF-mutated colorectal cancer. *Journal of Clinical Oncology, 33*(34), 4032−4038.

Kumar-Sinha, C., Kalyana-Sundaram, S., & Chinnaiyan, A. M. (2015). Landscape of gene fusions in epithelial cancers: Seq and ye shall find. *Genome Medicine, 7*, 129.

Lane, B. R., Bissonnette, J., Waldherr, T., Ritz-Holland, D., Chesla, D., Cottingham, S. L., ... Winn, M. (2015). Development of a center for personalized cancer care at a regional cancer center: Feasibility trial of an institutional tumor sequencing advisory board. *The Journal of Molecular Diagnostics, 17*(6), 695−704.

Larkin, J., Ascierto, P. A., Dréno, B., Atkinson, V., Liszkay, G., Maio, M., ... Ribas, A. (2014). Combined vemurafenib and cobimetinib in BRAF-mutated melanoma. *The New England Journal of Medicine, 371*(20), 1867−1876.

Le Tourneau, C., Delord, J.-P., Gonçalves, A., Gavoille, C., Dubot, C., Isambert, N., ... Paoletti, X. (2015). Molecularly targeted therapy based on tumour molecular profiling vs conventional therapy for advanced cancer (SHIVA): A multicentre, open-label, proof-of-concept, randomised, controlled phase 2 trial. *The Lancet Oncology, 16*(13), 1324−1334.

Levit, L. A., Kim, E. S., McAneny, B. L., Nadauld, L. D., Levit, K., Schenkel, C., & Schilsky, R. L. (2019). Implementing precision medicine in community-based oncology programs: Three models. *Journal of Oncology Practice, 15*(6), 325−329.

Li, M. M., Datto, M., Duncavage, E. J., Kulkarni, S., Lindeman, N. I., Roy, S., ... Nikiforova, M. N. (2017). Standards and guidelines for the interpretation and reporting of sequence variants in cancer: A joint consensus recommendation of the Association for Molecular Pathology, American Society of Clinical Oncology, and College of American Pathologists. *The Journal of Molecular Diagnostics, 19*(1), 4−23.

Long, G. V., Stroyakovskiy, D., Gogas, H., Levchenko, E., de Braud, F., Larkin, J., ... Flaherty, K. (2014). Combined BRAF and MEK inhibition vs BRAF inhibition alone in melanoma. *The New England Journal of Medicine, 371*(20), 1877−1888.

Luchini, C., Lawlor, R. T., Milella, M., & Scarpa, A. (2020). Molecular tumor boards in clinical practice. *Trends in Cancer.*

Maher, C. A., Kumar-Sinha, C., Cao, X., Kalyana-Sundaram, S., Han, B., Jing, X., ... Chinnaiyan, A. M. (2009). Transcriptome sequencing to detect gene fusions in cancer. *Nature, 458*(7234), 97−101.

Massard, C., Michiels, S., Ferté, C., Le Deley, M.-C., Lacroix, L., Hollebecque, A., ... Soria, J.-C. (2017). High-throughput genomics and clinical outcome in hard-to-treat advanced cancers: Results of the MOSCATO 01 trial. *Cancer Discovery, 7*(6), 586.

Mateo, J., Chakravarty, D., Dienstmann, R., Jezdic, S., Gonzalez-Perez, A., Lopez-Bigas, N., ... Pusztai, L. (2018). A framework to rank genomic alterations as targets for cancer precision medicine: The ESMO Scale for Clinical Actionability of molecular Targets (ESCAT). *Annals of Oncology, 29*(9), 1895−1902.

Meric-Bernstam, F., Brusco, L., Shaw, K., Horombe, C., Kopetz, S., Davies, M. A., ... Mills, G. B. (2015). Feasibility of large-scale genomic testing to facilitate enrollment onto genomically matched clinical trials. *Journal of Clinical Oncology, 33*(25), 2753−2762.

Mok, T. S., Wu, Y.-L., Ahn, M.-J., Garassino, M. C., Kim, H. R., Ramalingam, S. S., ... Papadimitrakopoulou, V. A. (2016). Osimertinib or platinum−pemetrexed in EGFR T790M−positive lung cancer. *New England Journal of Medicine, 376*(7), 629−640.

Network, N. C. C. (2020). *Soft Tissue Sarcoma (Version 2.2020).* NCCN Clinical Practive Guidelines in Oncology (NCCN Guidelines) 2020 August 1, 2020]; Available from https://www.nccn.org/professionals/physician_gls/pdf/sarcoma.pdf.

Network, N. C. C. (2020). *Cutaneous Melanoma (version 3.2020)*. NCCN Clinical Practice Guidelines in Oncology (NCCN Guidelines) 2020 [cited 2020 August 1, 2020]; Available from https://www.nccn.org/professionals/physician_gls/pdf/cutaneous_melanoma.pdf.

Network, N. C. C. (2020). *Non-Small Cell Lung Cancer (Version 6.2020)*. NCCN Clinical Practive Guidelines in Oncology (NCCN Guidelines) 2020 [cited 2020 August 1, 2020]; Available from https://www.nccn.org/professionals/physician_gls/pdf/nscl.pdf.

Norquist, B. M., Harrell, M. I., Brady, M. F., Walsh, T., Lee, M. K., Gulsuner, S., ... Birrer, M. J. (2016). Inherited mutations in women with ovarian carcinoma. *JAMA Oncology, 2*(4), 482−490.

Odogwu, L., Mathieu, L., Blumenthal, G., Larkins, E., Goldberg, K. B., Griffin, N., ... Pazdur, R. (2018). FDA approval summary: Dabrafenib and trametinib for the treatment of metastatic non-small cell lung cancers harboring BRAF V600E mutations. *The Oncologist, 23*(6), 740−745.

Oneal, P. A., Kwitkowski, V., Luo, L., Shen, Y. L., Subramaniam, S., Shord, S., ... Pazdur, R. (2018). FDA approval summary: Vemurafenib for the treatment of patients with Erdheim-Chester disease with the BRAFV600 mutation. *The Oncologist, 23*(12), 1520−1524.

Ozsolak, F., & Milos, P. M. (2011). RNA sequencing: Advances, challenges and opportunities. *Nature Reviews Genetics, 12*(2), 87−98.

Pagès, A., Foulon, S., Zou, Z., Lacroix, L., Lemare, F., de Baère, T., ... Bonastre, J. (2017). The cost of molecular-guided therapy in oncology: A prospective cost study alongside the MOSCATO trial. *Genetics in Medicine, 19*(6), 683−690.

Patterson, S. E., Liu, R., Statz, C. M., Durkin, D., Lakshminarayana, A., & Mockus, S. M. (2016). The clinical trial landscape in oncology and connectivity of somatic mutational profiles to targeted therapies. *Human Genomics, 10*(1), 4.

Pearlman, R., Frankel, W. L., Swanson, B., Zhao, W., Yilmaz, A., Miller, K., ... Hampel, H. (2017). Prevalence and spectrum of germline cancer susceptibility gene mutations among patients with early-onset colorectal cancer. *JAMA Oncology, 3*(4), 464−471.

Pennell, N. A., Mutebi, A., Zhou, Z.-Y., Ricculli, M. L., Tang, W., Wang, H., ... Otterson, G. A. (2019). Economic impact of next-generation sequencing vs single-gene testing to detect genomic alterations in metastatic non−small-cell lung cancer using a decision analytic model. *JCO Precision Oncology, 3*, 1−9.

Pishvaian, M. J., Blais, E. M., Bender, R. J., Rao, S., Boca, S. M., Chung, V., ... Madhavan, S. (2019). A virtual molecular tumor board to improve efficiency and scalability of delivering precision oncology to physicians and their patients. *JAMIA Open, 2*(4), 505−515.

Rao, S., Pitel, B., Wagner, A. H., Boca, S. M., McCoy, M., King, I., ... Madhavan, S. (2020). Collaborative, multidisciplinary evaluation of cancer variants through virtual molecular tumor boards informs local clinical practices. *JCO Clinical Cancer Informatics, 4*, 602−613.

Réda, M., Richard, C., Bertaut, A., Niogret, J., Collot, T., Fumet, J. D., ... Ghiringhelli, F. (2020). Implementation and use of whole exome sequencing for metastatic solid cancer. *EBioMedicine, 51*, 102624.

Rolfo, C., Manca, P., Salgado, R., Van Dam, P., Dendooven, A., Machado Coelho, A., ... Pauwels, P. (2018). Multidisciplinary molecular tumour board: A tool to improve clinical practice and selection accrual for clinical trials in patients with cancer. *ESMO Open, 3*(5), e000398.

Ross, D. S., Liu, B., Schram, A. M., Razavi, P., Lagana, S. M., Zhang, Y., ... Hechtman, J. F. (2020). Enrichment of kinase fusions in ESR1 wild-type, metastatic breast cancer revealed by a systematic analysis of 4854 patients. *Annals of Oncology, 31*(8), 991−1000.

Schwaederle, M., Parker, B. A., Schwab, R. B., Fanta, P. T., Boles, S. G., Daniels, G. A., . . . Kurzrock, R. (2014). Molecular tumor board: The University of California-San Diego Moores Cancer Center experience. *The Oncologist, 19*(6), 631−636.

Singer, F., Irmisch, A., Toussaint, N. C., Grob, L., Singer, J., Thurnherr, T., . . . Stekhoven, D. J. (2018). SwissMTB: Establishing comprehensive molecular cancer diagnostics in Swiss clinics. *BMC Medical Informatics and Decision Making, 18*(1), 89, 89.

Stetson, D., Ahmed, A., Xu, X., Nuttall, B. R. B., Lubinski, T. J., Johnson, J. H., Barrett, J. C., & Dougherty, B. A. (2019). Orthogonal comparison of four plasma NGS tests with tumor suggests technical factors are a major source of assay discordance. *JCO Precision Oncology* (3), 1−9.

Subbiah, V., Kreitman, R. J., Wainberg, Z. A., Cho, J. Y., Schellens, J. H. M., Soria, J. C., . . . Keam, B. (2017). Dabrafenib and trametinib treatment in patients with locally advanced or metastatic BRAF V600−mutant anaplastic thyroid cancer. *Journal of Clinical Oncology, 36* (1), 7−13.

Swisher, E. M., Lin, K. K., Oza, A. M., Scott, C. L., Giordano, H., Sun, J., . . . McNeish, A. (2017). Rucaparib in relapsed, platinum-sensitive high-grade ovarian carcinoma (ARIEL2 Part 1): An international, multicentre, open-label, phase 2 trial. *The Lancet Oncology, 18*(1), 75−87.

Tafe, L. J., Gorlov, I. P., de Abreu, F. B., Lefferts, J. A., Liu, X., Pettus, J. R., . . . Chamberlin, M. D. (2015). Implementation of a molecular tumor board: The impact on treatment decisions for 35 patients evaluated at Dartmouth-Hitchcock Medical Center. *The Oncologist, 20* (9), 1011−1018.

Tamborero, D., Rubio-Perez, C., Deu-Pons, J., Schroeder, M. P., Vivancos, A., Rovira, A., . . . Lopez-Bigas, N. (2018). Cancer Genome Interpreter annotates the biological and clinical relevance of tumor alterations. *Genome Medicine, 10*(1), 25.

Tamborero, D., Dienstmann, R., Rachid, M. H., Boekel, J., Baird, R., Braña, I., . . . Lehtiö, J. (2020). Support systems to guide clinical decision-making in precision oncology: The Cancer Core Europe Molecular Tumor Board Portal. *Nature Medicine, 26*(7), 992−994.

Tao, J., Eubank, M. H., Pamer, E., Cangemi, N. A., Drilon, A. E., Harding, J. J., . . . Hyman, D. M. (2018). PRECISE: A clinical-grade automated molecular eligibility screening and just-in-time (JIT) physician decision support solution for molecularly-selected oncology trials. *Journal of Clinical Oncology, 36*(15_suppl), 6507.

Tao, J. J., Eubank, M. H., Schram, A. M., Cangemi, N., Pamer, E., Rosen, E. Y., . . . Hyman, D. M. (2019). Real-world outcomes of an automated physician support system for genome-driven oncology. *JCO Precision Oncology, 3*, 1−13.

Tate, J. G., Bamford, S., Jubb, H. C., Sondka, Z., Beare, D. M., Bindal, N., . . . Forbes, S. A. (2018). COSMIC: The catalogue of somatic mutations in cancer. *Nucleic Acids Research, 47*(D1), D941−D947.

Torga, G., & Pienta, K. J. (2018). Patient-paired sample congruence between 2 commercial liquid biopsy tests. *JAMA Oncology, 4*(6), 868−870.

Trédan, O., Wang, Q., Pissaloux, D., Cassier, P., de la Fouchardière, A., Fayette, J., . . . Blay, J. Y. (2019). Molecular screening program to select molecular-based recommended therapies for metastatic cancer patients: Analysis from the ProfiLER trial. *Annals of Oncology, 30*(5), 757−765.

Twomey, J. D., Brahme, N. N., & Zhang, B. (2017). Drug-biomarker co-development in oncology - 20 years and counting. *Drug Resistance Updates: Reviews and Commentaries in Antimicrobial and Anticancer Chemotherapy, 30*, 48−62.

van der Velden, D. L., van Herpen, C. M. L., van Laarhoven, H. W. M., Smit, E. F., Groen, H. J. M., Willems, S. M., . . . Voest, E. E. (2017). Molecular tumor boards: Current practice and future needs. *Annals of Oncology, 28*(12), 3070−3075.

Wagner, A. H., Walsh, B., Mayfield, G., Tamborero, D., Sonkin, D., Krysiak, K., ... Margolin, A. A. (2020). A harmonized meta-knowledgebase of clinical interpretations of somatic genomic variants in cancer. *Nature Genetics, 52*(4), 448−457.

Willis, J., Lefterova, M. I., Artyomenko, A., Kasi, P. M., Nakamura, Y., Mody, K., ... Odegaard, J. I. (2019). Validation of microsatellite instability detection using a comprehensive plasma-based genotyping panel. *Clinical Cancer Research, 25*(23), 7035−7045.

Yao, Z., Yaeger, R., Rodrik-Outmezguine, V. S., Tao, A., Torres, N. M., Chang, M. T., ... Rosen, N. (2017). Tumours with class 3 BRAF mutants are sensitive to the inhibition of activated RAS. *Nature, 548*(7666), 234−238.

Yoshihara, K., Wang, Q., Torres-Garcia, W., Zheng, S., Vegesna, R., Kim, H., & Verhaak, R. G. (2015). The landscape and therapeutic relevance of cancer-associated transcript fusions. *Oncogene, 34*(37), 4845−4854.

Zill, O. A., Banks, K. C., Fairclough, S. R., Mortimer, S. A., Vowles, J. V., Mokhtari, R., ... Talasaz, A. (2018). The landscape of actionable genomic alterations in cell-free circulating tumor DNA from 21,807 advanced cancer patients. *Clinical Cancer Research, 24*(15), 3528−3538.

Chapter 5

Family health history and health risk assessment in oncology

Lori A. Orlando and R. Ryanne Wu

Duke Center for Applied Genomics & Precision Medicine, Duke Department of Medicine, Duke University, Durham, NC, United States

Introduction

In the continuum from health to disease there are several key transition periods. The first is from healthy to presymptomatic, where an individual still feels well and is asymptomatic but has developed a disease. An example of this health state is the beginning of colorectal cancer when the cancer is present but too small to cause any symptoms yet. The second is from presymptomatic to disease diagnosis and the third is from diagnosis to disease status, which can be either well controlled or uncontrolled. Health risk assessments (HRAs) are an essential component of the healthy period. Their purpose is to estimate an individual's risk for developing common diseases (see Table 5.1 for examples) allowing clinicians to tailor preventive care, screening, and testing to each individual's level of risk with the goal of keeping healthy people healthy. Personalized care plans developed with the aid of HRAs balance effectiveness and harms with risk, in a way that maximizes benefit and minimizes harm not only for each individual, but also when taken as a whole for the population as well. An example is the use of breast MRI for breast cancer screening. MRI has a greater sensitivity than other modalities such as mammogram for detecting breast cancer but results in frequent biopsies of benign lesions. To balance these risks and benefits, guidelines recommend only using breast MRI for screening in women with $\geq 20\%$ lifetime risk of breast cancer (National Comprehensive Cancer Network, 2018). Unfortunately, HRAs are not widely used in primary care, where they would be most effective, due to a number of constraints. This chapter discusses how HRAs were developed, their key aspects, and what needs to occur in order to integrate them into primary care settings.

Genomic and Precision Medicine. DOI: https://doi.org/10.1016/B978-0-12-800684-9.00021-6

TABLE 5.1 Examples of cancer conditions for which family health history–based health risk assessment is useful.

	Risk algorithm based on family health history only	Risk algorithms include family health history
Hereditary breast and ovarian cancer	×	
Hereditary nonpolyposis colon cancer (Lynch syndrome)	×	
Melanoma		×
Prostate cancer		×
Renal cell cancer		×
Lung cancer		×
Familial adenomatous polyposis	×	
Pancreatic cancer		×

In the beginning

In 1948, Joseph Mountain, the Assistant Surgeon General, initiated the Framingham Heart Study, an innovative longitudinal study arising from the field of epidemiology. The goal, as devised by the director, Thomas Dawber, was to closely follow a group of individuals living in Framingham, Massachusetts, collecting as much data as possible over the course of many years in order to develop a risk prediction model for heart disease (Dawber, Meadors, & Moore, 1951). This was the first time the phrase "factor of risk," more commonly termed risk factor today, was introduced (Kannel, Dawber, Kagan, Revotskie, & Stokes, 1961). Despite initial skepticism among both the research and medical communities, the trial was successful beyond expectations and the field of HRA was born. In 2009, when Clay Christensen coined the term "Precision Medicine," he defined it as precisely predicting a medical outcome by combining a variety of data into rules (Christensen, Grossman, & Hwang, 2009). By this definition, HRAs are simply the application of precision medicine to those who are healthy.

Today most HRAs include the following components: data collection (either through a web-based or paper questionnaire), risk calculation, and report of risk results. This last component, the report, may or may not provide guidance about how to manage your risk. Some are exceptionally detailed and even indicate how much your risk can be lowered by initiating one or more recommended preventive actions, while others merely indicate that you are at increased risk for the specified condition. For the first component, data collection, the data collected vary depending upon which conditions are included in the risk assessment, but at a minimum they all include: demographics, lifestyle, personal health history, *family health history*, and biometrics (such as blood pressure, weight, and cholesterol). Other types of data, such as genetic/genomic and individual preferences, are just now starting to be incorporated into some risk assessment models and have the potential to not only refine the accuracy of risk calculations, but also improve shared decision making with medical providers (Glasgow et al., 2011; Krist et al., 2013).

Why family health history is central to HRAs

Family health history is an unassuming and often overlooked, but essential data element in HRAs. For many conditions, family health history is the strongest predictor of disease risk and for some, such as hereditary cancer syndromes, it is the only predictor (and thus the only component of the HRA) (see Table 5.1). An example of the impact of family health history on disease risk is prostate cancer, where having a father or brother with the disease, more than doubles the individual's risk (Kiciński, Vangronsveld, & Nawrot, 2011). In some cases, excluding a family health history can lead to missing those at highest risk for developing a condition. For example, risk assessments for lung cancer ask about environmental exposures (such as smoking and asbestos), but do not ask about family history; however, a family history of lung cancer can double the risk of developing lung cancer in smokers (Coté et al., 2012). Renal cell carcinoma, a tumor of the kidney, is another example. Almost all risk assessments include smoking, alcohol, and exercise, and some include family members with renal cell carcinoma, but most do not ask about a family history of other cancers even though renal cell carcinoma is part of the constellation of cancers that can occur in two hereditary cancer syndromes, Lynch and Von Hippel-Lindau (Chow, Dong, & Devesa, 2010). While those with hereditary cancer syndromes are only a small proportion of those developing renal cell carcinoma, they are the ones at the highest risk of developing disease.

In addition to being highly predictive, family health history also serves as the basis for a number of evidence-based guidelines that not only indicate the level of disease risk associated with a given combination of affected relatives, but also actions to take to manage risk. For example, the National Comprehensive Cancer Network's guidelines for breast and ovarian cancer recommend BRCA testing if an individuals' first degree relative (parents or child) developed breast cancer at age 45 or younger (Daly et al., 2020). Another example is colorectal cancer screening. If an individual has a relative with colorectal cancer or advanced adenomatous polyps, then screening is recommended to start earlier than for the general population and to be done more frequently (National Comprehensive Cancer Network, 2021).

Thus family health history is the only data element in HRAs that is both highly predictive and actionable in combination with other data elements and by itself. Unfortunately, family health history is often hard to obtain. Individuals often do not know much about their relatives' health and what they do know is often piecemeal or may be inaccurate (Quereshi et al., 2007). This leads to the problem that one of the most informative data elements in HRAs is also one of the more difficult to collect.

An implementation crisis

Despite the acclaim surrounding the publication of the Framingham Heart Study results, there was little movement in the field of HRA until 1980 when the Center for Disease Control (CDC) released a publicly available HRA tool (Yoon, Scheuner, Jorgensen, & Khoury, 2009). Incidentally, 1980 was also a time when employers and insurers were being faced with rapidly increasing health care costs. In their search for a way to manage these costs, they turned to HRAs (Breslow, Fielding, Herrman, & Wilbur, 1990). To explore the impact of this resource, Prudential funded updates to the CDC's tool, which ultimately showed that use of an HRA tool in the workplace could lower company health care expenditures, as well as reduce absenteeism and increase productivity (DeFriese & Fielding, 1990; Loeppke et al., 2008). These results and Prudential's takeover of the program in 1986 led to rapid uptake among US employers and insurance companies; however, uptake continued to be anemic in the healthcare setting (Goetzel et al., 2011).

Explanations for why implementation in the healthcare system failed to take root include: the disconnect between public health and health care, increasing demands on primary care providers, and a perverse incentive system that rewarded interventions over maintaining health (Goetzel et al., 2011; McGinnis, Williams-Russo, & Knickman, 2002). The combination of these factors encouraged the development of a healthcare system incapable of responding to the needs of the healthy segment of the population, quickly leading to a negative feedback loop dominated by sick patients getting sicker, less time to manage risk among healthy patients, and ultimately healthy patients getting sick (Stange, Zyzanski, & Jaen, 1998). In this environment it is easy to see how adoption of HRA in clinical practice was slow.

Fortunately, recent studies have highlighted these findings and their significant impact on the US health care system. In particular, the Mirror, Mirror studies performed by the Commonwealth Fund to assess health care quality and cost in 11 international healthcare systems between 2004 and 2021, not only ranked the United States last in quality and highest in expenditures; but showed little improvement over the 17-year period (Schneider et al., 2021). In addition, the Affordable Care Act enacted in 2010 has emphasized the need for improvements in quality of care, maintaining health, and lowering costs. HRAs are neatly aligned with these objectives and are now viewed as a useful tool for redesigning healthcare systems (Goetzel et al., 2011). That being said, there are still a number of practical barriers to overcome before implementation in primary care can become widespread: ease of use for providers, ease of use for patients, quality of the data-entered into the HRA (particularly family health history data), and evaluation of HRAs' potential to improve quality of care in primary care populations. Each of these are describe in detail below.

Will providers use it?

Providers (primary care providers in particular) are frequently overloaded by the number of tasks to achieve within the constraints of the healthcare visit, and with visit times averaging 15−18 min for most appointments many lower priority and/or complex tasks often lose out to higher acuity concerns (Shaw, Davis, Fleischer, & Feldman, 2014; Tai-Seale, McGuire, & Zhang, 2007). Because HRA data collection, risk calculation, and evidence synthesis are complex and time consuming (particularly for family health history data), it often poses a significant challenge for integration into normal work flow (Rich et al., 2004; Sweet, Bradley, & Westman, 2002a, 2002b). In particular, risk algorithms are often complex, requiring a computer to calculate, however, the calculators are typically scattered across the Internet and not integrated into electronic medical record systems. In addition, the sheer magnitude of the literature available makes synthesizing an actionable risk management plan difficult and efforts to initiate provider education around these topics have fallen flat for many of the same reasons that implementing HRAs have (Barrison, Smith, Oviedo, Heeren, & Schroy, 2003; Schroy et al., 2005).

One solution to these complex and interrelated barriers is to leverage the burgeoning field of health information technology (IT). Patient-facing web- or computer-based HRA tools can eliminate the data collection component from the physician's office, moving it to the patient's home, and provide risk calculations and actionable risk management plans to the provider at the point-of-care. Similarly, mobile health technologies (mHealth) are beginning to demonstrate that they can be used to facilitate risk-related data collection (environmental, behavioral, psychological, and biological) and communication between patients and their healthcare system (Silva, Rodrigues, de la Torre Díez, López-Coronado, & Saleem, 2015). To date a number of family health history−based HRA tools have been built with just such capabilities (Ginsburg, Wu, & Orlando, 2019; Welch et al., 2018). Uptake of these tools has been anecdotally promising. There have been limited evaluations of physician experience with and uptake of HRA tools. In the two studies that did evaluate this, primary care providers indicated that the HRA tool was easy to incorporate into workflow, improved their quality of care, made their practice easier, and enhanced their understanding of the importance of family health history (Emery et al., 2007; Wu et al., 2013). These results suggest that with the right combination of features, electronic HRA tools can gain acceptance by busy primary care clinicians.

Will the patients use it?

With the expansion of web-access, health systems' greater emphasis on patient engagement, and the increasing demands on medical providers' time, there has been a movement toward increased utilization of patient-derived data for HRAs. Some have raised concern that patients do not have the IT know-how to use these tools or the inclination to do so. But current trends would argue that that is not the case. Millions of patients use the Internet on a daily basis to better understand their health (Fox & Duggan, 2021; Hesse, Moser, & Rutten, 2010) and even among minorities and those with household incomes of $25,000–$50,000 per year Internet use is over 77% (Martin, 2021). In addition, individuals recognize the importance of family health history with over 90% of Americans surveyed reporting that they felt family health history was important to their personal health (Centers for Disease Control and Prevention, 2004; Foland & Burke, 2014). When HRAs are placed on health systems' patient portals significant uptake occurs. For example, when Health Heritage was launched on the Northshore University Health System's health portal in May 2014, 500 patients completed the assessment within the first 30 days (unpublished data). This was without any coordinated efforts by the health system to alert patients to the availability of the tool.

There are also concerns that such tools discriminate against those of lower socio-economic status who may not have the IT literacy or access required to complete such tools. In fact, the reverse is true—HRA tools can potentially reduce health disparities. Algorithms to assess risk and provide recommendations do not hold the inherent biases that providers may subconsciously bring to the patient encounter. While early family health history studies largely enrolled Caucasian middle-class participants, the focus is now on engaging and evaluating their performance in more diverse populations (Wang et al., 2015; Wu et al., 2019). In two recent studies, minorities and those of lower socio-economic status reported that the tools were easy to use (91%−94%) and understand (97%) irrespective of education level or ethnicity (Wang et al., 2015; Wu et al., 2013).

When patients do use IT tools for HRA, reactions are overwhelmingly positive. In five studies, participants reported being generally satisfied with their experience (77%−90%), that completing the tool did not cause persistent anxiety (96%), and that they would recommend it to others (83%−99%)

(Cohn et al., 2010; Fung et al., 2021; Qureshi, Standen, Hapgood, & Hayes, 2001; Wang et al., 2015; Wu et al., 2013). In addition to having a positive experience using the tool itself, patients incur the additional benefits of: raised awareness about their risk for disease and the actions they can take to mitigate their risk, and greater engagement in their health with the potential for an enhanced shared-decision-making experience with their providers (Glasgow et al., 2011; Krist et al., 2013) (see Fig. 5.1 for HRA screenshot).

What about patient-entered data?

If we move HRA data collection out of the provider encounter, are the data entered by patients reliable? In particular, there are concerns about patients' ability to accurately collect and interpret their family's health history. Questions arise around differentiation of a primary cancer versus metastatic cancer, different types of cardiovascular disease such as coronary artery disease versus arrhythmias, and diseases with names that sound similar such as cirrhosis and psoriasis. In a systematic review of family history questionnaires, four studies examined the agreement between patient-entered data and the presumed gold standard of genetic counselor acquired data with evidence of strong agreement between them (70%−100%) (Reid, Walter, Brisbane, &, Emery, 2009). In fact there is evidence that patient-entered data

FIGURE 5.1   MeTree HRA screen shot.

are significantly superior to what is collected in routine practice within primary care (Cohn et al., 2010; Frezzo, Rubinstein, Dunham, & Ormond, 2003; Reid et al., 2009; Sweet et al., 2002a). In our own experience, less than 4% of patients' medical records contained high-quality family history information documented for even one relative (Powell et al., 2013). In comparison, with use of MeTree >99% of pedigrees had at least one relative with high-quality family history information and over 50% of pedigrees had at least 50% of relatives with high-quality information (Wu et al., 2014).

Certainly, the context within which family health history is collected will affect its validity. When patients are offered education and the opportunity to discuss their family health history with relatives, significant improvements in accuracy can be seen. When patients are informed of the key components of a quality family history and use that information to guide discussions with family members, almost half will provide new or updated information and 16% will have a change in recommendations for disease risk management following conversations with family members (Beadles et al., 2014).

Does it make a difference?

Knowing that providers and patients will use a HRA and that patient-reported data are reliable, does use of such tools make a difference in outcomes? The value of HRAs is that there are now a considerable number of guidelines that tailor recommended risk management strategies to an individual's risk for disease. Strategies for higher-risk individuals often have a greater sensitivity than strategies recommended for population-based screening but are also typically associated with higher costs or adverse events rates that warrant limiting them to those that are most likely to benefit. This tailoring of risk strategies to risk level helps to balance benefits and harms at the individual level and when assessed at the population level is better than using one-size fits all recommendations. To show that HRAs are in fact able to increase uptake of these guidelines and improve individual/population health they need to be able to: improve identification of individuals at increased disease risk, increase the match between risk level and risk management strategy, and improve patient acceptance of management recommendations above what is currently occurring in routine care.

Surveys of providers have shown that physicians frequently over- or under-estimate risk and do not feel confident in their ability to assess risk (Acton et al., 2000; Baldwin et al., 2014; Barrison et al., 2003; Bellcross et al., 2011; Dhar et al., 2011; Gramling, Nash, Siren, Eaton, & Culpepper, 2004). Fortunately several studies have shown that family health history−based HRAs can accurately estimate risk and identify a significant number of patients who were not identified during routine visits with their providers (Cohn et al., 2010; Frezzo et al., 2003; Qureshi et al., 2012; Rubinstein et al., 2011). For example, The Family Healthware Trial found that 82% of participants were at strong or moderate risk for at least one of six conditions (O'Neill et al., 2009); the Health Heritage trial found 42% to be at increased risk (Cohn et al., 2010); and the IGNITE study found 58% to be at increased risk (Orlando et al., 2020). The majority of these were not previously identified as being at high risk and were not receiving guideline-based care (Cohn et al., 2010; Emery et al., 2007; Rubinstein et al., 2011).

While evidence is clear that using HRAs increases identification of at-risk individuals, less has been done to evaluate the impact on provider and patient behavior as a result of this risk information. In the authors' experience with MeTree, use of the tool decreased over-utilization of high-risk services among average risk individuals by 81% while increasing appropriate use among high-risk individuals (Orlando et al., in press). In another study, the implementation of a breast cancer risk assessment tool, HughesRiskApps, resulted in significant increases in referrals to genetic counseling and genetic testing (Ozanne et al., 2009). Other studies have also confirmed that understanding of family history increases appropriate utilization of cancer screening services (Codori, Petersen, Miglioretti, & Boyd, 2001; Murabito et al., 2001). Further work needs to be done in this area to better understand the impact of HRAs on patient and provider behavior as it relates to screening (Qureshi et al., 2009).

Conclusion

HRAs, a precision medicine tool intended to be used by well patients, hold promise for improving both individual health and population health by accurately estimating risk for disease and improving the match between risk level and risk management strategy above what is currently occurring in clinical care. These tools may also enhance patient experience by increasing patient activation and shared decision-making, thereby incorporating patient values into the patient-provider encounter and improving adherence to recommendations. Their accuracy and effectiveness may be further enhanced by combining genetic and genomic data about disease risk into risk algorithms. While uptake in the US healthcare system has been slow to start, it is beginning to gain traction as the evidence builds around their benefit and innovations in health IT permit more seamless integration into clinical workflow. It is one of the few genomic and personalized medicine tools that is ready for immediate translation into clinical care (especially primary care) and as such can lead the way for translation of other genomic and personalized medicine in these fields by increasing clinicians' familiarity with these types of tools and their benefits.

References

Acton, R. T., Burst, N. M., Casebeer, L., et al. (2000). Knowledge, attitudes, and behaviors of Alabama's primary care physicians regarding cancer genetics. *Academic Medicine: Journal of the Association of American Medical Colleges, 75*(8), 850–852.

Baldwin, L. M., Trivers, K. F., Andrilla, C. H., et al. (2014). Accuracy of ovarian and colon cancer risk assessments by U.S. physicians. *Journal of General Internal Medicine.*

Barrison, A. F., Smith, C., Oviedo, J., Heeren, T., & Schroy, P. C., 3rd (2003). Colorectal cancer screening and familial risk: A survey of internal medicine residents' knowledge and practice patterns. *The American Journal of Gastroenterology, 98*(6), 1410–1416.

Beadles, C. A., Ryanne, Wu. R., Himmel, T., et al. (2014). Providing patient education: Impact on quantity and quality of family health history collection. *Familial Cancer.*

Bellcross, C. A., Kolor, K., Goddard, K. A., Coates, R. J., Reyes, M., & Khoury, M. J. (2011). Awareness and utilization of BRCA1/2 testing among U.S. primary care physicians. *American Journal of Preventive Medicine, 40*(1), 61–66.

Breslow, L., Fielding, J., Herrman, A. A., & Wilbur, C. S. (1990). Worksite health promotion: Its evolution and the Johnson & Johnson experience. *Preventive Medicine, 19*(1), 13–21.

Centers for Disease Control and Prevention. (2004). Awareness of family health history as a risk factor for disease–United States, 2004. *Morbidity and Mortality Weekly Report, 53*(44), 1044–1047.

Chow, W. H., Dong, L. M., & Devesa, S. S. (2010). Epidemiology and risk factors for kidney cancer. *Nature Reviews Urology, 7*(5), 245–257.

Christensen, C. M., Grossman, J. H., & Hwang, J. (2009). *The innovator's prescription: A disruptive solution for health care.* New York: McGraw-Hill.

Codori, A. M., Petersen, G. M., Miglioretti, D. L., & Boyd, P. (2001). Health beliefs and endoscopic screening for colorectal cancer: Potential for cancer prevention. *Preventive Medicine, 33*(2 Pt 1), 128–136.

Cohn, W. F., Ropka, M. E., Pelletier, S. L., et al. (2010). Health Heritage(c) a web-based tool for the collection and assessment of family health history: Initial user experience and analytic validity. *Public Health Genomics, 13*(7–8), 477–491.

Coté, M. L., Liu, M., Bonassi, S., et al. (2012). Increased risk of lung cancer in individuals with a family history of the disease: A pooled analysis from the International Lung Cancer Consortium. *European Journal of Cancer (Oxford, England: 1990), 48*(13), 1957–1968.

Daly, M. B., Pilarski, R., Yurgelun, M. B., et al. (2020). NCCN guidelines insights: Genetic/familial high-risk assessment: Breast, ovarian, and pancreatic, version 1.2020: Featured updates to the NCCN Guidelines. *Journal of the National Comprehensive Cancer Network, 18*(4), 380–391.

Dawber, T. R., Meadors, G. F., & Moore, F. E., Jr. (1951). Epidemiological approaches to heart disease: The Framingham Study. *American Journal of Public Health and the Nation's Health, 41*(3), 279–281.

DeFriese, G. H., & Fielding, J. E. (1990). Health risk appraisal in the 1990s: Opportunities, challenges, and expectations. *Annual Review of Public Health, 11*(1), 401–418.

Dhar, S. U., Cooper, H. P., Wang, T., et al. (2011). Significant differences among physician specialties in management recommendations of BRCA1 mutation carriers. *Breast Cancer Research and Treatment, 129*(1), 221–227.

Emery, J., Morris, H., Goodchild, R., et al. (2007). The GRAIDS Trial: A cluster randomised controlled trial of computer decision support for the management of familial cancer risk in primary care. *British Journal of Cancer, 97*, 486.

Foland, J., & Burke, B. (2014). *Family health history data collection in Connecticut.* Hartford, CT: Connecticut Department of Public Health, Genomics Office.

Fox, S., & Duggan, M. (2013). Health Online 2013: A project of the Pew Research Center. https://www.pewresearch.org/internet/2013/01/15/health-online-2013/ (Accessed 11.08.21).

Frezzo, T. M., Rubinstein, W. S., Dunham, D., & Ormond, K. E. (2003). The genetic family history as a risk assessment tool in internal medicine. *Genetics in Medicine: Official Journal of the American College of Medical Genetics, 5*(2), 84−91.

Fung, S. M., Wu, R. R., Myers, R. A., et al. (2021). Clinical implementation of an oncology-specific family health history risk assessment tool. *Hereditary Cancer in Clinical Practice, 19*(1), 20.

Ginsburg, G. S., Wu, R. R., & Orlando, L. A. (2019). Family health history: Underused for actionable risk assessment. *Lancet, 394*(10198), 596−603.

Glasgow, R. E., Dickinson, P., Fisher, L., et al. (2011). Use of RE-AIM to develop a multimedia facilitation tool for the patient-centered medical home. *Implementation Science, 6*, 118.

Goetzel, R. Z., Staley, P., Ogden, L., et al. (2011). *A framework for patient-centered health risk assessments−providing health promotion and disease prevention services to Medicare beneficiaries.* Atlanta, GA: U.S. Department of Health and Human Services; Centers for Disease Control and Prevention.

Gramling, R., Nash, J., Siren, K., Eaton, C., & Culpepper, L. (2004). Family physician self-efficacy with screening for inherited cancer risk. *Annals of Family Medicine, 2*(2), 130−132.

Hesse, B. W., Moser, R. P., & Rutten, L. J. (2010). Surveys of physicians and electronic health information. *New England Journal of Medicine, 362*(9), 859−860.

Kannel, W. B., Dawber, T. R., Kagan, A., Revotskie, N., & Stokes, J., III (1961). Factors of risk in the development of coronary heart disease−six year follow-up experience. The Framingham Study. *Annals of Internal Medicine, 55*, 33−50.

Kiciński, M., Vangronsveld, J., & Nawrot, T. S. (2011). An epidemiological reappraisal of the familial aggregation of prostate cancer: A *meta*-analysis. *PLoS One, 6*(10), e27130.

Krist, A. H., Glenn, B. A., Glasgow, R. E., et al. (2013). Designing a valid randomized pragmatic primary care implementation trial: The my own health report (MOHR) project. *Implementation Science, 8*, 73.

Loeppke, R., Nicholson, S., Taitel, M., Sweeney, M., Haufle, V., & Kessler, R. C. (2008). The impact of an integrated population health enhancement and disease management program on employee health risk, health conditions, and productivity. *Population Health Management, 11*(6), 287−296.

Martin, M. (2021). Computer and internet use in the United States, 2018. In: United States Census Bureau, editor.

McGinnis, J. M., Williams-Russo, P., & Knickman, J. R. (2002). The case for more active policy attention to health promotion. *Health Affairs, 21*(2), 78−93.

Murabito, J. M., Evans, J. C., Larson, M. G., et al. (2001). Family breast cancer history and mammography: Framingham Offspring Study. *American Journal of Epidemiology, 154*(10), 916−923.

National Comprehensive Cancer Network. (2018). NCCN Clincial Practice Guidelines in Oncology: Genetic/Familial High Risk Assessment: Breast and Ovarian. https://www.nccn.org/professionals/physician_gls/pdf/genetics_screening.pdf.

National Comprehensive Cancer Network. (2021). NCCN Clinical Practice Guidelines in Oncology: Colorectal Cancer Screening. https://www.nccn.org/professionals/physician_gls/pdf/colorectal_screening.pdf (Accessed 10.08.21).

O'Neill, S. M., Rubinstein, W. S., Wang, C., et al. (2009). Familial risk for common diseases in primary care: The family healthware impact trial. *American Journal of Preventive Medicine, 36*(6), 506−514.

Orlando, L., Wu, R. R., Myers, R. A., et al. (in press). Clinical utility of a web-enabled risk assessment and clinical decision support program. *Genetics in Medicine.*

Orlando, L. A., Wu, R. R., Myers, R. A., et al. (2020). At the intersection of precision medicine and population health: An implementation-effectiveness study of family health history based systematic risk assessment in primary care. *BMC Health Services Research, 20*(1), 1015.

Ozanne, E. M., Loberg, A., Hughes, S., et al. (2009). Identification and management of women at high risk for hereditary breast/ovarian cancer syndrome. *The Breast Journal, 15*(2), 155–162.

Powell, K. P., Christianson, C. A., Hahn, S. E., et al. (2013). Collection of family health history for assessment of chronic disease risk in primary care. *North Carolina Medical Journal, 74* (4), 279–286.

Quereshi, N. W. B., Santaguida, P., Carroll, J., Allanson, J., Culebro, C., Brouwers, M., & Raina, P. (2007). Collection and use of cancer family history in primary Care. AHRQ Publication No. 08-E001. http://www.ahrq.gov/clinic/tp/famhisttp.htm.2015.

Qureshi, N., Armstrong, S., Dhiman, P., et al. (2012). Effect of adding systematic family history enquiry to cardiovascular disease risk assessment in primary care: A matched-pair, cluster randomized trial. *Annals of Internal Medicine, 156*(4), 253–262.

Qureshi, N., Standen, P. J., Hapgood, R., & Hayes, J. (2001). A randomized controlled trial to assess the psychological impact of a family history screening questionnaire in general practice. *Family Practice, 18*(1), 78–83.

Qureshi, N., Wilson, B., Santaguida, P., et al. (2009). Family history and improving health. *Evidence Report/Technology Assessment, 186*, 1–135.

Reid, G. T., Walter, F. M., Brisbane, J. M., & Emery, J. D. (2009). Family history questionnaires designed for clinical use: A systematic review. *Public Health Genomics, 12*(2), 73–83.

Rich, E. C., Burke, W., Heaton, C. J., et al. (2004). Reconsidering the family history in primary care. *Journal of General Internal Medicine, 19*(3), 273–280.

Rubinstein, W. S., Acheson, L. S., O'Neill, S. M., et al. (2011). Clinical utility of family history for cancer screening and referral in primary care: A report from the Family Healthware Impact Trial. *Genetics in Medicine: Official Journal of the American College of Medical Genetics, 13*(11), 956–965.

Schneider, E., Shah, A., Doty, M. M., Tikkanen, R., Fields, K., & Willaims II, R. D. (2021). Mirror, Mirror 2021: Reflecting poorly healthcare in the U.S. compared to other high-income countries.

Schroy, P. C., III, Glick, J. T., Geller, A. C., Jackson, A., Heeren, T., & Prout, M. (2005). A novel educational strategy to enhance internal medicine residents' familial colorectal cancer knowledge and risk assessment skills. *The American Journal of Gastroenterology, 100*(3), 677–684.

Shaw, M. K., Davis, S. A., Fleischer, A. B., & Feldman, S. R. (2014). The duration of office visits in the United States, 1993 to 2010. *The American Journal of Managed Care, 20*(10), 820–826.

Silva, B. M. C., Rodrigues, J. J. P. C., de la Torre Díez, I., López-Coronado, M., & Saleem, K. (2015). Mobile-health: A review of current state in 2015. *Journal of Biomedical Informatics, 56*, 265–272.

Stange, K. C., Zyzanski, S. J., Jaen, C. R., et al. (1998). Illuminating the 'black box'. A description of 4454 patient visits to 138 family physicians. *The Journal of Family Practice, 46*(5), 377–389.

Sweet, K. M., Bradley, T. L., & Westman, J. A. (2002a). Identification and referral of families at high risk for cancer susceptibility. *Journal of Clinical Oncology: Official Journal of the American Society of Clinical Oncology, 20*(2), 528–537.

Sweet, K. M., Bradley, T. L., & Westman, J. A. (2002b). Identification and referral of families at high risk for cancer susceptibility. *Journal of Clinical Oncology, 20*(2), 528−537.

Tai-Seale, M., McGuire, T. G., & Zhang, W. (2007). Time allocation in primary care office visits. *Health Services Research, 42*(5), 1871−1894.

Wang, C., Bickmore, T., Bowen, D. J., et al. (2015). Acceptability and feasibility of a virtual counselor (VICKY) to collect family health histories. *Genetics in Medicine: Official Journal of the American College of Medical Genetics, 17*(10), 822−830.

Welch, B. M., Wiley, K., Pflieger, L., et al. (2018). Review and comparison of electronic patient-facing family health history tools. *Journal of Genetic Counseling.*

Wu, R. R., Himmel, T. L., Buchanan, A. H., et al. (2014). Quality of family history collection with use of a patient facing family history assessment tool. *BMC Family Practice, 15*(1), 31.

Wu, R. R., Myers, R. A., Buchanan, A. H., et al. (2019). Effect of sociodemographic factors on uptake of a patient-facing information technology family health history risk assessment platform. *Applied Clinical Informatics, 10*(2), 180−188.

Wu, R. R., Orlando, L. A., Himmel, T. L., et al. (2013). Patient and primary care provider experience using a family health history collection, risk stratification, and clinical decision support tool: A type 2 hybrid controlled implementation-effectiveness trial. *BMC Family Practice, 14*(1), 111.

Yoon, P. W., Scheuner, M. T., Jorgensen, C., & Khoury, M. J. (2009). Developing Family Healthware, a family history screening tool to prevent common chronic diseases. *Preventing Chronic Disease, 6*(1), A33.

Chapter 6

Lymphomas: molecular subsets and advances in therapeutics

Sandeep Dave

Division of Hematologic Malignancies and Cellular Therapy, Department of Medicine,
Duke University Medical Center, Durham, NC, United States

Introduction

Lymphomas are a heterogeneous group of malignancies that arise from lymphocytes and have associated unique oncogenic drivers that can be targeted by precision medicine approaches. Since the first report of lymphoma by Thomas Hodgkin in 1832 (Hodgkin, 1832), lymphomas have been grouped into either Hodgkin's lymphoma or non-Hodgkin lymphoma (NHL) and the most recent classifications for pathology diagnosis list dozens of lymphoma subtypes (Swerdlow, Campo, & Pileri, 2016). In 2020 an estimated 77,240 cases of NHL were diagnosed causing 19,940 deaths in the United States alone (National Cancer Institute: Surveillance Epidemiology, and End Results Program). Advances in technology have revealed multiple additional subtypes of NHL. Of these subtypes, diffuse large B-cell lymphoma (DLBCL) accounts for the largest percentage of cases. Lymphomas are generally treatable with chemotherapy. Even so, the majority of patients with DLBCL and a number of other lymphoma subtypes will eventually succumb to their disease. This chapter will begin with an overview of lymphoma and discuss the influence of genomics on diagnosis, prognosis, and treatment, with emphasis on the most common entities including DLBCL, follicular lymphoma, mantle cell lymphoma (MCL), Burkitt's lymphoma, and Hodgkin's lymphoma.

Diffuse large B-cell lymphoma

DLBCL is the most common subtype of NHL with an annual incidence in the United States of >20,000 cases (National Cancer Institute: Surveillance Epidemiology, and End Results, Program). Less than 50% of cases can be cured with initial combination chemotherapy and immunotherapy with rituximab, which is currently standard of care (Coiffier, Lepage, & Briere, 2002;

Genomic and Precision Medicine. DOI: https://doi.org/10.1016/B978-0-12-800684-9.00016-2
85

Feugier, Van Hoof, & Sebban, 2005) and many of these patients relapse after or cannot tolerate salvage chemotherapy with high-dose chemotherapy and stem cell support (Philip, Guglielmi, & Hagenbeek, 1995; Shipp, Abeloff, & Antman,1999). Thus effective and well-tolerated novel strategies in relapsed/ refractory DLBCL are urgently needed.

A major impediment to identifying new therapeutics in lymphoma is the inherent heterogeneity of the disease. DLBCL, the most common form of lymphoma, can be subdivided into at least three subtypes based on gene expression profiles [activated B cell (ABC), germinal center B cell (GCB), and primary mediastinal B cell (PMBL)] (Alizadeh, Eisen, & Davis, 2000; Rosenwald, Wright, & Chan, 2002); other classification schemes based on molecular/genomic phenotypes have been reported (Monti, Savage, & Kutok, 2005) (see Fig. 6.1 for overview of heterogeneity of lymphomas including DLBCL). In addition, >100 recurrent oncogenic mutations in DLBCL underlie these broad classifications and only 10%−15% of DLBCL cases harbor the most frequent mutations (Lohr, Stojanov, & Lawrence, 2012; Morin, Mendez-Lago, & Mungall, 2011; Zhang, Grubor, & Love, 2013) (Fig. 6.1). Table 6.1 summarizes the most common molecular alterations in DLBCL and other lymphomas.

The malignant cells of the ABC subgroup of DLBCLs arise from post-GCBs that are poised to differentiate into plasma cells. The genes in the ABC DLBCL-defining gene expression signatures are associated with plasma-cell differentiation, with high secretion of proteins and immunoglobulins (Alizadeh, Eisen, & Davis, 2000). Genetic alterations associated with the ABC subtype include deletion of the known tumor suppressor INK4a/ARF, and trisomy 3 (Chapuy, Stewart, & Dunford, 2018; Lenz, Wright, & Dave, 2008; Lenz, Wright, & Emre, 2008; Pasqualucci & Dalla-Favera, 2018; Rosenwald, Wright, & Chan, 2002; Schmitz, Wright, & Huang, 2018). Unlike GCB DLBCLs, which express BCL6, the ABC subgroup of DLBCLs overexpresses IRF4, which normally functions to increase cell division after the presentation of an antigen, but becomes permanently active in the development of DLBCL (Alizadeh et al., 2000). The genes PRKCB1 and PDE4B were more highly expressed in ABC than GCB DLBCL (Staudt & Dave, 2005). A third distinct subtype of DLBCL has since emerged, called PMBL, which overlaps with the other subtypes with regard to expression of the LN signature and increased activity of NF-κB, and its signature had one-third of its genes in common with a gene expression signature reflecting Hodgkin's lymphoma (Rosenwald, Wright, & Leroy, 2003). There was also a clinical overlap with Hodgkin's disease, with most cases occurring in young women as mediastinal disease (Boleti & Johnson 2007; Brittig, Csanaky, Kecskés, & István, 1991), with better 5-year overall survival compared to DLBCL (64% vs. 46%) (Mottok, Hung, & Chavez, 2019; Steidl & Gascoyne 2011; Twa & Steidl 2015). The gene that was best able to distinguish PMBL from the other DLBCL subtypes was PDL2 (programmed death ligand 2), which is located on chromosome 9p and is also characteristic of Hodgkin's lymphoma (Rosenwald et al., 2003).

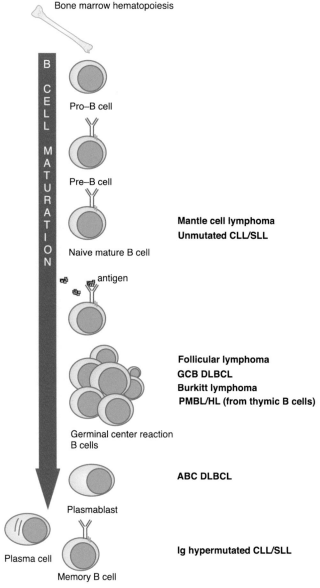

Bone marrow hematopoiesis

B CELL MATURATION

Pro–B cell

Pre–B cell

Naive mature B cell

Mantle cell lymphoma
Unmutated CLL/SLL

antigen

Follicular lymphoma
GCB DLBCL
Burkitt lymphoma
PMBL/HL (from thymic B cells)

Germinal center reaction
B cells

ABC DLBCL

Plasmablast

Plasma cell

Memory B cell

Ig hypermutated CLL/SLL

FIGURE 6.1 Schematic of B cell maturation and cell or origin for B cell lymphomas and plasma cell dyscrasias. B cell lymphomas are represented by heterogeneous entities with molecular phenotypes that mirror various stages of B cell maturation. Within defined entities such as DLBCL there are additional subgroups that can be identified along this spectrum.

The three defined subtypes of DLBCL differ in their survival outcomes and responses to therapy. Five-year overall survival for GCB, PMBL, and ABC has been reported to be 59%, 64%, and 31%, respectively (Lenz et al., 2008;

TABLE 6.1 Summary of lymphoma subtypes including their point of origin during B-cell differentiation, key oncogenic events, and associated gene expression signatures.

Subtype	Origin	Oncogenic events	Signatures
ABC DLBCL	Postgerminal center	Trisomy 3, *INK4alpha*, *ARF* deletion, BLIMP1, *XBP1*, NF-κB pathway, mutations in *CDKN2A*, *PIM1*, *MYD88*, *PTEN*	ABC
GCB DLBCL	Germinal center	t (14;18), *bcl-2*, *c-rel*, *LMO2*, *BCL*, mutations in *CREBBP*, *BCL2*, *EZH2*	Good prognosis: GCB, LN, MHC-2, stromal 1; bad prognosis: proliferation, stromal 2
Primary mediastinal B-cell lymphoma	Germinal center	*PDL2* on chromosome 9p, NF-κB, and JAK-STAT pathway, mutations in *CIITA*, *CD58*, *EP300*	30% overlap with Hodgkin's lymphoma
Follicular lymphoma	Germinal center	t (14;18), *bcl-2*, *BCL6*	Good prognosis: immune response 1; poor prognosis: immune response 2
Burkitt's lymphoma	Germinal center	*c-myc* translocation, mutation in *ID3*, *TCF3*, *GNA13* and activation of PI-3 kinase	mBL signature; composite of myc, subset GCB, and low-level NF-κB and MHC class 1
Hodgkin's lymphoma	Germinal center	*PDL2* on chromosome 9p in tandem with JAK2 amplification, NF-κB, *STAT6*, *SOCS1* loss	30% overlap with PMBL
MCL	Pregerminal center	t (11;14); cyclin D1; deletion of *INK4alpha/ ARF*, *TP53*, *ATM*, *RB* mutations	MCL proliferation

Scott, Mottok, & Ennishi, 2015). These survival differences were independent of the microarray type used to determine the gene expression profiles. In general, GCB DLBCL shows a better prognosis than ABC DLBCL, but considerable variability remains in the outcome, with 30% mortality for GCB patients at 2 years from diagnosis and around one-quarter of ABC DLBCL patients still alive more than 10 years from diagnosis. Prognostic gene expression signatures have been developed in DLBCL that reflect GCB cells, LN, proliferation, and

MHC class II, which are highly efficacious in prognosticating survival in these patients (Lenz et al., 2008). Similar to follicular lymphoma, the microenvironment and immune response are important in DLBCL. In addition to the previously described signatures, two gene expression signatures, identified as stromal 1 and stromal 2, were predictive of response to therapy with CHOP or R-CHOP (rituximab in combination with CHOP chemotherapy). The stromal 1 gene expression signature comprised genes of the extracellular matrix responsible for fibronectin, osteonectin, and collagen. In patient samples, the products of these genes were found in the fibrous strands between tumor cells as well as histiocytic cells in the DLBCL biopsies. This signature is believed to represent a monocytic immune response to the lymphoma cells. The stromal 2 signature includes genes for endothelial cells and blood vessel formation. This correlated clinically with high expression of the signature producing increased blood vessel density in patients. These data indicate that gene expression signatures in DLBCL are robust measures of the underlying biology of these tumors and have shed new light with regard to the observed clinical heterogeneity in these patients (Lenz et al., 2008).

Next-generation sequencing approaches have defined the genomic landscapes of lymphomas across the whole genome and furthermore have linked clinical outcomes to clustering of individual alterations. At least three relatively large studies evaluating the association of DLBCL genomic alteration associations with clinical outcomes across hundreds of patients have defined the molecular features of this entity at a fine level of detail (Chapuy et al., 2018; Reddy, Zhang, & Davis, 2017; Schmitz et al., 2018). As expected, mutations segregate according to previously GEP-define COO subtypes. Subtype clustering was used in two of the largest genomic studies of DLBCL and further defined molecular subgroups via this approach, whereas another study documented that the collaboration of individual mutations appears to drive clinical outcome and this effect is additive to clinical features that are used in clinical prognostic scores such as the international prognostic index (IPI).

Follicular lymphoma

Follicular lymphoma is the second most frequent NHL, and is generally an indolent disease, with a median survival of 10 years (Batlevi, Sha, & Alperovich, 2020; Freedman & Jacobsen, 2020; Solal-Céligny, Roy, & Colombat, 2004). The malignant cells of this disease arise from GCB cells and are frequently characterized by translocation of BCL2, rendering it under the control of immunoglobulin gene promoters. Patients display a wide range of survival from under 1 year to more than 20 years and survival can be predicted with host and tumor-related features (Solal-Céligny et al., 2004). Two gene expression signatures have been identified that together strongly predicted survival in a synergistic manner: immune response 1 (good prognosis) and 2 (poor prognosis). The immune response 1 gene expression signature included genes encoding T-cell markers (e.g., CD8 and STAT4). Immune response 2 included genes expressed in macrophages and/or dendritic cells (e.g., CD68, TLR5, and C3AR1). The survival of 191 follicular lymphoma patients was found to be significantly different when patients were grouped into quartiles based on their expression of these two gene expression signatures: 13.6, 11.1, 10.8, and 3.9 years. Further evaluation after sorting by flow cytometry confirmed that both immune response signatures were expressed predominantly in the nonmalignant, CD19-negative cells. Furthermore, compared to the gene expression profiles of normal hematopoietic cells, genes from immune responses 1 and 2 were over-expressed in T cells and monocytes, but not in B cells. Taken together, these data suggest that the prognostically important immune response gene expression signatures are derived from nonmalignant cells and highlight the interaction of the patient's immune system and the malignant cells as an important factor in tumor progression (Dave, Wright, & Tan, 2004).

Similar to DLBLC, studies utilizing next-generation sequencing to define the genomic landscape of follicular lymphoma have been completed by several groups (Bouska, Zhang, & Gong, 2017; Krysiak, Gomez, & White, 2017; Morin, Johnson, & Severson, 2010; Morin et al., 2011; Okosun, Bödör, & Wang, 2014; Zamò et al., 2018). Overall, significant overlap in the mutational profiles of FL and GCB DLBCL exists; however, FL cases have frequent KMT2D (Morin et al., 2011; Ortega-Molina, Boss, & Canela, 2015; Zhang, Dominguez-Sola, & Hussein, 2015) and EZH2 mutations (Béguelin, Teater, & Meydan, 2020) that may deliver an epigenetic "hit" that is important for driving FL pathogenesis in tandem with BCL2 mutations or other alterations. Genomic studies to understand features across the whole genome or whole exome landscape of FL (as has been reported with DLBCL) in context of clinical outcomes are ongoing, although a 7-gene predictor used in association with clinical features to improve on the identification of high- and low-risk FL patients (m7FLIPI) has been published (Pastore, Jurinovic, & Kridel, 2015) as have other predictive models (Huet, Tesson, & Jais, 2018); the ideal clinical utility of these tools is unclear (Bolen, Mattiello, & Herold, 2021).

Burkitt lymphoma

Burkitt lymphoma (BL) is an aggressive lymphoma with high cure rates after treatment with methotrexate/cytarabine (90% children, 70% adults) and high mortality without treatment (Crombie & LaCasce, 2021; Evens, Danilov, & Jagadeesh, 2021; Graham & Lynch 2021; Olszewski, Jakobsen, & Collins, 2021). The need for accurate diagnosis, particularly differentiating BL and DLBCL, is crucial given the intense nature of BL chemotherapy compared to the standard CHOP-based regimens for DLBCL. Diagnostic accuracy for

BLs remains low, with an estimated interobserver agreement of approximately 50% among experts diagnosing the disease (Dave, Fu, & Wright, 2006).

Gene expression profiling has demonstrated an improvement to diagnostic accuracy. Hummel and colleagues defined biologic BL as cases that showed increased expression of their molecular Burkitt lymphoma (mBL) signature, containing 58 genes. Cases expressing the mBL signature had an overall good outcome, with 5-year overall survival of 75% (Hummel, Bentink, & Berger, 2006). A parallel study (Dave et al., 2006) defined a classifier with three components: high MYC target gene expression, increased expression of a subset of GCB genes, and low expression of both MHC class I and NF-κB target genes. Both these studies indicate the feasibility and accuracy of gene expression profiling for rendering this clinically important distinction, which is difficult using existing methods in pathology.

As in DLBCL and FL, next generation sequencing (NGS) approaches have profiled BL at the exome, transcriptome, and whole genome level. These studies have revealed novel mutations in ID3 and TCF3, which are transcription factors that drive PI-3 kinase signaling in BL cells (Love, Sun, & Jima, 2012; Schmitz, Young, & Ceribelli, 2012). BL tumors also frequently have mutations in focal adhesion signaling mediators such as GNA13 and RHOA and these mutations likely effect a loss of tumor suppressor function in nascent malignant cells and thus driving the formation of BL (Healy, Nugent, & Rempel, 2016; Muppidi, Schmitz, & Green, 2014). BL diagnosis and treatment may benefit from the discovery of these mutations as ID3/TCF3 mutations can aid in distinguishing DLBCL from BL and PI-3 kinase signaling may be a druggable pathway with currently available pharmacologic agents.

Mantle cell lymphoma

MCL accounts for 6% of lymphomas and has a disproportionately higher mortality rate. It arises from naïve B cells that have not yet undergone the germinal center reaction. Characteristic genetic events include t (11;14), INK4a/ARF deletion, TP53 deletion, and ATM deletion, in addition to increased expression of BMI-1 (Alkan, Schnitzer, Thompson, Moscinski, & Ross, 1995; Cortelazzo, Ponzoni, Ferreri, & Dreyling, 2020; de Boer, Schuuring, & Dreef, 1995; Raffeld & Jaffe, 1991; Raffeld, Sander, Yano, & Jaffe, 1992). In MCL, a gene expression signature indicating proliferation was associated with poorer survival, with a survival difference of 6.7 years (vs. 0.8 years) based on the level of expression (Hoster, Rosenwald, & Berger, 2016; Rosenwald, Wright, & Wiestner, 2003). Although cyclin D1 over-expression is a hallmark of the tumor, an additional subset of cyclin-D1-negative MCL patients were identified using gene expression profiling. These patients are believed to over-express cyclin D2 or D3 in the place of cyclin D1 with similar down-stream consequences (Herens et al., 2008; Kurita, Takeuchi, & Kobayashi, 2016; Seto, 2013; Stefancikova, Moulis, & Fabian, 2009).

Leveraging genomics for precision medicine in lymphoma clinical care

A growing number of small molecule inhibitors and biologic therapeutics are undergoing evaluation in clinical trials and several have become standard of care agents. Agents targeting B-cell lineage markers, BCR receptor or other oncogenic signaling pathways, and epigenetic modifiers are among promising drugs in the field and hold the promise of improving patient outcomes through precision medicine.

Targeting B-cell lineage markers has been one of the first precision medicine approaches to be applicable to lymphomas. Lymphomas expressing CD20 and other B-cell lineage markers can be targeted effective with antibody therapies such as rituximab. Rituximab was found to significantly improve the complete response rate when added to CHOP chemotherapy in DLBCL (CR/Cru 75% with rituximab vs. 63%) and provided a significant survival benefit (0.64 RR of death, $P = 0.007$) in a large prospective randomized study that led to its approval (Coiffier, Lepage, & Briere, 2002). Other approved CD20-targeted agents include obinutuzumab, which showed a progression-free survival benefit when added to frontline follicular lymphoma chemotherapy-based treatment compared to rituximab and chemotherapy in the GALLIUM study and in a randomized trial in rituximab-refractory FL showed a survival benefit with the addition of obinutuzumab to bendamustine chemotherapy compared to bendamustine alone (Hiddemann, Barbui, & Canales, 2018). Bendamustine in combination with rituximab and polatuzumab vedotin—a CD79B-targeted monoclonal antibody drug conjugate—has received FDA-approval as third line therapy with DLBCL after a clinical trial documented a 40% response rate with PFS of 9.5 m, which appeared higher than the combination of bendamustine with rituximab alone (Sehn, Herrera, & Flowers, 2020). Tafasitamab, a CD19-directed monoclonal antibody, also was approved recently in combination with lenalidomide after results from the L-MIND study, which enrolled relapsed/refractory DLBCL patients after a median of two lines of therapy; in this study there was a 43% CR rate and an additional 14% partial responders in an unselected patient population (Salles, Duell, & González Barca, 2020). Additional monoclonal antibodies, antibody drug-conjugates, antigen-directed toxin conjugates, and bispecific antibodies are under investigation in numerous clinical trials in B-cell lymphomas.

Malignant B cells may also be effectively targeted with cellular therapies. Autologous and allogeneic cell therapy approaches with chimeric antigen receptor modified (CAR) T- or NK-cells are actively being used as standard of care agents or being investigated in clinic trials. Cell constructs including axicabtagene ciloleucel (Neelapu, Locke, & Bartlett, 2017) (which is a CD19-directed autologous CAR T construct utilizing a CD28 costimulatory construct) and tisagenlecleucel (which contains a 4−1BB costimulatory construct) have both become standard of care as third line treatments for DLBCL. In the ZUMA-1 trial, which enrolled chemorefractory DLBCL patients and utilized axicabtagene ciloleucel after lymphodepleting chemotherapy, the median survival was noted reached after 27.1 months of trial follow-up (Locke, Ghobadi, & Jacobson, 2019) and this vastly exceeds estimates for survival of chemo-refractory patients treated in the era prior to availability of CAR T cells (Crump, Neelapu, & Farooq, 2017). Future strategies for immunotherapy of B-cell malignancies will likely focus on not only improving the efficacy of such strategies but also the availability and toxicity of immunotherapies as the currently available CAR T strategies are complicated and associated with relatively high rates of cytokine release syndrome and neurologic toxicity associated with these therapies. New approaches including allogeneic products and NK cell−based CAR-modified therapies hold additional promise in the field and there is a burgeoning field investigating the use of bispecific antibodies targeting B-cell lineage antigens including CD19 and CD20 across NHL subtypes.

Other studies have sought to leverage recent discoveries regarding biologic subgroups defined by gene expression or genetic alterations to more accurately diagnosis and treatment lymphomas. Lymphoma pathologic diagnostic algorithms routinely integrate assays for MYC, BCL2, and other rearrangements to define subgroups such as high-grade B-cell lymphoma with MYC and BCL2 or BCL6 (or other rearrangements) and these assays are essential to differentiate this subgroup from DLBCL for diagnosis [the WHO 2016 classification of lymphoid malignancies classifies high-grade B-cell lymphoma with MYC and other rearrangements as an entity separate from DLBCL (Swerdlow et al., 2016)], prognostic and treatment considerations. Assessment of COO phenotype is commonly carried out using IHC algorithms that can identify activated B-cell type DLBCL versus germinal center subtypes with relatively high accuracy rates. Higher rates of accuracy in determining COO type relative to IHC (with Bayesian algorithms utilizing gene expression arrays or RNA sequencing being the gold standard) can be achieved with Nanostring technology and more precise COO determination may have important prognostic implications clinically (Scott et al., 2015) and may predict more accurately the response novel agents having differential activity in ABC DLBCL such as lenalidomide (Czuczman, Trněný, & Davies, 2017).

Precision medicine approaches utilizing genomic assays to predict response to targeted therapies have been studied in a number of clinical trials and translational studies.

B-cell receptor pathway small molecule inhibitors have become important therapies now approved in chronic lymphocytic leukemia/small lymphocytic lymphoma (CLL/SLL) (approved agents including ibrutinib and acalabrutinib), MCL (zanubrutinib, acalabrutinib, and ibrutinib), and Waldenstrom macroglobulinemia (ibrutinib); genomic features of these entities can be used to assess resistance features to these agents in these entities (Czuczman et al., 2017). In CLL/SLL acquired mutations in BTK and PLCG are found in 80% or greater of patients that acquire resistance to ibrutinib and there are often multiple clonal evolution events found in tissue or blood samples taken from patients that acquire resistance to ibrutinib (Ahn, Underbayev, & Albitar, 2017; Woyach, Furman, & Liu, 2014). Ibrutinib resistance appears to occur via somewhat different mechanism(s) in MCL and there is an evidence that noncanonical NF-κB signaling may bypass BTK receptor signaling inhibited by ibrutinib thus producing resistance (Rahal, Frick, & Romero, 2014). Resistance to ibrutinib appears to be mediated by an entirely different mechanism in WM. Waldenstrom's macroglobulinemia cases exhibit a high percentage of MYD88 L265P mutations (in more than 95% of cases) (Treon, Xu, & Yang, 2012; Treon et al., 2018) and in a prospective study of WM patients with symptomatic disease after one or more lines of therapy treated with ibrutinib monotherapy, profiles of MYD88 and CXCR4 mutations appeared to be associated with response rate. Rates of response were highest in patients with MYD88^{L265P} and wild-type CXCR4 (100% overall response rate and 91.2% major response rate compared to 71.4% ORR and 28.6% major response rate in MYD88^{WT} CXCR4^{WHIM} cases) (Treon, Tripsas, & Meid, 2015). Based on these data, ibrutinib became the first FDA-approved mediation for lymphoplasmacytic lymphoma/Waldenstrom macroglobulinemia and testing for MYD88 mutations with reflex testing for CXCR4 mutations is now standard of care for the diagnosis and assignment of treatment in this entity.

Studies have also sought to define targetable molecular alterations in common entities such as DLBCL. The DLBCL subtype having an activated B-cell phenotype (ABC DLBCL) is often characterized by molecular alterations in BCR signaling with downstream JAK-STAT and NF-κB activation and alterations in these pathways may be identified with immunohistochemistry, gene expression, or multiplexed gene mutation assays. Such assays have been studied as "gatekeepers" in trials utilizing novel agents with the goal of improving outcomes across the spectrum of DLBCL, particular in the front-line setting. Front-line prospective DLBCL trials studying the addition of novel agents to standard rituximab-CHOP have been carried out with agents including bortezomib, lenalidomide, and ibrutinib. At least two large studies evaluating the addition to bortezomib to a standard rituximab-CHOP backbone have been completed. The randomized phase II PYRAMID trial utilized the Hans IHC algorithm to identify non-GCB DLBCL cases prior to study enrollment and found no difference (and higher rates of toxicity) with the addition of bortezomib to chemoimmunotherapy in this subgroup with regard to response rates and overall survival (Leonard, Kolibaba, & Reeves, 2017). Outcomes were also similar with the addition of bortezomib to rituximab-CHOP in a different front-line study that utilized next-generation sequencing—based gene expression and mutation analysis in that no improvement in progression free survival was noted with the addition of bortezomib to frontline DLBCL therapy (Davies, Cummin, & Barrans, 2019). Lenalidomide has additionally been studied in multiple trials enrolling both treatment-naïve and relapsed/refractory DLBCL patients. In relapsed/refractory DLBCL, lenalidomide appears to have preferential activity in ABC DLBCL and this effect appeared to be higher when COO subtype is defined by GEP methods such as Nanostring technology as opposed to IHC algorithms; a phase II/III study showed response rates of 45.5% (CI: 16.7−76.6) with lenalidomide versus 18.8% (CI: 4.0−45.6) with investigator choice chemoimmunotherapy in ABC DLBCL defined by Nanostring GEP assay where response rates of 28.6% (CI: 13.2−48.7) with lenalidomide and 11.5% (CI 2.4−30.2) with investigators choice chemoimmunotherapy were found in COO type was determined by IHC analysis (Czuczman et al., 2017). Based on these results, the phase III ROBUST study utilized GEP to define ABC DLBCL and patients were randomized to rituximab-CHOP with either lenalidomide or placebo. This study did not meet its specific PFS improvement endpoint, although there was a trend toward improvement in lymphoma-specific outcomes with the addition of lenalidomide in this population (Nowakowski, Chiappella, & Gascoyne, 2021). Other studies have evaluated similar approaches with lenalidomide as well as ibrutinib (Wilson, Young, & Schmitz, 2015) and as of yet there have been no changes to the standard of care chemoimmunotherapy approach in DLBCL.

Other genetic alterations in lymphoma have been validated as precision medicine targets or are under active study as such. EZH2 mutations are found in at least 20% of FL cases and function as gain of mutations that function to trimethylate lysine 27 on histone 3 (H3K27). H3K27me3 fostered by EZH2-activating mutations can be effectively inhibited with small molecule agents. One of these, tazemetostat gained US FDA approval for patients with FL after two lines of therapy after a phase II trial documented an ORR of 69% (12% CRR) in EZH2-mutated FL (the overall response rate in FL with wild-type EZH2 was 38%) (Morschhauser, Tilly, & Chaidos, 2020), making this the first therapy to be granted approval based upon presence of an actionable mutation across lymphomas.

More work is required to leverage our current understanding of the genomic landscapes of lymphomas. Other pathways of interest include epigenetic alterations as well as JAK-STAT activation and other pathways in T-cell lymphomas (Van Arnam, Lim, & Elenitoba-Johnson, 2018). These entities generally exhibit unacceptable mortality rates with currently available therapies and hopefully novel combinations of molecularly targeted agents can improve upon these dismal outcomes.

Summary

Lymphomas are a heterogeneous group of malignancies with respect to both clinical behavior and genetic alterations. The application of genomics has shed new light on these complex tumors and enabled a better understanding of the molecular heterogeneity that underlies the clinical heterogeneity. Advances include the identification of subgroups with distinct genomics and responses to therapy within the previously broad category of lymphoma, and the identification of prognostic and diagnostic gene expression signatures as well as other lesions identified through high-throughput NGS approaches. The clinical translation of this work will improve multiple aspects of care including diagnosis, counseling of patients regarding prognosis, identification of patients most likely to benefit from therapy, and the design of clinical trials to further develop the current knowledge base and there are significant opportunities exist to leverage precision medicine approaches in these diseases to lengthen and improve the lives of patients.

References

Ahn, I. E., Underbayev, C., Albitar, A., Herman, S. E. M., Tian, X., Maric, I., ... Wiestner, A. (2017). Clonal evolution leading to ibrutinib resistance in chronic lymphocytic leukemia. *Blood*, *129*(11), 1469−1479. Available from https://doi.org/10.1182/blood-2016-06-719294.

Alizadeh, A. A., Eisen, M. B., Davis, R. E., Ma, C., Lossos, I. S., Rosenwald, A., ... Staudt, L. M. (2000). Distinct types of diffuse large B-cell lymphoma identified by gene expression profiling. *Nature*, *403*(6769), 503−511. Available from https://doi.org/10.1038/35000501.

Alkan, S., Schnitzer, B., Thompson, J. L., Moscinski, L. C., & Ross, C. W. (1995). Cyclin D1 protein expression in mantle cell lymphoma. *Annals of Oncology: Official Journal of the European Society for Medical Oncology*, *6*(6), 567−570. Available from https://doi.org/10.1093/oxfordjournals.annonc.a059245.

Batlevi, C. L., Sha, F., Alperovich, A., Ni, A., Smith, K., Ying, Z., ... Younes, A. (2020). Follicular lymphoma in the modern era: Survival, treatment outcomes, and identification of high-risk subgroups. *Blood Cancer Journal*, *10*(7), 74. Available from https://doi.org/10.1038/s41408-020-00340-z.

Béguelin, W., Teater, M., Meydan, C., Hoehn, K. B., Phillip, J. M., Soshnev, A. A., ... Melnick, A. M. (2020). Mutant EZH2 induces a pre-malignant lymphoma niche by reprogramming the immune response. *Cancer Cell*, *37*(5), 655−673. Available from https://doi.org/10.1016/j.ccell.2020.04.004, e11. (In eng).

Bolen, C. R., Mattiello, F., Herold, M., Hiddemann, W., Huet, S., Klapper, W., ... Venstrom, J. M. (2021). Treatment dependence of prognostic gene expression signatures in de novo follicular lymphoma. *Blood*. Available from https://doi.org/10.1182/blood.2020008119.

Boleti, E., & Johnson, P. W. (2007). Primary mediastinal B-cell lymphoma. *Hematological Oncology*, *25*(4), 157−163. Available from https://doi.org/10.1002/hon.818.

Bouska, A., Zhang, W., Gong, Q., Iqbal, J., Scuto, A., Vose, J., ... Chan, W. C. (2017). Combined copy number and mutation analysis identifies oncogenic pathways associated with transformation of follicular lymphoma. *Leukemia: Official Journal of the Leukemia Society of America, Leukemia Research Fund, UK*, *31*(1), 83−91. Available from https://doi.org/10.1038/leu.2016.175.

Brittig, F., Csanaky, G., Kecskés, L., & István, L. (1991). Primary (B-cell) mediastinal lymphoma. *Orvosi Hetilap*, *132*(29), 1599−1602.

Chapuy, B., Stewart, C., Dunford, A. J., Kim, J., Kamburov, A., Redd, R. A., ... Shipp, M. A. (2018). Molecular subtypes of diffuse large B cell lymphoma are associated with distinct pathogenic mechanisms and outcomes. *Nature Medicine*, *24*(5), 679−690. Available from https://doi.org/10.1038/s41591-018-0016-8.

Coiffier, B., Lepage, E., Briere, J., Herbrecht, R., Tilly, H., Bouabdallah, R., ... Gisselbrecht, C. (2002). CHOP chemotherapy plus rituximab compared with CHOP alone in elderly patients with diffuse large-B-cell lymphoma. *The New England Journal of Medicine*, *346*(4), 235−242. Available from http://www.ncbi.nlm.nih.gov/htbin-post/Entrez/query?db = m&form = 6&dopt = r&uid = 11807147. Available from http://content.nejm.org/cgi/content/full/346/4/235. Available from http://content.nejm.org/cgi/content/abstract/346/4/235.

Cortelazzo, S., Ponzoni, M., Ferreri, A. J. M., & Dreyling, M. (2020). Mantle cell lymphoma. *Critical Reviews in Oncology/Hematology*, *153*, 103038. Available from https://doi.org/10.1016/j.critrevonc.2020.103038.

Crombie, J., & LaCasce, A. (2021). The treatment of Burkitt lymphoma in adults. *Blood*, *137*(6), 743−750. Available from https://doi.org/10.1182/blood.2019004099.

Crump, M., Neelapu, S. S., Farooq, U., Van Den Neste, E., Kuruvilla, J., Westin, J., ... Gisselbrecht, C. (2017). Outcomes in refractory diffuse large B-cell lymphoma: Results from the international SCHOLAR-1 study. *Blood*, *130*(16), 1800−1808. Available from https://doi.org/10.1182/blood-2017-03-769620.

Czuczman, M. S., Trněný, M., Davies, A., Rule, S., Linton, K. M., Wagner-Johnston, N., ... Lewis, I. D. (2017). A phase 2/3 multicenter, randomized, open-label study to compare the efficacy and safety of lenalidomide vs investigator's choice in patients with relapsed or refractory diffuse large B-cell lymphoma. *Clinical Cancer Research: An Official Journal of the American Association for Cancer Research*, *23*(15), 4127−4137. Available from https://doi.org/10.1158/1078-0432.Ccr-16-2818.

Dave, S. S., Fu, K., Wright, G. W., Lam, L. T., Kluin, P., Boerma, E.-J., ... Lymphoma/Leukemia Molecular Profiling Project. (2006). Molecular diagnosis of Burkitt's lymphoma. *The New England Journal of Medicine*, *354*(23), 2431−2442. Available from https://doi.org/10.1056/NEJMoa055759.

Dave, S. S., Wright, G., Tan, B., Rosenwald, A., Gascoyne, R. D., Chan, W. C., ... Staudt, L. M. (2004). Prediction of survival in follicular lymphoma based on molecular features of tumor-infiltrating immune cells. *The New England Journal of Medicine*, *351*(21), 2159−2169. Available from https://doi.org/10.1056/NEJMoa041869.

Davies, A., Cummin, T. E., Barrans, S., Maishman, T., Mamot, C., Novak, U., ... Johnson, P. W. M. (2019). Gene-expression profiling of bortezomib added to standard chemoimmunotherapy for diffuse large B-cell lymphoma (REMoDL-B): An open-label, randomised, phase 3 trial. *The Lancet Oncology*, *20*(5), 649−662. Available from https://doi.org/10.1016/s1470-2045(18)30935-5.

de Boer, C. J., Schuuring, E., Dreef, E., Peters, G., Bartek, J., Kluin, P. M., & van Krieken, J. H. (1995). Cyclin D1 protein analysis in the diagnosis of mantle cell lymphoma. *Blood*, *86*(7), 2715−2723.

Evens, A. M., Danilov, A., Jagadeesh, D., Sperling, A., Kim, S.-H., Vaca, R., ... Alderuccio, J. P. (2021). Burkitt lymphoma in the modern era: Real-world outcomes and prognostication across 30 US cancer centers. *Blood*, *137*(3), 374−386. Available from https://doi.org/10.1182/blood.2020006926.

Feugier, P., Van Hoof, A., Sebban, C., Solal-Celigny, P., Bouabdallah, R., Fermé, C., ... Coiffier, B. (2005). Long-term results of the R-CHOP study in the treatment of elderly patients with diffuse large B-cell lymphoma: A study by the Groupe d'Etude des Lymphomes de l'Adulte. *Journal of Clinical Oncology: Official Journal of the American Society of Clinical Oncology*, *23*(18), 4117−4126. Available from https://doi.org/10.1200/jco.2005.09.131.

Freedman, A., & Jacobsen, E. (2020). Follicular lymphoma: 2020 update on diagnosis and management. *American Journal of Hematology*, *95*(3), 316−327. Available from https://doi.org/10.1002/ajh.25696.

Graham, B. S., & Lynch, D. T. (2021). *Burkitt lymphoma*. Treasure Island, FL: StatPearls Publishing LLC.

Healy, J. A., Nugent, A., Rempel, R. E., Moffitt, A. B, Davis, N. S., Jiang, X., ... Dave, S. S. (2016). GNA13 loss in germinal center B cells leads to impaired apoptosis and promotes lymphoma in vivo. *Blood*, *127*(22), 2723−2731. Available from https://doi.org/10.1182/blood-2015-07-659938.

Herens, C., Lambert, F., Quintanilla-Martinez, L., Bisig, B., Deusings, C., & de Leval, L. (2008). Cyclin D1-negative mantle cell lymphoma with cryptic t (12;14) (p13;q32) and cyclin D2 overexpression. *Blood*, *111*(3), 1745−1746. Available from https://doi.org/10.1182/blood-2007-10-120824.

Hiddemann, W., Barbui, A. M., Canales, M. A., Cannell, P. K., Collins, G. P., Dürig, J., ... Marcus, R. E. (2018). Immunochemotherapy with obinutuzumab or rituximab for previously untreated follicular lymphoma in the GALLIUM study: Influence of chemotherapy on efficacy and safety. *Journal of Clinical Oncology: Official Journal of the American Society of Clinical Oncology*, *36*(23), 2395−2404. Available from https://doi.org/10.1200/jco.2017.76.8960.

Hodgkin. (1832). On some morbid appearances of the absorbent glands and spleen. *Medico-Chirurgical Transacions*, *17*, 68−114. Available from https://doi.org/10.1177/095952873201700106.

Hoster, E., Rosenwald, A., Berger, F., Bernd, H.-W., Hartmann, S., Loddenkemper, C., ... Klapper, W. (2016). Prognostic value of Ki-67 index, cytology, and growth pattern in mantle-cell lymphoma: Results from randomized trials of the European Mantle cell lymphoma network. *Journal of Clinical Oncology: Official Journal of the American Society of Clinical Oncology*, *34*(12), 1386−1394. Available from https://doi.org/10.1200/jco.2015.63.8387.

Huet, S., Tesson, B., Jais, J. P., Feldman, A. L., Magnano, L., Thomas, E., ... Salles, G. (2018). A gene-expression profiling score for prediction of outcome in patients with follicular lymphoma: a retrospective training and validation analysis in three international cohorts. *The Lancet Oncology*, *19*(4), 549−561. Available from https://doi.org/10.1016/s1470-2045(18)30102-5.

Hummel, M., Bentink, S., Berger, H., Klapper, W., Wessendorf, S., Barth, T. F. E., ... Molecular Mechanisms in Malignant Lymphomas Network Project of the Deutsche, K. (2006). A biologic definition of Burkitt's lymphoma from transcriptional and genomic profiling. *The New England Journal of Medicine*, *354*(23), 2419−2430. Available from https://doi.org/10.1056/NEJMoa055351.

Krysiak, K., Gomez, F., White, B. S., Matlock, M., Miller, C. A., Trani, L., ... Fehniger, T. A. (2017). Recurrent somatic mutations affecting B-cell receptor signaling pathway genes in follicular lymphoma. *Blood*, *129*(4), 473−483. Available from https://doi.org/10.1182/blood-2016-07-729954.

Kurita, D., Takeuchi, K., Kobayashi, S., Hojo, A., Uchino, Y., Sakagami, M., ... Takei, M. (2016). A cyclin D1-negative mantle cell lymphoma with an IGL-CCND2 translocation that relapsed with blastoid morphology and aggressive clinical behavior. *Virchows Archiv: An International Journal of Pathology*, *469*(4), 471−476. Available from https://doi.org/10.1007/s00428-016-1995-9.

Lenz, G., Wright, G., Dave, S. S., Xiao, W., Powell, J., Zhao, H., ... Lymphoma/Leukemia Molecular Profiling Project. (2008). Stromal gene signatures in large-B-cell lymphomas. *The New England Journal of Medicine*, *359*(22), 2313−2323. Available from https://doi.org/10.1056/NEJMoa0802885.

Lenz, G., Wright, G. W., Emre, N. C., Kohlhammer, H., Dave, S. S., Davis, E. R., ... Staudt, L. M. (2008). Molecular subtypes of diffuse large B-cell lymphoma arise by distinct genetic pathways. *Proceedings of the National Academy of Sciences of the United States of America*, *105*(36), 13520−13525. Available from https://doi.org/10.1073/pnas.0804295105.

Leonard, J. P., Kolibaba, K. S., Reeves, J. A., Tulpule, A., Flinn, I. W., Kolevska, T., ... de Vos, S. (2017). Randomized phase II study of R-CHOP with or without bortezomib in previously untreated patients with non-germinal center B-cell-like diffuse large B-cell lymphoma. *Journal of Clinical Oncology: Official Journal of the American Society of Clinical Oncology*, *35*(31), 3538−3546. Available from https://doi.org/10.1200/jco.2017.73.2784.

Locke, F. L., Ghobadi, A., Jacobson, C. A., Miklos, D. B., Lekakis, L. J., Oluwole, O. O., ... Neelapu, S. S. (2019). Long-term safety and activity of axicabtagene ciloleucel in refractory large B-cell lymphoma (ZUMA-1): A single-arm, multicentre, phase 1−2 trial. *The Lancet Oncology*, *20*(1), 31−42. Available from https://doi.org/10.1016/s1470-2045(18)30864-7.

Lohr, J. G., Stojanov, P., Lawrence, M. S., Auclair, D., Chapuy, B., Sougnez, C., ... Golub, T. R. (2012). Discovery and prioritization of somatic mutations in diffuse large B-cell lymphoma (DLBCL) by whole-exome sequencing. *Proceedings of the National Academy of Sciences of the United States of America*, *109*(10), 3879−3884. Available from https://doi.org/10.1073/pnas.1121343109.

Love, C., Sun, Z., Jima, D., Li, G., Zhang, J., Miles, R., ... Dave, S. S. (2012). The genetic landscape of mutations in Burkitt lymphoma. *Nature Genetics*, *44*(12), 1321−1325. Available from https://doi.org/10.1038/ng.2468.

Monti, S., Savage, K. J., Kutok, J. L., Feuerhake, F., Kurtin, P., Mihm, M., ... Shipp, M. A. (2005). Molecular profiling of diffuse large B-cell lymphoma identifies robust subtypes including one characterized by host inflammatory response. *Blood*, *105*(5), 1851−1861. Available from https://doi.org/10.1182/blood-2004-07-2947.

Morin, R. D., Johnson, N. A., Severson, T. M., Mungall, A. J., An, J., Goya, R., ... Marra, M. A. (2010). Somatic mutations altering EZH2 (Tyr641) in follicular and diffuse large B-cell lymphomas of germinal-center origin. *Nature Genetics*, *42*(2), 181−185. Available from https://doi.org/10.1038/ng.518.

Morin, R. D., Mendez-Lago, M., Mungall, A. J., Goya, R., Mungall, K. L., Corbett, R. D., ... Marra, M. M. (2011). Frequent mutation of histone-modifying genes in non-Hodgkin lymphoma. *Nature*, *476*(7360), 298−303. Available from https://doi.org/10.1038/nature10351.

Morschhauser, F., Tilly, H., Chaidos, A., McKay, P., Phillips, T., Assouline, S., ... Salles, G. (2020). Tazemetostat for patients with relapsed or refractory follicular lymphoma: An open-label, single-arm, multicentre, phase 2 trial. *The Lancet Oncology*, *21*(11), 1433−1442. Available from https://doi.org/10.1016/s1470-2045(20)30441-1.

Mottok, A., Hung, S. S., Chavez, E. A., Woolcock, B., Telenius, A., Chong, L. C., ... Steidl, C. (2019). Integrative genomic analysis identifies key pathogenic mechanisms in primary mediastinal large B-cell lymphoma. *Blood*, *134*(10), 802−813. Available from https://doi.org/10.1182/blood.2019001126.

Muppidi, J. R., Schmitz, R., Green, J. A., Xiao, W., Larsen, A. B., Braun, S. E., ... Cyster, J. G. (2014). Loss of signalling via Gα13 in germinal centre B-cell-derived lymphoma. *Nature*, *516*(7530), 254−258. Available from https://doi.org/10.1038/nature13765.

National Cancer Institute: Surveillance Epidemiology, and End Results Program. Cancer Stat Facts: Non-Hodgkin Lymphoma. https://seer.cancer.gov/statfacts/html/nhl.html.

Neelapu, S. S., Locke, F. L., Bartlett, N. L., Lekakis, L. J., Miklos, D. B., Jacobson, C. A., ... Go, W. Y. (2017). Axicabtagene ciloleucel CAR T-cell therapy in refractory large B-cell lymphoma. *The New England Journal of Medicine*, *377*(26), 2531−2544. Available from https://doi.org/10.1056/NEJMoa1707447.

Nowakowski, G. S., Chiappella, A., Gascoyne, R. D., Scott, D. W., Zhang, Q., Jurczak, W., . . . Vitolo, U. (2021). ROBUST: A phase III study of lenalidomide plus R-CHOP vs placebo plus R-CHOP in previously untreated patients with ABC-type diffuse large B-cell lymphoma. *Journal of Clinical Oncology: Official Journal of the American Society of Clinical Oncology*. Available from https://doi.org/10.1200/jco.20.01366, Jco2001366.

Okosun, J., Bödör, C., Wang, J., Araf, S., Yang, C.-Y., Pan, C., . . . Fitzgibbon, J. (2014). Integrated genomic analysis identifies recurrent mutations and evolution patterns driving the initiation and progression of follicular lymphoma. *Nature Genetics*, *46*(2), 176−181. Available from https://doi.org/10.1038/ng.2856.

Olszewski, A. J., Jakobsen, L. H., Collins, G. P., Cwynarski, K., Bachanova, V., Blum, K. A., . . . Evens, A. M. (2021). Burkitt lymphoma international prognostic index. *Journal of Clinical Oncology: Official Journal of the American Society of Clinical Oncology*. Available from https://doi.org/10.1200/jco.20.03288, Jco2003288. (In eng).

Ortega-Molina, A., Boss, I. W., Canela, A., Pan, H., Jiang, Y., Zhao, C., . . . Wendel, H. G. Heng 5, Yanwen 2 5, Chunying Zhao. (2015). The histone lysine methyltransferase KMT2D sustains a gene expression program that represses B cell lymphoma development. *Nature Medicine*, *21*(10), 1199−1208. Available from https://doi.org/10.1038/nm.3943.

Pasqualucci, L., & Dalla-Favera, R. (2018). Genetics of diffuse large B-cell lymphoma. *Blood*, *131*(21), 2307−2319. Available from https://doi.org/10.1182/blood-2017-11-764332.

Pastore, A., Jurinovic, V., Kridel, R., Hoster, E., Staiger, A. M., Szczepanowski, M., . . . Weigert, O. (2015). Integration of gene mutations in risk prognostication for patients receiving first-line immunochemotherapy for follicular lymphoma: A retrospective analysis of a prospective clinical trial and validation in a population-based registry. *The Lancet Oncology*, *16*(9), 1111−1122. Available from https://doi.org/10.1016/s1470-2045(15)00169-2.

Philip, T., Guglielmi, C., Hagenbeek, A., Somers, R., Van der Lelie, H., Bron, D., et al. (1995). Autologous bone marrow transplantation as compared with salvage chemotherapy in relapses of chemotherapy-sensitive non-Hodgkin's lymphoma. *The New England Journal of Medicine*, *333*(23), 1540−1545. Available from https://doi.org/10.1056/nejm199512073332305.

Raffeld, M., & Jaffe, E. S. (1991). bcl-1, t (11;14), and mantle cell-derived lymphomas. *Blood*, *78*(2), 259−263.

Raffeld, M., Sander, C. A., Yano, T., & Jaffe, E. S. (1992). Mantle cell lymphoma: An update. *Leukemia & Lymphoma*, *8*(3), 161−166. Available from https://doi.org/10.3109/10428199209054902.

Rahal, R., Frick, M., Romero, R., Korn, J. M., Kridel, R., Chan, F. C., . . . Stegmeier, F. (2014). Pharmacological and genomic profiling identifies NF-κB-targeted treatment strategies for mantle cell lymphoma. *Nature Medicine*, *20*(1), 87−92. Available from https://doi.org/10.1038/nm.3435.

Reddy, A., Zhang, J., Davis, N. S., Moffitt, A. B., Love, C. L., Waldrop, A., . . . Dave, S. S. (2017). Genetic and functional drivers of diffuse large B cell lymphoma. *Cell*, *171*(2), 481−494. Available from https://doi.org/10.1016/j.cell.2017.09.027, e15. (In eng).

Rosenwald, A., Wright, G., Chan, W. C., Connors, J. M., Campo, E., Fisher, R. I., . . . Lymphoma/Leukemia Molecular Profiling Project. (2002). The use of molecular profiling to predict survival after chemotherapy for diffuse large-B-cell lymphoma. *The New England Journal of Medicine*, *346*(25), 1937−1947. Available from https://doi.org/10.1056/NEJMoa012914.

Rosenwald, A., Wright, G., Leroy, K., Yu, X., Gaulard, P., Gascoyne, R. D., . . . Staudt, L. M. (2003). Molecular diagnosis of primary mediastinal B cell lymphoma identifies a clinically favorable subgroup of diffuse large B cell lymphoma related to Hodgkin lymphoma. *The Journal of Experimental Medicine*, *198*(6), 851−862. Available from https://doi.org/ 10.1084/jem.20031074.

Rosenwald, A., Wright, G., Wiestner, A., Chan, W. C., Connors, J. M., Campo, E., . . . Staudt, L. M. (2003). The proliferation gene expression signature is a quantitative integrator of oncogenic events that predicts survival in mantle cell lymphoma. *Cancer Cell*, *3*(2), 185−197. Available from https://doi.org/10.1016/s1535-6108(03)00028-x.

Salles, G., Duell, J., González Barca, E., Tournilhac, O., Jurczak, W., Liberati, A. M., . . . Maddocks, K. (2020). Tafasitamab plus lenalidomide in relapsed or refractory diffuse large B-cell lymphoma (L-MIND): A multicentre, prospective, single-arm, phase 2 study. *The Lancet Oncology*, *21*(7), 978−988. Available from https://doi.org/10.1016/s1470-2045(20) 30225-4.

Schmitz, R., Wright, G. W., Huang, D. W., Johnson, C. A., Phelan, J. D., Wang, J. Q., . . . Staudt, L. M. (2018). Genetics and pathogenesis of diffuse large B-cell lymphoma. *The New England Journal of Medicine*, *378*(15), 1396−1407. Available from https://doi.org/10.1056/ NEJMoa1801445.

Schmitz, R., Young, R. M., Ceribelli, M., Jhavar, S., Xiao, W., Zhang, M., . . . Staudt, L. M. (2012). Burkitt lymphoma pathogenesis and therapeutic targets from structural and functional genomics. *Nature*, *490*(7418), 116−120. Available from https://doi.org/10.1038/ nature11378.

Scott, D. W., Mottok, A., Ennishi, D., Wright, G. W., Farinha, P., Ben-Neriah, S., . . . Gascoyne, R. D. (2015). Prognostic significance of diffuse large B-cell lymphoma cell of origin determined by digital gene expression in formalin-fixed paraffin-embedded tissue biopsies. *Journal of Clinical Oncology: Official Journal of the American Society of Clinical Oncology*, *33*(26), 2848−2856. Available from https://doi.org/10.1200/jco.2014.60.2383.

Sehn, L. H., Herrera, A. F., Flowers, C. R., Kamdar, M. K., McMillan, A., Hertzberg, M., . . . Matasar, M. J. (2020). Polatuzumab vedotin in relapsed or refractory diffuse large B-cell lymphoma. *Journal of Clinical Oncology: Official Journal of the American Society of Clinical Oncology*, *38*(2), 155−165. Available from https://doi.org/10.1200/jco.19.00172.

Seto, M. (2013). Cyclin D1-negative mantle cell lymphoma. *Blood*, *121*(8), 1249−1250. Available from https://doi.org/10.1182/blood-2013-01-475954.

Shipp, M. A., Abeloff, M. D., Antman, K. H., Carroll, G., Hagenbeek, A., Loeffler, M., . . . Coiffier, B. (1999). International consensus conference on high-dose therapy with hematopoietic stem cell transplantation in aggressive non-Hodgkin's lymphomas: Report of the jury. *Journal of Clinical Oncology: Official Journal of the American Society of Clinical Oncology*, *17*(1), 423−429. Available from http://www.ncbi.nlm.nih.gov/htbin-post/Entrez/ query?db = m&form = 6&dopt = r&uid = 10458261.

Solal-Céligny, P., Roy, P., Colombat, P., White, J., Armitage, J. O., Arranz-Saez, R., . . . Montserrat, E. (2004). Follicular lymphoma international prognostic index. *Blood*, *104*(5), 1258−1265. Available from https://doi.org/10.1182/blood-2003-12-4434.

Staudt, L. M., & Dave, S. (2005). The biology of human lymphoid malignancies revealed by gene expression profiling. *Advances in Immunology*, *87*, 163−208. Available from https:// doi.org/10.1016/s0065-2776(05)87005-1.

Stefancikova, L., Moulis, M., Fabian, P., Falkova, I., Vasova, I., Kren, L., . . . Smardova, J. (2009). Complex analysis of cyclin D1 expression in mantle cell lymphoma: Two cyclin

D1-negative cases detected. *Journal of Clinical Pathology, 62*(10), 948−950. Available from https://doi.org/10.1136/jcp.2008.063701.

Steidl, C., & Gascoyne, R. D. (2011). The molecular pathogenesis of primary mediastinal large B-cell lymphoma. *Blood, 118*(10), 2659−2669. Available from https://doi.org/10.1182/blood-2011-05-326538.

Swerdlow, S. H., Campo, E., Pileri, S. A., Harris, N. L., Stein, H., Siebert, R., . . . Jaffe, E. S. (2016). The 2016 revision of the World Health Organization classification of lymphoid neoplasms. *Blood, 127*(20), 2375−2390. Available from https://doi.org/10.1182/blood-2016-01-643569.

Treon, S. P., Tripsas, C. K., Meid, K., Warren, D., Varma, G., Green, R., . . . Advani, R. H. (2015). Ibrutinib in previously treated Waldenström's macroglobulinemia. *The New England Journal of Medicine, 372*(15), 1430−1440. Available from https://doi.org/10.1056/NEJMoa1501548.

Treon, S. P., Xu, L., Liu, X., Hunter, Z. R., Yang, G., & Castillo, J. J. (2018). Genomic landscape of Waldenström macroglobulinemia. *Hematology/Oncology Clinics of North America, 32*(5), 745−752. Available from https://doi.org/10.1016/j.hoc.2018.05.003.

Treon, S. P., Xu, L., Yang, G., Zhou, Y., Liu, X., Cao, Y., . . . Hunter, Z. R. (2012). MYD88 L265P somatic mutation in Waldenström's macroglobulinemia. *The New England Journal of Medicine, 367*(9), 826−833. Available from https://doi.org/10.1056/NEJMoa1200710.

Twa, D. D., & Steidl, C. (2015). Structural genomic alterations in primary mediastinal large B-cell lymphoma. *Leukemia & Lymphoma, 56*(8), 2239−2250. Available from https://doi.org/10.3109/10428194.2014.985673.

Van Arnam, J. S., Lim, M. S., & Elenitoba-Johnson, K. S. J. (2018). Novel insights into the pathogenesis of T-cell lymphomas. *Blood, 131*(21), 2320−2330. Available from https://doi.org/10.1182/blood-2017-11-764357.

Wilson, W. H., Young, R. M., Schmitz, R., Yang, Y., Pittaluga, S., Wright, G., . . . Staudt, L. M. (2015). Targeting B cell receptor signaling with ibrutinib in diffuse large B cell lymphoma. *Nature Medicine, 21*(8), 922−926. Available from https://doi.org/10.1038/nm.3884.

Woyach, J. A., Furman, R. R., Liu, T. M., Ozer, H. G., Zapatka, M., Ruppert, A. S., . . . Zhang, J. (2014). Resistance mechanisms for the Bruton's tyrosine kinase inhibitor ibrutinib. *The New England Journal of Medicine, 370*(24), 2286−2294. Available from https://doi.org/10.1056/NEJMoa1400029.

Zamò, A., Pischimarov, J., Horn, H., Ott, G., Rosenwald, A., & Leich, E. (2018). The exomic landscape of t (14;18)-negative diffuse follicular lymphoma with 1p36 deletion. *British Journal of Haematology, 180*(3), 391−394. Available from https://doi.org/10.1111/bjh.15041.

Zhang, J., Dominguez-Sola, D., Hussein, S., Lee, J.-E., Holmes, A. B., Bansal, M., . . . Pasqualucci, L. (2015). Disruption of KMT2D perturbs germinal center B cell development and promotes lymphomagenesis. *Nature Medicine, 21*(10), 1190−1198. Available from https://doi.org/10.1038/nm.3940.

Zhang, J., Grubor, V., Love, C. L., Banerjee, A., Richards, K. L., Mieczkowski, P. A., . . . Dave, S. S. (2013). Genetic heterogeneity of diffuse large B-cell lymphoma. *Proceedings of the National Academy of Sciences of the United States of America.* Available from https://doi.org/10.1073/pnas.1205299110. (In Eng). 1205299110 [pii].

Chapter 7

Precision medicine in myeloid neoplasms/acute leukemias

Sendhilnathan Hari Ramalingam
Duke University School of Medicine, Durham, North Carolina, United States

Introduction

The 2016 revision to the World Health Organization (WHO) recognizes dozens of entities under the classification of myeloid neoplasms and acute leukemias (Arber, Orazi, & Hasserjian, 2016). These entities exhibit significant molecular heterogeneity even while being driven by interconnected/related biological pathways. High-throughput genomic techniques have defined the drivers of these entities, which have increased our understanding of how genomics determine a disease's clinical behavior. As in other malignancies, such studies have suggested therapeutic targets and precision medicine approaches that have revolutionized the care of patients with diseases such as chronic myeloid leukemia (CML) (Rabian, Lengline, & Rea, 2019; Saglio, Fava, & Gale, 2019) and acute myeloid leukemia (AML) (Chung & Ma, 2017; Cucchi, Polak, & Ossenkoppele, 2021). For example, the discovery of genomic alterations like *FLT3* internal tandem duplication or *IDH1/2* has afforded targeted approaches using commercially available therapies that improve outcome by inhibiting oncogenic signaling.

Spectrum of disease(s)

Trailblazing investigators in the field of genetics and genomics established the concept of recurrent chromosomal rearrangements in cancer via the study of karyotype banding in the early 1970s. The identification of recurrent alterations including t (8;22) (subsequently found to involve *RUNX1-RUNX1T1* fusion (Miyoshi, Kozu, & Shimizu, 1993)), as well as the Philadelphia chromosome and t (15;17) (Rowley, Golomb, & Dougherty, 1977) (producing PML-RARA fusion (Goddard, Borrow, & Solomon, 1992)) by Janet Rowley and colleagues (Rowley, 1973a,b; Rowley & Potter, 1976) revolutionized our understanding of cancer biology by introducing the concept of the genetic basis of cancer; these

Genomic and Precision Medicine. DOI: https://doi.org/10.1016/B978-0-12-800684-9.00002-2

entities were the first malignancies to be defined by pathognomonic genetic abnormalities. The realization that cancer is a process driven by genomic/epigenomic alterations has led to exploration of the genomic landscapes across the spectrum of neoplastic disease (Fig. 7.1).

These efforts have been aided by the geometric gains in the throughput technology used to probe the genomes and epigenomes of cancers. AML was among the first cancers where the somatic genomic landscape was described at the whole genome level (Mardis, Ding, & Dooling, 2009). Furthermore, epigenetic profiling in AML was one of the earliest examples of the use of epigenetic markers to define disease subtypes (Figueroa, Lugthart, & Li, 2010) and this has led to an appreciation of both novel disease subgroups and the factors that govern hematopoiesis as many of the mutations in myelodysplastic syndrome (MDS)/AML alter the machinery that orchestrate normal blood cell production in the bone marrow. Manipulation of the methylation state with hypomethylating agents such as decitabine (Kantarjian, Issa, & Rosenfeld, 2006; Lübbert, Suciu, & Baila, 2011) and azacytidine (Fenaux, Mufti, & Hellstrom-Lindberg, 2009) is the backbone of treatment for MDS. These agents improve survival and are seldom used in other cancer entities. All these discoveries have led to an ever-increasing number of accepted discrete pathologic entities recognized by the WHO, reflecting the improved understanding of genomic heterogeneity in these diseases.

The most recent WHO category of myeloid neoplasms/acute leukemias incorporates a heterogeneous group of entities ranging from myeloproliferative neoplasms (MPNs), myeloid/lymphoid neoplasms defined by recurrent gene rearrangements including *PDGFRA*, *PDGFRB*, *FGFR1*, or *PCM1-JAK2*, MDS, MDS/MPN "overlap" syndromes, as well as acute myeloid and lymphoblastic leukemia/lymphoma. Refer to Table 7.1 for an overview of

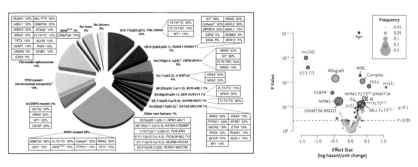

FIGURE 7.1 **Genomic alterations and correlation with associations with overall survival in acute myeloid leukemia.** Left panel shows categories of AML characterized by most common categories as defined by recurrent genomic alterations regarding co-occurring somatic mutations (from ELN guidelines (Döhner et al., 2017)). Right panel shows association between clinical features or genomic alterations by hazard and effect size derived from 1540 patients with AML (positive scale on effect size associated with worse relative overall survival) (Papaemmanuil, Gerstung, & Bullinger, 2016).

TABLE 7.1 World Health Organization revised 2016 classification of myeloid neoplasm and leukemias by category (Arber et al., 2016) listing selected molecular alterations and precision medicine targets (Arber et al., 2016).

WHO myeloid neoplasm and acute leukemia classification	Key molecular alteration	Precision medicine target(s)
Myeloproliferative neoplasms (MPN)		
Chronic myeloid leukemia (CML), *BCR-ABL1*+	t(9;22), *BCR-ABL1* fusion; TKI resistance mutaitons	Imatinib and other tyrosine kinase inhibitors
Chronic neutrophilic leukemia (CNL)	*CSF3R*	
Polycythemia vera (PV)	JAK2 mutations	
Primary myelofibrosis (PMF)	JAK2, CALR, MPL mutations	
PMF, prefibrotic/early stage	JAK2, CALR, MPL mutations	
PMF, overt fibrotic stage	JAK2, CALR, MPL mutations	
Essential thrombocythemia (ET)	JAK2, CALR, MPL mutations	
Chronic eosinophilic leukemia, not otherwise specified (NOS)	Multiple	
MPN, unclassifiable		
Mastocytosis	KIT alterations	Imatinib (D816 codon mutations predict resistance); midostaurin
Myeloid/lymphoid neoplasms with eosinophilia and rearrangement of PDGFRA, PDGFRB, *or* FGFR1, *or with* PCM1-JAK2		
Myeloid/lymphoid neoplasms with *PDGFRA* rearrangement	*PDGFRA* fusion	Imatinib and other tyrosine kinase inhibitors
Myeloid/lymphoid neoplasms with *PDGFRB* rearrangement	*PDGFRB* fusion	Imatinib and other tyrosine kinase inhibitors
Myeloid/lymphoid neoplasms with *FGFR1* rearrangement	*FGFR1* fusion	Imatinib and other tyrosine kinase inhibitors
Provisional entity: Myeloid/ lymphoid neoplasms with PCM1-JAK2	*PCM1-JAK2* fusion	Ruxolitinib and other JAK-STAT inhibitors

(Continued)

TABLE 7.1 (Continued)

WHO myeloid neoplasm and acute leukemia classification	Key molecular alteration	Precision medicine target(s)
Myelodysplastic/myeloproliferative neoplasms (MDS/MPN)		
Chronic myelomonocytic leukemia (CMML)		
Atypical chronic myeloid leukemia (aCML), *BCR-ABL1⁻*		
Juvenile myelomonocytic leukemia (JMML)		
MDS/MPN with ring sideroblasts and thrombocytosis (MDS/MPN-RS-T)	SF3B1 mutation and ring sideroblast formation	Luspatercept (approved for cases with >15% ringed sideroblasts)
MDS/MPN, unclassifiable	Multiple	
Myelodysplastic syndromes (MDS)		
MDS with single lineage dysplasia	Multiple	
MDS with ring sideroblasts (MDS-RS)	*SF3B1* mutations	Luspatercept (approved for cases with >15% ringed sideroblasts)
MDS-RS and single lineage dysplasia		
MDS-RS and multilineage dysplasia		
MDS with multilineage dysplasia		
MDS with excess blasts		
MDS with isolated del (5q)	Multiple	
MDS, unclassifiable		
Provisional entity: Refractory cytopenia of childhood		
Myeloid neoplasms with germ line predisposition		
Acute myeloid leukemia (AML) and related neoplasms		
AML with recurrent genetic abnormalities		

(Continued)

TABLE 7.1 (Continued)

WHO myeloid neoplasm and acute leukemia classification	Key molecular alteration	Precision medicine target(s)
AML with t (8;21) (q22;q22.1); *RUNX1-RUNX1T1*	Multiple	
AML with inv (16) (p13.1q22) or t (16;16) (p13.1;q22); *CBFB-MYH11*	Multiple	
APL with *PML-RARA*	*PML-RARA* fusions	All trans-retinoic acid (ATRA), arsenic trioxide
AML with t (9;11) (p21.3; q23.3); *MLLT3-KMT2A*	*KMT2A* fusion	
AML with t (6;9) (p23;q34.1); *DEK-NUP214*		
AML with inv (3) (q21.3q26.2) or t (3;3) (q21.3;q26.2); *GATA2, MECOM*		
AML (megakaryoblastic) with t (1;22) (p13.3;q13.3); *RBM15-MKL1*		
Provisional entity: AML with BCR-ABL1		
AML with mutated *NPM1*	*NPM1* exon 12 mutation	Menin inhibitors, relatively good prognosis
AML with biallelic mutations of *CEBPA*		
Provisional entity: AML with mutated RUNX1		
AML with myelodysplasia-related changes		
Therapy-related myeloid neoplasms	Multiple	
AML, NOS	*FLT3* ITD, *FLT3* TKD, *IDH1/2* mutations	Midostaurin, gilteritinib (*FLT-3*), ivosidenib (*IDH1*), enasidenib (*IDH2*)
AML with minimal differentiation	*FLT3* and *IDH1/2* mutations occur in multiple subtypes	
AML without maturation		
AML with maturation		

(Continued)

TABLE 7.1 (Continued)

WHO myeloid neoplasm and acute leukemia classification	Key molecular alteration	Precision medicine target(s)
Acute myelomonocytic leukemia		
Acute monoblastic/monocytic leukemia		
Pure erythroid leukemia		
Acute megakaryoblastic leukemia		
Acute basophilic leukemia		
Acute panmyelosis with myelofibrosis		
Myeloid sarcoma		
Myeloid proliferations related to Down syndrome		
Transient abnormal myelopoiesis (TAM)		
Myeloid leukemia associated with Down syndrome		
Blastic plasmacytoid dendritic cell neoplasm		
Acute leukemias of ambiguous lineage		
Acute undifferentiated leukemia	Multiple	
Mixed phenotype acute leukemia (MPAL) with t (9;22) (q34.1;q11.2); *BCR-ABL1*	*BCR-ABL1* fusion	Imatinib and other tyrosinde kinase inhibitors
MPAL with t (v;11q23.3); *KMT2A* rearranged	*KMT2A* fusions with multiple partners	DOTL1 inhibitors, Menin targeted agents
MPAL, B/myeloid, NOS	Multiple	
MPAL, T/myeloid, NOS	Multiple	
B-lymphoblastic leukemia/lymphoma		
B-lymphoblastic leukemia/ lymphoma, NOS	Multiple	
B-lymphoblastic leukemia/ lymphoma with recurrent genetic abnormalities	Multiple	

(Continued)

TABLE 7.1 (Continued)

WHO myeloid neoplasm and acute leukemia classification	Key molecular alteration	Precision medicine target(s)
B-lymphoblastic leukemia/lymphoma with t (9;22) (q34.1;q11.2);*BCR-ABL1*	*BCR-ABL1* fusion	Imatinib and other tyrosinde kinase inhibitors
B-lymphoblastic leukemia/lymphoma with t (v;11q23.3); *KMT2A* rearranged	*KMT2A* fusions with multiple partners	Menin inhibitors
B-lymphoblastic leukemia/lymphoma with t (12;21) (p13.2;q22.1); *ETV6-RUNX1*	Multiple	
B-lymphoblastic leukemia/lymphoma with hyperdiploidy	Multiple	
B-lymphoblastic leukemia/lymphoma with hypodiploidy	Multiple	
B-lymphoblastic leukemia/lymphoma with t (5;14) (q31.1;q32.3) *IL3-IGH*	Multiple	
B-lymphoblastic leukemia/lymphoma with t (1;19) (q23; p13.3);*TCF3-PBX1*	Multiple	
Provisional entity: B-lymphoblastic leukemia/lymphoma, BCR-ABL1−like	JAK-STAT activation	Ruxolitinib and other JAK-STAT inhibitors
Provisional entity: B-lymphoblastic leukemia/lymphoma with iAMP21	Multiple	
T-lymphoblastic leukemia/lymphoma		
Provisional entity: Early T-cell precursor lymphoblastic leukemia	Multiple	
Provisional entity: Natural killer (NK) cell lymphoblastic leukemia/lymphoma	Multiple	

these entities as well as the genomic markers and therapeutic targets across the spectrum of myeloid neoplasms/acute leukemias.

Many of these entities have been previously characterized by the presence of disease defining genomic alterations identifiable at the cytogenetic level (e.g., t (9;22) in CML) and for many years diagnosis and prognostication of myeloid entities have relied on the use of cytogenetics analysis. Such karyotype analysis is part of the standard of care, and the long-standing use of chromosomal analysis is relatively unique compared to other malignancies. This is likely a reflection of the relatively low burden of genomic alterations in myeloid neoplasms/acute leukemias and the importance of single gene fusions as oncologic drivers. Next-generation sequencing approaches have rapidly built upon decades of investigation utilizing karyotype, fluorescence in situ hybridization (FISH), array comparative genomic hybridization (CGH), and gene expression analysis. In MPNs, advances in RNA sequencing technologies and computational approaches recently led to the discovery of *CALR* frameshift mutations in a significant proportion of *JAK2* mutation negative MPNs (Klampfl, Gisslinger, & Harutyunyan, 2013; Nangalia, Massie, & Baxter, 2013). This finding refined the diagnostic approach to MPNs: currently multiplexed gene sequencing panels are frequently used to assay the presence of either *JAK2*, *CALR*, or *MPL* as oncogenic drivers in MPNs. Additionally, variation in patterns of *CALR* mutations have prognostic implications (Rumi, Pietra, & Ferretti, 2014; Rumi, Pietra, & Pascutto, 2014; Tefferi, Lasho, & Finke, 2014; Tefferi, Lasho, & Tischer, 2014). The discovery of mutations in *IDH1* and *IDH2*, whereby the altered protein product (isocitrate dehydrogenase) produces a novel metabolite, 2-hydroxyglutarate (2HG), has become revolutionary to the treatment of AML. 2HG is a competitive inhibitor of a-ketoglutarate dioxygenases (a-KG) including histone demethylases and TET family 5-methylcytosine (5mC) hydroxylases; *IDH1/2* mutations thus lead to downstream loss of function of these epigenetic modifiers as an oncogenic mechanism, a process that is readily targetable with small molecule inhibitors like ivosidenib and enasidenib. These agents have rapidly become part of the standard of care in AML. Even rare entities such as chronic neutrophilic leukemia and histiocytic neoplasms have recently benefited from genomic discovery studies. Both of these entities have activating *CSF1R* (Durham, Lopez Rodrigo, & Picarsic, 2019) mutations and histiocytic neoplasms have additional MAP-kinase activating mutations as well as *RET* and *ALK* rearrangements; dramatic responses have been reported to targeted inhibitors of these pathways (Diamond, Durham, & Haroche, 2016; Diamond, Durham, & Ulaner, 2019; Gounder, Solit, & Tap, 2018; Lee, Gasilina, & Roychoudhury, 2017).

The expanding availability of clinically validated high-throughput next generation sequencing may eventually supplant the use of chromosomal banding and even targeted FISH assays to define disease-associated genomic alterations at the cytogenetic level by instead providing timely assessment of alterations at the whole genome level. An experience published by investigators at Washington University found that whole genome-based sequencing in MDS/AML patients could identify cytogenetic alterations that were not revealed with conventional karyotyping, and in a timely manner (as quickly as 14 days). Other reports confirm the utility of this approach for clinical prognostication and therapy selection. NGS approaches in MDS/AML may thus replace the long-standing standard of testing for chromosomal alterations in these diseases.

The expansion of whole exome/genome sequence approaches to large population cohorts has also defined premalignant states analogs to monoclonal gammopathy of unknown significance as it relates to plasma cell dyscrasias or adenomatous polyps as they relate to the development of colorectal adenocarcinoma. Reanalysis of two large cohorts involving 1000 s of peripheral blood samples collected from healthy individuals characterized by targeted sequencing revealed the presence of recurrent somatic mutations characteristic of myeloid neoplasms. Ostensibly, this finding represents the development of clonal hematopoiesis (CH; often termed clonal hematopoiesis of indeterminate potential, abbreviated CHiP). Persons with CHiP are at increased risk of developing myeloid neoplasms upon follow-up and the burden of mutations indicating CHiP as determined by the variant allele frequency (VAF) appears positively correlated to the risk of development of a myeloid neoplasm over long-term follow up. These findings indicate that a myeloid premalignant state likely precedes the development of frank myeloid neoplasms. An additional moniker—clonal cytopenias of unknown/unclear significance (CCUS)—has been coined to indicate patients with cytopenias that cannot be classified as having a myeloid neoplasm such as MDS but in the setting of cytopenia(s) may be at increased risk of developing a myeloid neoplasm driven by the discovered CH. CH has interestingly been linked to increased overall mortality and worse cardiovascular outcomes; it is theorized that CH results in production of abnormal monocytes or other immune effectors that cause alterations in the regulation of vascular endothelial function and may increase atherosclerotic mediators. CHiP is also encountered on genomic assays performed to guide therapy in patients with advanced solid tumors where patients with CH appear to have worse outcomes perhaps in part related to high rates of hematologic toxicity during cytotoxic therapy. Guidelines for following patients found to have CHiP or CCUS incidentally are unclear, but clinics dedicated to the monitoring of the clinical course of these patients have been formed at many institutions and these patients may benefit from multidisciplinary care delivered by hematologists and preventive cardiologists.

Thus there are multiple instances whereby multiplexed targeted clinical grade gene sequencing panels can be utilized to provide novel options for patients with hematologic malignancies. Some of the most exciting clinical advances in myeloid neoplasms in terms of advancing therapeutics matched to the underlying tumor genomic makeup have occurred in AML and so the remainder of this review will focus on progress in our understanding of the genomic landscape and implementation of precision therapeutics in AML.

Approach to prognosis and treatment recommendations in acute myeloid leukemia

AML is the most common type of acute leukemia with an annual incidence in the United States of >20,000; AML accounts for 10,000 deaths annually (Sasaki, Ravandi, & Kadia, 2021; Siegel, Miller, & Jemal, 2020). Generally, two-thirds of patients will have disease relapse following initial chemotherapy, dependent largely upon age and other prognostic factors. For greater than 30 years, there were no improvements in treatment of AML beyond standard induction programs utilizing 7 days of cytarabine infusion combined with anthracycline chemotherapy (the so- called "7 + 3" program) followed by consolidation using a program of high dose cytarabine treatments and for those with either relapse or high risk of relapse, allogeneic transplant. Patients not fit for intensive chemotherapy have traditionally had very poor outcomes with alternative low-intensity therapies, although the recently described combination of hypomethylator therapy (azacytidine or decitabine) in combination with venetoclax produces improved complete response rates with acceptable toxicity in patients that cannot tolerate more intensive strategies.

Genomics has been leveraged to improve the crude precision medicine approaches long used in AML. AML prognostication utilizing pretreatment karyotyping has been established for decades and estimates prognosis and the risk of frontline treatment failure with standard induction therapies such as "7 + 3" with consolidative high dose cytarabine. Patients with complete response to therapy have traditionally been stratified to observation or to consideration of "consolidative" allogeneic bone marrow transplant: in those who have high-risk genomic alterations and adequate medical fitness, allogeneic transplant improves survival. Conversely, patients with favorable-risk genomic alterations can be assigned to clinical observation alone after initial treatment/high dose cytarabine consolidation without exposure to the risk of transplant-related morbidity/mortality. More recently, massively parallel deep sequencing has led to the development of assays that can define dynamically the presence of minimal residual disease (MRD) in patients for whom allogeneic stem cell transplantation is planned. Hourigan and colleagues analyzed frozen whole blood specimens from the Blood and Marrow Transplant Clinical Trials Network (BMT CTN) 0901 trial in which 190 AML patients in morphologic complete remission undergoing allogeneic transplant were randomized (1:1) to preparative regimens with either a myeloablative or reduced intensity (Hourigan, Dillon, & Gui, 2020). Utilizing ultra-deep error-corrected sequencing techniques capable of detection of AML-associated somatic mutations to a VAF of 0.001 or greater, each patient was assayed for MRD. With the exception of residual mutations in *ASXL1*, *DNMT3A*, and *TET2* that may have represented residual CH, MRD positivity was associated with worse survival in patients that were assigned to the reduced intensity conditioning (RIC) arm. It is interesting that residual CH after initial intensive induction therapy appears not to hold the same significance as the finding of somatic mutations marking persistent clones bearing oncogenic mutations such as *FLT3-ITD* or *IDH1/2*. MRD as defined by *NPM1* mutation (Ivey, Hills, & Simpson, 2016) has also been shown to be a negative prognostic marker in the setting of AML treatment. These assays can quickly move to clinical practice and appear to improve prognostication beyond standard cytogenetics and morphologic assessment of blasts, with important implications pertaining to how clinicians decide to recommend allogeneic transplant and what preparative regimen to use.

Recent genome sequencing efforts have identified recurrent genetic mutations targetable by novel small molecule therapies. One, FMS-like tyrosine kinase 3 (FLT-3), is expressed in 90% of myeloid blasts, with mutations in *FLT3* observed in ∼30% of AML patients. *FLT3* sequencing is now recommended in all new AML patients as both a prognostic marker and to direct the addition of targeted therapy to standard therapy. FLT-3 ITD cases with high allele burden (> 0.5 ratio of the area of the curve of *FLT3* ITD relative to *FLT3* wild type as detected by small quantitative sequencing) have an adverse prognosis relative to cases with lower allelic ratios and this determination of high risk has been listed as a standard of care in the 2017 European LeukemiaNet (ELN) guidelines (Döhner, Estey, & Grimwade, 2017), among others. For those who are medically fit, allogeneic transplantation has been recommended as standard of care in FLT3 ITD altered AML with the goal of overcoming the adverse prognostic implications of FLT-3 ITD alterations. This approach improves survival but at the cost of the morbidity associated with allogeneic transplantation.

Multiple small molecule inhibitors of FLT-3 exist, and these agents are classified into multiple categories according to kinase specificity and site of action on the FLT-3 structure itself. Sorafenib, a nonspecific multikinase inhibitor, has been studied in multiple Phase I/II studies in FLT3-altered AML. In a trial in young (age < 65 years) AML patients evaluating toxicity and response rates with sorafenib combined with standard chemotherapy (idarubicin and cytarabine) complete remission was achieved in 14/15 patients with *FLT3* mutated AML which appeared favorable compared to previous experience in that patient population. The SORAML trial addressed a similar question in a placebo-controlled randomized approach. Early results and longer term follow up show an event free and leukemia progression-free survival benefit with the addition of sorafenib to standard chemotherapy. This combination is associated with more toxicity as expected; interestingly there appeared to be benefit in *FLT3* wild-type patients as well since the trial did not limit enrollment to patients with FLT3 alterations. Sorafenib is being studied in multiple trials including as a maintenance agent after allogeneic transplant but sorafenib's role in the standard care of MDS/AML patients is unclear.

Other FLT-3-targeted agents have shown efficacy in *FLT3*-altered AML. Midostaurin and gilteritinib are standard of care agents approved by the FDA for the care of these patients. The Phase III RATIFY trial (Levis, 2017; Stone, Mandrekar, & Sanford, 2017) randomized 717 AML patients with either FLT-3 ITD or FLT-3 TKD (with stratification of alteration type and allelic ratio) to treatment with standard induction and consolidation with the addition of either placebo or midostaurin added during induction and consolidation therapy as well as a schedule of midostaurin maintenance. The addition of midostaurin resulted in improvements in both event free and overall survival compared to standard chemotherapy and this combination is approved for use in the United States. Interestingly, subset analysis suggests that the magnitude of benefit in RATIFY was higher in *FLT3* TKD altered patients compared to those with *FLT3* ITD but this approach is nevertheless used in both scenarios. The role of allogeneic transplant for AML defined as high risk due to FLT-3 ITD with high allelic ratio (Bazarbachi, Bug, & Baron, 2020; Poiré, Labopin, & Polge, 2018) has not necessarily been supplanted by the use of a midostaurin-containing regimen and the role of consolidation therapy still requires investigation.

In relapsed/refractory *FLT3*-altered AML, outcomes have historically been unacceptably poor with any treatment and multiple targeted agents have been studied. Gilteritinib, a selective FLT-3 inhibitor (Dhillon, 2019; Levis & Perl, 2020), is the only approved agent in this setting based on a randomized trial showing a complete response rate of 34% in gilteritnib-treated FLT-3-altered relapsed/refractory AML patients as compared to a 15.3% rate in salvage chemotherapy-treated patients (Perl, Martinelli, & Cortes, 2019). This trial showed improvement in symptoms with gilteritinib treatment compared to chemotherapy even given still low CR rates and short event free survival in the overall study group. Multiple other agents are under investigation both as single agents or in combination with other agents such as hypomethylators (Cooper, Kindwall-Keller, & Craig, 2015; Gilteritinib plus azacitidine combination..., 2021; Tomlinson, Gallogly, & Kane, 2020) (azacytidine/decitabine), venetoclax, and others.

Mutations in *IDH1* and *IDH2* are also highly actionable alterations in AML. *IDH1/2* mutations occur in approximately 1/3 of AML cases and may co-occur with multiple other alterations (Marcucci, Maharry, & Wu, 2010; Paschka, Schlenk, & Gaidzik, 2010; Patel, Ravandi, & Ma, 2011). Given this overlap with other molecular alterations in various subtypes of AML, *IDH1/2* mutations do not define a molecular subtype of AML but are nonetheless targetable as a therapeutic. Because both *IDH1* and *IDH2* mutations result in an oncogene product with novel enzymatic function that produces an epigenetically active metabolite (2HG (Lu, Ward, & Kapoor, 2012; Su, Dong, & Li, 2018)), mutant IDH1/2 protein—analogs to the BCR-ABL1 fusion protein—represents a novel therapeutic target with theoretically less on-target toxicity. Indeed, multiple inhibitors have been described from high-throughput screens that selectively inhibit mutant IDH1/2 enzymatic function, lower 2HG levels and result in decreased tumor cell growth and/or AML cell differentiation.

Enasidenib, an oral, selective inhibitor of mutant IDH2 (Stein, 2018; Yen, Travins, & Wang, 2017) was the first described IDH1/2 inhibitor to be reported as active in MDS/AML. In a Phase I/II trialstudying the efficacy and toxicity of enasidenib in relapsed/refractory *IDH2*-mutated AML, enasidenib was well tolerated and the dose limiting toxicity was not reached in the dose escalation phase (Stein, DiNardo, & Pollyea, 2017). The overall response rate in this study was 40.3% with a median response of 5.8 months and 19.3% of patients attained complete remission; in patients reaching CR the median overall survival was 19.7 months, which is very favorable compared to salvage therapy. Interestingly, one of the key severe adverse events appreciated in this study was the development of Grade 3/4 differentiation syndrome in 7% of patients. This phenomenon is analogs to the differentiation encountered with the use of ATRA in acute promyelocytic leukemia and can be treated with corticosteroids and leukoreduction along with supportive care (Fathi, DiNardo, & Kline, 2018; Sanz, Fenaux, & Tallman, 2019; Stahl & Tallman, 2019). Ivosidenib, a selective oral inhibitor of mutant IDH1 (DiNardo, Stein, & de Botton, 2018; Popovici-Muller, Lemieux, & Artin, 2018), has been studied in an analogs manner to enasidenib and exhibits similar response rates in IDH1 mutated relapsed/refractory AML. Studies of both agents have led to the approval of enasidenib and ivosidenib in *IDH2*- and *IDH1*-mutated AML, respectively, in the United States and Europe. Ongoing trials are looking into these agents in combination with both standard and novel therapies; these approaches hold significant promise for improving the outcomes across the spectrum of IDH1/2-mutated AML.

Application of multiplexed gene panels to assess the presence of recurrent mutations in myeloid neoplasms is now routinely done in a longitudinal manner in the clinical care of these patients. This has made possible translational research resulting in a preliminary understanding of the mechanisms of resistance to standard cytotoxic chemotherapy as well as novel therapies, particularly in AML. AMLs appear to undergo distinct patterns of clonal evolution as they become resistant to chemotherapy or targeted therapy with disappearance of clones with targetable mutations or the emergence and expansion of clones bearing molecular mediators of resistance. Acquired mutations in IDH2 (*trans*- or *cis*-dimer interface mutations) have been described as a mechanism of resistance to allosteric IDH inhibitors (Intlekofer, Shih, & Wang, 2018). In the initial Phase II experience with enasidenib, mutations in *NRAS*, *FLT3* (-ITD or -TKD), or cases with more numerous mutations in individual genes were associated with low-response rates to IDH2 inhibition. *KRAS*, *NRAS*, and other MAP kinase alterations have been associated with treatment resistance in other entities including ALL, and the RAS mutations have been traditionally thought to be "undruggable" targets. Similarly, gilteritinib resistant relapsed/refractory AML cases are enriched in MAP kinase alterations. Presumably, these mutations are capable of mediating resistance to ivosidenib and other FLT-3 inhibitors such as midostaurin. With the approval of hypomethylator/venetoclax combination therapy in elderly/unfit patients with AML (DiNardo, Pratz, & Pullarkat, 2019; DiNardo, Jonas, & Pullarkat, 2020), analogs studies have analyzed the prognosis of AML molecular subgroups with hypomethylator/venetoclax treatment including in the "real world" frontline setting (DiNardo, Tiong, & Quaglieri, 2020; Winters, Gutman, & Purev, 2019). While complete responses occur with this combination in about 2/3 of patients across AML subtypes, CR rates in AML subtypes classified by karyotype or gene mutation vary drastically (from >90% in *IDH1/2*-mutated cases to <40% in *TP53*-altered AML) (DiNardo et al., 2020). Thus clinically relevant genomic assays can help guide prognostication and potentially selection of therapy, particularly regarding treatment intensity.

Therapeutic matching pathways leveraging genomic assays and advances in targeted precision therapies at diagnosis may improve outcomes in AML. The Beat AML Master Trial (Burd, Levine, & Ruppert, 2020) enrolled newly diagnosed patients >59 years old; diagnostic samples were subjected to multiplexed sequencing assays with the goal of identifying actionable genetic alterations within 7 days. Patients were offered substudy enrollment with a matched therapeutic combination if a potentially actionable mutation was discovered (in some groups, novel therapy such as CD47 antibody blockade was added to standard therapy where there was not a clearly actionable alteration). Of 395 patients enrolled, 56.7% enrolled on a substudy and the median overall survival was 12.8 months in patients receiving matched genomics-guided therapy versus 3.9 months in the cohort receiving standard of care agents. This study thus proves the feasibility and utility of applying precision medicine in older patients with AML where outcomes have been traditionally grim. Studies are ever ongoing to determine effective novel combinations and it is likely that high-throughput genomics and precision cancer medicine will continue to improve outcomes across myeloid neoplasms and leukemias.

Summary

Investigation of the genomic landscapes in myeloid neoplasms/acute leukemias has produced many paradigm-changing discoveries. Advances in technology have improved our ability to understand the molecular subtypes of these diseases at ever-increasing resolution and continue to produce exciting discoveries in the field of translational genomics. Such findings provide opportunities for basic scientists to elucidate the biologic underpinnings of these diseases. Efforts thus far have resulted in discoveries that have changed profoundly our approach to treatment and the resulting patient outcomes; however, we still need further development of genomic assays and corresponding therapeutic advances to continue improving outcomes in myeloid neoplasms and acute leukemias.

References

Arber, D. A., Orazi, A., Hasserjian, R., et al. (2016). The 2016 revision to the World Health Organization classification of myeloid neoplasms and acute leukemia. *Blood, 127*(20), 2391−2405. Available from https://doi.org/10.1182/blood-2016-03-643544.

Bazarbachi, A., Bug, G., Baron, F., et al. (2020). Clinical practice recommendation on hematopoietic stem cell transplantation for acute myeloid leukemia patients with FLT3-internal tandem duplication: A position statement from the Acute Leukemia Working Party of the European Society for Blood and Marrow Transplantation. *Haematologica, 105*(6), 1507−1516. Available from https://doi.org/10.3324/haematol. 2019.243410.

Burd, A., Levine, R. L., Ruppert, A. S., et al. (2020). Precision medicine treatment in acute myeloid leukemia using prospective genomic profiling: Feasibility and preliminary efficacy of the Beat AML Master Trial. *Nature Medicine, 26*(12), 1852−1858. Available from https://doi.org/10.1038/s41591-020-1089-8.

Chung, C., & Ma, H. (2017). Driving toward precision medicine for acute leukemias: Are we there yet? *Pharmacotherapy, 37*(9), 1052−1072. Available from https://doi.org/10.1002/phar.1977.

Cooper, B. W., Kindwall-Keller, T. L., Craig, M. D., et al. (2015). A phase I study of midostaurin and azacitidine in relapsed and elderly AML patients. *Clinical Lymphoma, Myeloma & Leukemia, 15*(7), 428−432. Available from https://doi.org/10.1016/j.clml.2015.02.017, e2.

Cucchi, D. G. J., Polak, T. B., Ossenkoppele, G. J., et al. (2021). Two decades of targeted therapies in acute myeloid leukemia. *Leukemia, 35*(3), 651−660. Available from https://doi.org/10.1038/s41375-021-01164-x.

Dhillon, S. (2019). Gilteritinib: First global approval. *Drugs, 79*(3), 331−339. Available from https://doi.org/10.1007/s40265-019-1062-3.

Diamond, E. L., Durham, B. H., Haroche, J., et al. (2016). Diverse and targetable kinase alterations drive histiocytic neoplasms. *Cancer Discovery, 6*(2), 154−165. Available from https://doi.org/10.1158/2159-8290.Cd-15-0913.

Diamond, E. L., Durham, B. H., Ulaner, G. A., et al. (2019). Efficacy of MEK inhibition in patients with histiocytic neoplasms. *Nature, 567*(7749), 521−524. Available from https://doi.org/10.1038/s41586-019-1012-y.

DiNardo, C. D., Jonas, B. A., Pullarkat, V., et al. (2020). Azacitidine and venetoclax in previously untreated acute myeloid leukemia. *The New England Journal of Medicine, 383*(7), 617−629. Available from https://doi.org/10.1056/NEJMoa2012971.

DiNardo, C. D., Pratz, K., Pullarkat, V., et al. (2019). Venetoclax combined with decitabine or azacitidine in treatment-naive, elderly patients with acute myeloid leukemia. *Blood, 133*(1), 7−17. Available from https://doi.org/10.1182/blood-2018-08-868752.

DiNardo, C. D., Stein, E. M., de Botton, S., et al. (2018). Durable remissions with ivosidenib in IDH1-mutated relapsed or refractory AML. *The New England Journal of Medicine, 378*(25), 2386−2398. Available from https://doi.org/10.1056/NEJMoa1716984.

DiNardo, C. D., Tiong, I. S., Quaglieri, A., et al. (2020). Molecular patterns of response and treatment failure after frontline venetoclax combinations in older patients with AML. *Blood, 135*(11), 791−803. Available from https://doi.org/10.1182/blood.2019003988.

Döhner, H., Estey, E., Grimwade, D., et al. (2017). Diagnosis and management of AML in adults: 2017 ELN recommendations from an international expert panel. *Blood, 129*(4), 424−447. Available from https://doi.org/10.1182/blood-2016-08-733196.

Durham, B. H., Lopez Rodrigo, E., Picarsic, J., et al. (2019). Activating mutations in CSF1R and additional receptor tyrosine kinases in histiocytic neoplasms. *Nature Medicine, 25*(12), 1839–1842. Available from https://doi.org/10.1038/s41591-019-0653-6.

Fathi, A. T., DiNardo, C. D., Kline, I., et al. (2018). Differentiation syndrome associated with enasidenib, a selective inhibitor of mutant isocitrate dehydrogenase 2: Analysis of a Phase 1/2 study. *JAMA Oncology, 4*(8), 1106–1110. Available from https://doi.org/10.1001/jamaoncol.2017.4695.

Fenaux, P., Mufti, G. J., Hellstrom-Lindberg, E., et al. (2009). Efficacy of azacitidine compared with that of conventional care regimens in the treatment of higher-risk myelodysplastic syndromes: A randomised, open-label, phase III study. *The Lancet Oncology, 10*(3), 223–232. Available from https://doi.org/10.1016/s1470-2045(09)70003-8.

Figueroa, M. E., Lugthart, S., Li, Y., et al. (2010). DNA methylation signatures identify biologically distinct subtypes in acute myeloid leukemia. *Cancer Cell, 17*(1), 13–27. Available from https://doi.org/10.1016/j.ccr.2009.11.020.

Gilteritinib plus azacitidine combination shows promise in newly diagnosed FLT3-mutated AML. (2021). *The Oncologist, 26*(Suppl. 1), S10. Available from https://doi.org/10.1002/onco.13652.

Goddard, A. D., Borrow, J., & Solomon, E. (1992). A previously uncharacterized gene, PML, is fused to the retinoic acid receptor alpha gene in acute promyelocytic leukaemia. *Leukaemia, 6*(Suppl. 3), 117s–119s.

Gounder, M. M., Solit, D. B., & Tap, W. D. (2018). Trametinib in histiocytic sarcoma with an activating MAP2K1 (MEK1) mutation. *The New England Journal of Medicine, 378*(20), 1945–1947. Available from https://doi.org/10.1056/NEJMc1511490.

Hourigan, C. S., Dillon, L. W., Gui, G., et al. (2020). Impact of conditioning intensity of allogeneic transplantation for acute myeloid leukemia with genomic evidence of residual disease. *Journal of Clinical Oncology, 38*(12), 1273–1283. Available from https://doi.org/10.1200/jco.19.03011.

Intlekofer, A. M., Shih, A. H., Wang, B., et al. (2018). Acquired resistance to IDH inhibition through trans or cis dimer-interface mutations. *Nature, 559*(7712), 125–129. Available from https://doi.org/10.1038/s41586-018-0251-7.

Ivey, A., Hills, R. K., Simpson, M. A., et al. (2016). Assessment of minimal residual disease in standard-risk AML. *The New England Journal of Medicine, 374*(5), 422–433. Available from https://doi.org/10.1056/NEJMoa1507471.

Kantarjian, H., Issa, J. P., Rosenfeld, C. S., et al. (2006). Decitabine improves patient outcomes in myelodysplastic syndromes: Results of a phase III randomized study. *Cancer, 106*(8), 1794–1803. Available from https://doi.org/10.1002/cncr.21792.

Klampfl, T., Gisslinger, H., Harutyunyan, A. S., et al. (2013). Somatic mutations of calreticulin in myeloproliferative neoplasms. *The New England Journal of Medicine, 369*(25), 2379–2390. Available from https://doi.org/10.1056/NEJMoa1311347.

Lee, L. H., Gasilina, A., Roychoudhury, J., et al. (2017). Real-time genomic profiling of histiocytoses identifies early-kinase domain BRAF alterations while improving treatment outcomes. *JCI Insight, 2*(3), e89473. Available from https://doi.org/10.1172/jci.insight.89473.

Levis, M. (2017). Midostaurin approved for FLT3-mutated AML. *Blood, 129*(26), 3403–3406. Available from https://doi.org/10.1182/blood-2017-05-782292.

Levis, M., & Perl, A. E. (2020). Gilteritinib: Potent targeting of FLT3 mutations in AML. *Blood Advances, 4*(6), 1178–1191. Available from https://doi.org/10.1182/bloodadvances.2019000174.

Lu, C., Ward, P. S., Kapoor, G. S., et al. (2012). IDH mutation impairs histone demethylation and results in a block to cell differentiation. *Nature, 483*(7390), 474−478. Available from https://doi.org/10.1038/nature10860.

Lübbert, M., Suciu, S., Baila, L., et al. (2011). Low-dose decitabine vs best supportive care in elderly patients with intermediate- or high-risk myelodysplastic syndrome (MDS) ineligible for intensive chemotherapy: Final results of the randomized phase III study of the European Organisation for Research and Treatment of Cancer Leukemia Group and the German MDS Study Group. *Journal of Clinical Oncology, 29*(15), 1987−1996. Available from https://doi.org/10.1200/jco.2010.30.9245.

Marcucci, G., Maharry, K., Wu, Y. Z., et al. (2010). IDH1 and IDH2 gene mutations identify novel molecular subsets within de novo cytogenetically normal acute myeloid leukemia: A Cancer and Leukemia Group B study. *Journal of Clinical Oncology, 28*(14), 2348−2355. Available from https://doi.org/10.1200/jco.2009.27.3730.

Mardis, E. R., Ding, L., Dooling, D. J., et al. (2009). Recurring mutations found by sequencing an acute myeloid leukemia genome. *The New England Journal of Medicine, 361*(11), 1058−1066. Available from https://doi.org/10.1056/NEJMoa0903840.

Miyoshi, H., Kozu, T., Shimizu, K., et al. (1993). The t (8;21) translocation in acute myeloid leukemia results in production of an AML1-MTG8 fusion transcript. *The EMBO Journal, 12* (7), 2715−2721.

Nangalia, J., Massie, C. E., Baxter, E. J., et al. (2013). Somatic CALR mutations in myeloproliferative neoplasms with nonmutated JAK2. *The New England Journal of Medicine, 369*(25), 2391−2405. Available from https://doi.org/10.1056/NEJMoa1312542.

Papaemmanuil, E., Gerstung, M., Bullinger, L., et al. (2016). Genomic classification and prognosis in acute myeloid leukemia. *The New England Journal of Medicine, 374*(23), 2209−2221. Available from https://doi.org/10.1056/NEJMoa1516192.

Paschka, P., Schlenk, R. F., Gaidzik, V. I., et al. (2010). IDH1 and IDH2 mutations are frequent genetic alterations in acute myeloid leukemia and confer adverse prognosis in cytogenetically normal acute myeloid leukemia with NPM1 mutation without FLT3 internal tandem duplication. *Journal of Clinical Oncology, 28*(22), 3636−3643. Available from https://doi.org/10.1200/jco.2010.28.3762.

Patel, K. P., Ravandi, F., Ma, D., et al. (2011). Acute myeloid leukemia with IDH1 or IDH2 mutation: Frequency and clinicopathologic features. *American Journal of Clinical Pathology, 135*(1), 35−45. Available from https://doi.org/10.1309/ajcpd7nr2rmnqdvf.

Perl, A. E., Martinelli, G., Cortes, J. E., et al. (2019). Gilteritinib or chemotherapy for relapsed or refractory FLT3-mutated AML. *The New England Journal of Medicine, 381*(18), 1728−1740. Available from https://doi.org/10.1056/NEJMoa1902688.

Poiré, X., Labopin, M., Polge, E., et al. (2018). Allogeneic stem cell transplantation benefits for patients ≥ 60 years with acute myeloid leukemia and FLT3 internal tandem duplication: A study from the Acute Leukemia Working Party of the European Society for Blood and Marrow Transplantation. *Haematologica, 103*(2), 256−265. Available from https://doi.org/10.3324/haematol.2017.178251.

Popovici-Muller, J., Lemieux, R. M., Artin, E., et al. (2018). Discovery of AG-120 (ivosidenib): A first-in-class mutant IDH1 inhibitor for the treatment of IDH1 mutant cancers. *ACS Medical Chemical Letters, 9*(4), 300−305. Available from https://doi.org/10.1021/acsmedchemlett.7b00421.

Rabian, F., Lengline, E., & Rea, D. (2019). Towards a personalized treatment of patients with chronic myeloid leukemia. *Current Hematologic Malignancy Reports, 14*(6), 492−500. Available from https://doi.org/10.1007/s11899-019-00546-4.

Rowley, J. D. (1973b). Chromosomal patterns in myelocytic leukemia. *The New England Journal of Medicine*, *289*(4), 220−221. Available from https://doi.org/10.1056/nejm197307262890421.

Rowley, J. D. (1973a). Letter: A new consistent chromosomal abnormality in chronic myelogenous leukaemia identified by quinacrine fluorescence and Giemsa staining. *Nature*, *243* (5405), 290−293. Available from https://doi.org/10.1038/243290a0.

Rowley, J. D., Golomb, H. M., & Dougherty, C. (1977). 15/17 translocation, a consistent chromosomal change in acute promyelocytic leukaemia. *Lancet (London, England)*, *1*(8010), 549−550. Available from https://doi.org/10.1016/s0140-6736(77)91415-5.

Rowley, J. D., & Potter, D. (1976). Chromosomal banding patterns in acute nonlymphocytic leukemia. *Blood*, *47*(5), 705−721.

Rumi, E., Pietra, D., Ferretti, V., et al. (2014). JAK2 or CALR mutation status defines subtypes of essential thrombocythemia with substantially different clinical course and outcomes. *Blood*, *123*(10), 1544−1551. Available from https://doi.org/10.1182/blood-2013-11-539098.

Rumi, E., Pietra, D., Pascutto, C., et al. (2014). Clinical effect of driver mutations of JAK2, CALR, or MPL in primary myelofibrosis. *Blood*, *124*(7), 1062−1069. Available from https://doi.org/10.1182/blood-2014-05-578435.

Saglio, G., Fava, C., & Gale, R. P. (2019). Precision tyrosine kinase inhibitor dosing in chronic myeloid leukemia? *Haematologica*, *104*(5), 862−864. Available from https://doi.org/10.3324/haematol.2018.214445.

Sanz, M. A., Fenaux, P., Tallman, M. S., et al. (2019). Management of acute promyelocytic leukemia: Updated recommendations from an expert panel of the European LeukemiaNet. *Blood*, *133*(15), 1630−1643. Available from https://doi.org/10.1182/blood-2019-01-894980.

Sasaki, K., Ravandi, F., Kadia, T. M., et al. (2021). De novo acute myeloid leukemia: A population-based study of outcome in the United States based on the Surveillance, Epidemiology, and End Results (SEER) database, 1980 to 2017. *Cancer*, *127*(12), 2049−2061. Available from https://doi.org/10.1002/cncr.33458.

Siegel, R. L., Miller, K. D., & Jemal, A. (2020). Cancer statistics, 2020. *CA: A Cancer Journal for Clinicians*, *70*(1), 7−30. Available from https://doi.org/10.3322/caac.21590.

Stahl, M., & Tallman, M. S. (2019). Differentiation syndrome in acute promyelocytic leukaemia. *British Journal of Haematology*, *187*(2), 157−162. Available from https://doi.org/10.1111/bjh.16151.

Stein, E. M. (2018). Enasidenib, a targeted inhibitor of mutant IDH2 proteins for treatment of relapsed or refractory acute myeloid leukemia. *Future Oncology (London, England)*, *14*(1), 23−40. Available from https://doi.org/10.2217/fon-2017-0392.

Stein, E. M., DiNardo, C. D., Pollyea, D. A., et al. (2017). Enasidenib in mutant IDH2 relapsed or refractory acute myeloid leukemia. *Blood*, *130*(6), 722−731. Available from https://doi.org/10.1182/blood-2017-04-779405.

Stone, R. M., Mandrekar, S. J., Sanford, B. L., et al. (2017). Midostaurin plus chemotherapy for acute myeloid leukemia with a FLT3 mutation. *The New England Journal of Medicine*, *377* (5), 454−464. Available from https://doi.org/10.1056/NEJMoa1614359.

Su, R., Dong, L., Li, C., et al. (2018). R-2HG exhibits anti-tumor activity by targeting FTO/m (6)A/MYC/CEBPA signaling. *Cell*, *172*(1−2), 90−105.e23. Available from https://doi.org/10.1016/j.cell.2017.11.031.

Tefferi, A., Lasho, T. L., Finke, C. M., et al. (2014). CALR vs JAK2 vs MPL-mutated or triple-negative myelofibrosis: Clinical, cytogenetic and molecular comparisons. *Leukemia*, *28*(7), 1472−1477. Available from https://doi.org/10.1038/leu.2014.3.

Tefferi, A., Lasho, T. L., Tischer, A., et al. (2014). The prognostic advantage of calreticulin mutations in myelofibrosis might be confined to type 1 or type 1-like CALR variants. *Blood*, *124*(15), 2465−2466. Available from https://doi.org/10.1182/blood-2014-07-588426.

Tomlinson, B. K., Gallogly, M. M., Kane, D. M., et al. (2020). A phase II study of midostaurin and 5-azacitidine for untreated elderly and unfit patients with FLT3 wild-type acute myelogenous leukemia. *Clinical Lymphoma, Myeloma & Leukemia, 20*(4), 226−233.e1. Available from https://doi.org/10.1016/j.clml.2019.10.018.

Winters, A. C., Gutman, J. A., Purev, E., et al. (2019). Real-world experience of venetoclax with azacitidine for untreated patients with acute myeloid leukemia. *Blood Advances, 3*(20), 2911−2919. Available from https://doi.org/10.1182/bloodadvances.2019000243.

Yen, K., Travins, J., Wang, F., et al. (2017). AG-221, a first-in-class therapy targeting acute myeloid leukemia harboring oncogenic IDH2 mutations. *Cancer Discovery, 7*(5), 478−493. Available from https://doi.org/10.1158/2159-8290.Cd-16-1034.

Chapter 8

Lung cancer

James Isaacs[1] and Jeffrey Clarke[2]
[1]*Department of Medicine, Duke University Medical Center, Durham, NC, United States,*
[2]*Division of Medical Oncology, Department of Medical Oncology, Duke University, Durham, NC, United States*

Introduction

Lung cancer is the leading cause of cancer-related death in both men and women in the United States with an estimated 135,720 deaths in 2020, accounting for 22.4% of all cancer-related deaths (National Cancer Institute, 2020). Lung cancer is divided into nonsmall-cell lung cancer (NSCLC), comprising 80% of cases and small-cell lung cancer (SCLC) accounting for 15% of cases (National Cancer Institute, 2020). NSCLC is further subdivided into histologic subclassifications of adenocarcinoma, squamous cell carcinoma, and less commonly large cell carcinoma. Historically outcomes for lung cancer have been dismal with 5-year survival rates of $\sim 20\%$ (National Cancer Institute, 2020). This reflects the late stage at diagnosis as only 17% of patients are diagnosed with localized (Stage I or II) disease, and over half of patients have metastatic disease at the time of diagnosis (National Cancer Institute, 2020).

The past two decades have seen significant improvements in lung cancer outcomes. Surgical resection or chemoradiation can be curative in earlier stage disease so efforts at enhancing detection at a population level have been pursued (Patz, Goodman, & Bepler, 2000). In the metastatic setting, initial trials in the 1990s suggested that chemotherapy provides modest survival benefits (Group NM-AC, 2008). Now, advancements in basic science techniques including genomic and immune profiling have revealed significant heterogeneity within NSCLC and understanding these differences for each patient can allow personalization of therapy (Swanton & Govindan, 2016). In metastatic adenocarcinoma, the identification of driver mutations has allowed for improved patient survival when patients are given targeted therapies (Kris, Johnson, & Berry, 2014; Singal, Miller, & Agarwala, 2019). Patients across all subtypes of metastatic lung cancer can also benefit from immune checkpoint blockade (ICB) with a portion of patients achieving durable responses (Antonia, Borghaei, & Ramalingam, 2019). Biomarkers can select for patients

Genomic and Precision Medicine. DOI: https://doi.org/10.1016/B978-0-12-800684-9.00006-X

123

more likely to respond to ICB and there is now an extensive effort to personalize therapy as the interplay of genomics and the immune microenvironment are explored. Indeed, population level data suggest that mortality for NSCLC fell 6.3% annually from 2013 to 2016, with benefit coming from both reduced incidence and advancements in therapy (Howlader, Forjaz, & Mooradian, 2020). To continue to improve lung cancer–specific mortality, personalized medicine and genomics can impact therapy by both increasing early detection of lung cancer and improving therapies in the metastatic setting.

Lung cancer risk and screening

It is well established that cigarette smoking is the major risk factor for lung cancer as $\sim 80\%$ of cases involve current or former smokers (Siegel, Jacobs, & Newton, 2015). This has spurred efforts to screen high-risk patients for lung cancer. Two trials, one in the United States and one in Europe, have established the benefit of annual or biannual low-dose CT scanning in selected patients, which reduced lung cancer mortality by $\sim 20\%$ (de Koning, van der Aalst, & de Jong, 2020; The National Lung Screening Trial Research Team, 2011). These trials enrolled patients on the basis of smoking history—for example the United States trial enrolled patients who were between the ages of 55 and 74 with a cigarette smoking history of at least 30 pack years, and if a former smoker had quit within the previous 15 years.

Age and smoking history alone may be limited in selecting high-risk patients, as it is noted that $<20\%$ of patients with extensive smoking history will actually develop lung cancer (Wang, Liu, & Yuan, 2017). Several studies have examined adding demographic features such as age, sex, race, body mass index, family history, education, and COPD status (Katki, Kovalchik, Berg, Cheung, & Chaturvedi, 2016; Kovalchik, Tammemagi, & Berg, 2013; Tammemägi, Katki, & Hocking, 2013). These retrospective analyses have shown that the use of demographic features to select patients for CT screening improves both the sensitivity and specificity of screening programs. However, despite these efforts there is no prospective data to support using a demographic-based risk screening tool. Current clinical practice guidelines use age and smoking history alone for screening recommendations (Mazzone, Silvestri, & Patel, 2018; Moyer, 2014).

In addition to demographic factors, several lines of epidemiolocal data suggest that genetic factors may predispose certain smokers to a higher risk of lung cancer. Analysis of the international lung cancer consortium across 24,380 lung cancer cases and 23,305 controls showed that individuals with a first-degree relative with lung cancer had a 1.51-fold increase of lung cancer, adjusted for smoking and other risk factors (Coté, Liu, & Bonassi, 2012). Population-based studies have similarly concluded that family history increases smoking-related cancer risk (Jonsson, Thorsteinsdottir, & Gudbjartsson, 2004). Several cohort studies have also suggested ethnic differences in susceptibility to lung cancer—related smoking, with African-American smokers having higher incidence of lung cancer than Caucasian smokers when controlled for other demographic factors (Coté, Kardia, Wenzlaff, Ruckdeschel, & Schwartz, 2005; Haiman, Stram, & Wilkens, 2006). These data have spurred genome-wide association studies (GWAS) seeking to identify specific genetic links to lung cancer risks in smokers. The GWAS have identified specific loci including 5p15, 6p21, and 15q25 that contain genes of interest related to nicotine addiction and metabolism, response to DNA damage and oxidant stress, and telomerase expression (Bossé & Amos, 2018; Timofeeva, Hung, & Rafnar, 2012). While these genetic polymorphisms are prevalent and associated with increased risk of lung cancer, the penetrance remains low for any individual gene contributing to lung cancer incidence, limiting clinical applicability (Musolf, Simpson, & de Andrade, 2017). One case—control study examined adding the 5p15, 6p21, and 15q25 polymorphisms to a model including smoking history, age, and sex to predict incident lung cancer. They found that adding "high-risk" genotypes to this model had minimal clinical impact on ability to predict risk for lung cancer (Weissfeld, Lin, & Lin, 2015). Thus currently germline genetic testing is not clinically used to stratify people at high risk of smoking-related lung cancer.

With declining rates of smoking, increasing attention has been paid to underlying genetic risk for nonsmokers. The case—control and population-based studies discussed above also show increased risk of lung cancer in nonsmokers with a family history of lung cancer (Coté et al., 2012). This suggests that genetic risk is not entirely related to smoking susceptibility. Genetic linkage studies examining families with high risks of lung cancer have identified a locus at 6q that conferred high risk to never and light smokers (Amos, Pinney, & Li, 2010). Single gene germline mutations are also found in never smokers including $EGFR^{T790M}$ (Yu, Arcila, & Harlan Fleischut, 2014) and $ERBB2$ (Yamamoto, Higasa, & Sakaguchi, 2014). However, despite these high penetrance mutations, the prevalence on a population level is low, limiting the practical application to population-based lung cancer screening efforts.

Diagnosis and molecular characteristics

Diagnosis of lung cancer requires tissue biopsy and histologic examination to differentiate between NSCLC (including histologic subtypes of adenocarcinoma and squamous cell carcinoma) and SCLC. Advancements in genetic sequencing have allowed more detailed classification of these subtypes. Squamous cell NSCLC and SCLC are strongly associated with cigarette smoking and have significant genomic instability. Adenocarcinomas are also predominately driven by smoking, but up to 20% occur in never smokers, and these cancers can have specific genetic mutations that may drive the cancer and can be treated with targeted therapies. The Cancer Genome Atlas (TCGA) has profiled many subtypes of cancers, performing whole genome sequencing, copy number analysis, gene expression profiling through RNA sequencing, and DNA methylation analysis (Table 8.1). The publication of TCGA for squamous cell lung cancer, lung adenocarcinoma, and SCLC has allowed important insights into the molecular underpinnings of these subtypes (Hammerman, Lawrence, & Voet, 2012).

Squamous cell carcinoma

TCGA analyzed 178 primary lung squamous cell carcinomas of which 96% of patients had a history of tobacco use (Hammerman et al., 2012). Analysis demonstrated significant genomic instability with a mean of 360 exonic mutations, 323 altered copy number segments, and 165 genomic rearrangements per tumor. The genomic alterations most commonly targeted pathways involving cell cycle control, response to oxidative stress, apoptotic signaling, and squamous cell differentiation. *TP53* mutations were detected in 81% of the samples and *CDKN2A*—a key regulator of cell cycling—was inactivated in 72% of cases by several different mechanisms (including epigenetic silencing-methylation, mutations, exon skipping, or homozygous deletions). Pathway analysis suggested that fibroblast growth factor receptors (FGFR) was the most commonly altered receptor tyrosine kinase (RTK; in 12% of cases), the PI3K pathway (PTEN, PIK3CA, and AKT) contained alterations in 47% of cases, and alterations in the oxidative stress response (KEAP1 and NFE2L2) occurred in 34% of cases (Hammerman et al., 2012).

Adenocarcinoma

Lung adenocarcinomas are associated with specific driver alterations in RTKs and the mitogen-activated protein kinase (MAPK) pathway (Imielinski, Berger Alice, Hammerman, & Peter, 2012). There are also marked differences in genetic makeup between patients with a smoking history and never smokers as patients with a smoking history have been found to have significantly higher mutational burdens. Additionally, adenocarcinoma from patients with smoking history is characterized by high cytosine to adenine transversion rate and this can be used as a genomic signature of smoking (Govindan, Ding, & Griffith, 2012). *KRAS* mutations are significantly enriched in former smokers whereas alterations in *EGFR, ALK, ROS1*, and *RET* are enriched in never smokers. TP53 and KEAP1 mutations were enriched in tumors without MAPK or RTK driver mutations (Collisson, Campbell, & Brooks, 2014). These findings have served as a basis for molecular-targeted therapies predominantly for patients with adenocarcinoma and no smoking history

TABLE 8.1 Genomic alterations detected in pathways in the human genome project analysis.

Adenocarcinoma		Squamous cell carcinoma		Small-cell carcinoma	
Alteration	Prevalence (%)	Alteration	Prevalence (%)	Alteration	Prevalence (%)
RTK/RAS/RAF Pathway	76	p53	90	p53	100
Cell cycle regulators	64	Cell cycle regulators	72	Cell cycle regulators	100
p53 pathway	63	PI(3)K-mTOR pathway	47	NOTCH	25
Chromatin regulation and RNA splicing	49	Squamous differentiation genes	44		
PI(3)K-mTOR pathway	25	Oxidative stress pathways	34		
Oxidative stress pathways	22	RTK/RAS/RAF Pathway	26		

Small cell

One hundred and ten small-cell carcinomas have also been analyzed in TCGA. This analysis showed universal bi-allelic inactivation of both TP53 and RB1, suggesting that these mutations are required for the development of SCLC. Additional frequent mutations were found in NOTCH family genes (25% of cases) (George, Lim, & Jang, 2015).

Early stage disease

Even with molecular advancements in the characterization of lung cancer (described above), the prognosis and treatment of patients with lung cancer is largely dependent on staging. Patients with early stage, nonmetastatic disease can be managed for curative intent with combinations of surgery, radiation, or chemotherapy. There have been multiple studies attempting to use gene signatures to predict patients with early stage surgically resected NSCLC who are at higher risk of recurrence and would benefit the most from adjuvant chemotherapy (Beer, Kardia, & Huang, 2002; Chen, Yu, & Chen, 2007; Guo, Wan, & Tosun, 2008; Kratz, He, & Van Den Eeden, 2012; Lau, Boutros, & Pintilie, 2007; Raponi, Zhang, & Yu, 2006). However, none of these have been prospectively validated or entered routine clinical practice.

Genetic testing in metastatic disease

While the use of precision genomics for treatment of early stage NSCLC is still investigational, there have already been dramatic improvements in outcomes for patients with metastatic NSCLC. As outlined above, adenocarcinomas (typically in never smokers) are more likely to be driven by specific somatic mutations in pathways that drive cancer progression and can be targeted by oral tyrosine kinase inhibitors (TKIs). In contrast, squamous cell NSCLC and SCLC are associated with smoking, leading to a higher degree of genetic instability and higher mutational burden (George et al., 2015; Hammerman et al., 2012). Thus, at this current time, there are no therapies for specific genomic alterations in squamous cell NSCLC or SCLC. For adenocarcinoma, there are now multiple somatic DNA alterations with effective targeted treatment options, including *EGFR, ALK, ROS1, BRAF, RET, MET*, and *NTRK. KRAS* and *HER2* (*ERBB2*) are additional driver mutations for which targeted therapies are currently under development (Fig. 8.1). Initially, panels of these mutations were performed using Sanger sequencing for mutations (*EGFR, KRAS*) and fluorescence in situ hybridization (FISH) for translocations (*ALK, ROS1*) (Barlesi, Mazieres, & Merlio, 2016; Kris et al., 2014). Newer methods utilize next generation DNA sequencing techniques

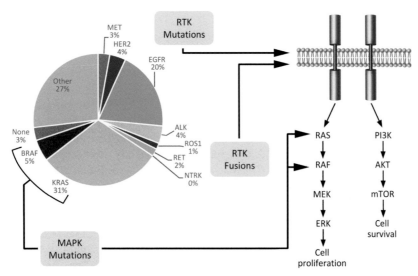

FIGURE 8.1 Molecular pathways with genomic driver alterations in NSCLC. Receptor tyrosine kinase mutations (EGFR, HER2, METex14), receptor tyrosine kinase fusions (ALK, ROS1, RET, NTRK), and MAPK mutations (KRAS, BRAF) with targeted therapies available for treatment are depicted. *Data on prevalence are from Foundation Medicine database as reported in Frampton, G. M., Ali, S. M., Rosenzweig, M., et al. (2015). Activation of MET via diverse exon 14 splicing alterations occurs in multiple tumor types and confers clinical sensitivity to MET inhibitors.* Cancer Discovery, 5, 850−859.

which can detect both mutations and translocations (Jordan, Kim, & Arcila, 2017; Singal et al., 2019).

Historically, genomic alterations were detected by testing tissue pathology specimens. A new technology complementary to tissue-based testing is the use of cell-free DNA (cfDNA) to detect circulating tumor DNA. Metastatic tumors will shed tumor DNA into circulation which can then be detected and analyzed for genomic alterations. This was initially tested in patients who were receiving treatment with EGFR TKIs, detecting T790M mutations in a group of patients with resistant disease (Oxnard, Paweletz, & Kuang, 2014; Oxnard, Thress, & Alden, 2016; Thompson, Yee, & Troxel, 2016). Several additional studies have examined the use of cfDNA assays for broad panels of actionable mutations for newly diagnosed NSCLC. This approach is appealing as there is commonly insufficient tissue to run next generation sequencing (NGS) testing on a biopsy sample (Li, Janku, & Jung, 2019; Zugazagoitia, Ramos, & Trigo, 2019). Additionally, due to tumor heterogeneity and sampling variability in biopsies, it is possible for cfDNA to detect mutations not picked up on initial tissue-based profiling (Aggarwal, Thompson, & Black, 2019). Finally, cfDNA offers faster turnaround time than traditional tissue-based testing (Leighl, Page, & Raymond, 2019).

EGFR mutations

Targeting the epidermal growth factor receptor (EGFR) in NSCLC is of one the most impressive examples of personalizing cancer care through genomic sequencing. There are now five FDA approved EGFR TKIs—erlotinib, gefitinib, afatinib, dacomitinib, and osimertinib (Table 8.2). EGFR is a cell surface RTK involved in intracellular signaling mediating prosurvival proliferative pathways through MAPK, PIK3/AKT, and JAK/STAT. As the molecular biology of these pathways became elucidated in the 1980s and 1990s evidence accumulated that EGFR is frequently over expressed in human cancers and in NSCLC (Hirsch, Varella-Garcia, & Bunn, 2003). This led to the rationale for the development of small molecules targeting the ATP-binding site of the EGFR tyrosine kinase—gefitinib and erlotinib. Early trials examined these agents in unselected patient populations, but it was then discovered that specific mutations in the EGFR tyrosine kinase domain led to increased signaling through EGFR and ultimately susceptibility to targeted therapy (Lynch, Bell, & Sordella, 2004; Paez, Jänne, & Lee, 2004).

Initial Phase III trials examined patients with NSCLC refractory to multiple lines of chemotherapy and compared erlotinib (Shepherd, Rodrigues Pereira, & Ciuleanu, 2005) or gefitinib (Thatcher, Chang, & Parikh, 2005) to best supportive care. The response rates were below 10% in both trials, with modest progression-free survival (PFS) and overall survival benefits in the erlotinib trial but no significant benefit in the gefitinib trial. Additional trials also found that combining gefitinib (Herbst, Giaccone, & Schiller, 2004) or erlotinib (Gatzemeier, Pluzanska, & Szczesna, 2007), with frontline chemotherapy regimens did not improve PFS compared to chemotherapy alone. These trials allowed a significant portion of patients with squamous histology and extensive smoking history. Subgroup analysis from these trials showed improved outcomes in patients with no smoking history, adenocarcinoma histology, and Asian ethnicity (which would select for patients with EGFR mutations). Translational studies sequenced the EGFR gene and found somatic mutations in the patients who had responded to gefitinib (Lynch et al., 2004; Paez et al., 2004). These mutations most often occur as exon 19 deletions or substitution mutations in exon 21 (L858R) which both serve to promote a constitutively active EGFR conformation, but also allow for competitive inhibition by EGFR TKIs. It was soon found that 10%−20% of Caucasian patients and 40%−60% of South-East Asian patients with lung adenocarcinoma have EGFR mutations (Hsu, Yang, Mok, & Loong, 2018). Trials that had enrolled clinically selected patients (adenocarcinoma, low smoking history) revealed PFS advantage for the first time in when comparing gefitinib to chemotherapy as a frontline treatment (Mok, Wu, & Thongprasert, 2009). Subsequent trials enrolling exclusively patients with EGFR exon 19 or 21 mutations established the standard of care of frontline TKI over chemotherapy in patients with EGFR driver mutations (Maemondo, Inoue, & Kobayashi, 2010; Mitsudomi, Morita, & Yatabe, 2010; Rosell, Carcereny, & Gervais, 2012).

TABLE 8.2 Pivotal clinical trials of FDA approved TKIs for first-line treatment of locally advanced or metastatic EGFR mutation positive adenocarcinoma.

Trial	Phase	Treatment	Number of patients	Response rate	Progression-free survival	Overall survival
IPASS, 2009	III	Gefitinib	609	43.00%	5.7 months	18.6 months
		Carboplatin + paclitaxel	608	32.30%	5.8 months	17.3 months
				$P < .001$	HR 0.72 (0.65–0.85) $P < .001$	HR 0.91 (0.76–1.10)
North-East Japan Study Group, 2010	III	Gefitinib	114	73.70%	10.8 months	30.5 months
		Carboplatin + paclitaxel	114	30.70%	5.4 months	23.6 months
				$P < .001$	HR 0.30 (0.22–0.41)	$P = .31$
West Japan Oncology Group, 2010	III	Gefitinib	86	62.10%	9.2 months	34.9 months
		Cisplatin + docetaxel	86	32.20%	6.3 months	27.3 months
				$P < .0001$	HR 0.489 (0.336–0.710)	HR 1.252 (0.883–1.775)
EURTAC, 2012	III	Erlotinib	86	58%	9.7 months	19.3 months
		Cisplatin + docetaxel or gemcitabine	87	15%	5.2 months	19.5 months
					HR 0.37 (0.25–0.54)	HR 1.04 (0.65–1.68)

(Continued)

TABLE 8.2 (Continued)

Trial	Phase	Treatment	Number of patients	Response rate	Progression-free survival	Overall survival
OPTIMAL, 2011	III	Erlotinib		83%	13.1 months	22.8 months
		Carboplatin + gemcitabine		36%	4.6 months	27.2 months
				$P < .0001$	HR 0.16 (0.10–0.26)	HR 1.19 (0.83–1.71)
LUX-6, 2014	III	Afatinib	242	66.90%	11 months	23.1 months
		Cisplatin + gemcitabine	122	23.00%	5.6 months	23.5 months
				$P < .0001$	HR 0.28 (0.20–0.39)	HR 0.93 (0.72–1.22)
LUX-7, 2016	IIB	Afatinib	160	70%	11 months	27.9 months
		Gefitinib	159	56%	10.9 months	24.5 months
				$P = .0083$	HR 0.73 (0.57–0.95)	HR 0.83 (0.58–1.17)
ARCHER, 2017	III	Dacomitinib	227	75%	14.7 months	34.1 months
		Gefitinib	225	75%	9.2 months	26.8 months
				$P = .423$	HR 0.59 (0.47–0.74)	$P = .044$
FLAURA, 2018	III	Osimertinib	279	80%	18.9 months	38.6 months
		Gefitinib or erlotinib	277	76%	10.2 months	31.8 months
				$P = .24$	$P < .001$	$P = .046$

HR, Hazard ratio.

Despite impressive clinical activity in the subgroup of patients with *EGFR* mutations, PFS remained <12 months with first-line TKIs gefitinib or erlotinib. Resistance mutations that developed while patients were receiving erlotinib or gefitinib were discovered, leading to another application of precision genetic testing. The most prevalent resistant mutation is $EGFR^{T790M}$, which occurs in up to 60% of patients who have progressed on first-line TKIs (Jänne, Yang, & Kim, 2015). First described in 2005 (Kobayashi, Boggon, & Dayaram, 2005), this mutation alters the affinity of first- and second-generation TKI through steric hindrance (Sos, Rode, & Heynck, 2010), preventing their inhibition of ATP binding. This led to the development of osimertinib, an irreversible third-generation EGFR TKI that was designed to overcome resistance to $EGFR^{T790M}$ mutations. It was first evaluated in comparison to standard chemotherapy as treatment after progression on first- or second-generation TKI in the AURA3 trial and showed a 71% response rate with PFS of 10.1 months compared to 4.4 months with chemotherapy (Mok, Wu, & Ahn, 2016). The FLAURA trial then studied osimertinib as frontline therapy in EGFR mutated NSCLC compared to gefitinib or erlotinib and showed a 6.8-month overall survival benefit (38.6 vs 31.8 months) despite 31% cross-over rate (Ramalingam, Vansteenkiste, & Planchard, 2019). Particularly notable was the enhanced central nervous system (CNS) activity, with a higher CNS response and prolonged CNS PFS with osimertinib in comparison to erlotinib or gefitinib (Reungwetwattana, Nakagawa, & Cho, 2018). This has established osimertinib as a new standard frontline therapy in EGFR mutant NSCLC.

Analysis of patients who have developed resistance to osimertinib revealed that the majority who maintained $EGFR^{T790M}$ mutations had developed an additional $EGFR^{C797S}$ mutation (Oxnard, Hu, & Mileham, 2018). Patients who progressed and no longer had detectable $EGFR^{T790M}$ mutations, or who progress on frontline osimertinib frequently demonstrate alterations leading to bypass-signaling pathways, including *MET* alterations (7%−24%) or gene fusions (such as *RET*, *BRAF*, or *NTRK* fusions, 1%−10%) (Piper-Vallillo, Sequist, & Piotrowska, 2020). Additionally, 2%−15% of patients have been shown to have histologic transformation to SCLC, and squamous cell transformation has also been described in patients treated with frontline osimertinib (Piper-Vallillo et al., 2020; Schoenfeld, Chan, & Kubota, 2020). For patients with *EGFR* mutation positive NSCLC that has progressed on osimertinib, several targeted therapies based on genomic alterations detected at progression are being investigated in clinical trials. The Phase I TATTON trial evaluated the addition of savolitinib to osimertinib in patients with *EGFR* mutation positive NSCLC who had developed *MET* amplification. Among 69 patients who had progressed on osimertinib prior to enrollment, the trial reported a 20% response rate and 7.9-month median duration of response (Sequist, Han, & Ahn, 2020). The SAVANNAH study (NCT03778229) will study a similar strategy in patients who have received first-line osimertinib and are found to have *MET* amplification upon disease progression. The Phase II ORCHARD study (NCT03944772) will assign patients to specific treatment arms based on the specific molecular resistance mechanism to osimertinib. In addition to using targeted therapy combinations at progression after treatment with EGFR TKIs, new frontline combinations of therapies are now being tested to try to achieve deeper treatment response and prevent the emergence of resistance mutations. The combination of gefitinib and chemotherapy has shown overall survival benefit when compared to gefitinib alone (Hosomi, Morita, & Sugawara, 2020; Noronha, Patil, & Joshi, 2020). The combination of erlotinib and VEGF inhibitor ramucirumab also showed PFS benefit over erlotinib alone (Nakagawa, Garon, & Seto, 2019). As the standard frontline treatment has shifted to osimertinib, trials evaluating osimertinib + chemotherapy (NCT04035486) or ramucirumab (NCT02789345) are now awaited.

Given the survival benefit in the metastatic setting, osimertinib is now being evaluated as adjuvant therapy following surgical resection of early stage NSCLC. In the randomized Phase III ADAURA trial, the group of patients with Stages II–IIIA adenocarcinoma receiving 3 years of adjuvant osimertinib following chemotherapy had a 24-month disease-free survival rate of 90% compared to 44% in patients given placebo (Hazard ratio (HR) 0.17, $P < .0001$) (Wu, Tsuboi, & He, 2020). While the magnitude of DFS benefit is substantial, it remains to be seen if this approach will improve overall survival, as gefitinib was previously shown to prolong DFS in the adjuvant setting but this did not translate to an overall survival benefit (Wu, Zhong, & Wang, 2020). Importantly, the most common site of initial relapse with adjuvant gefitinib was within the CNS, but the ADAURA trial has shown a CNS disease-free survival benefit with osimertinib (Wu et al., 2020; Xu, Xi, & Zhong, 2019). The role for adjuvant osimertinib in patients with locally advanced NSCLC treated with definitive chemoradiation is also being investigated in the LAURA (NCT03521154) trial.

Up to 10% of *EGFR* mutant NSCLC involves insertion mutations in exon 20. Unlike exon 19 deletions or exon 21 $EGFR^{L858R}$ mutations, patients with exon 20 mutant NSCLC have poor response to first- and second-generation EGFR TKIs (Yang, Sequist, & Geater, 2015). The insertions at exon 20 alter the conformation of a C-helix in EGFR, promoting constitutively active signaling activity. Although activating the EGFR tyrosine kinase, this conformational change does not impair ATP binding or lead to preferential binding of gefitinib in preclinical models (Yasuda, Park, & Yun, 2013). Poziotinib is an EGFR TKI with a smaller and more flexible conformation that is shown in preclinical models to inhibit *EGFR* exon 20 mutant NSCLC (Robichaux, Elamin, & Tan, 2018). In a Phase II trial, among 115 patients with previously treated *EGFR* exon 20 mutant NSCLC, poziotinib demonstrated a response rate of 14.8% (Le, Goldman, & Clarke, 2020). Toxicity may have limited consistent dosing, so alternate dosing schedules are now being explored in an ongoing trial (NCT03318939). Osimertinib at an increased dose of 160 mg daily has also shown a response rate of 25% among 21 patients with *EGFR* exon 20 mutant NSCLC (Piotrowska, Wang, Sequist, & Ramalingam, 2020). Finally, mobocertinib is a TKI specifically designed to target *EGFR* exon 20 insertions and has shown a 54% response rate among 26 previously treated patients in a Phase II trial (Janne, Neal, & Camidge, 2019). Mobocertinib is now being evaluated in a randomized Phase III against platinum-based chemotherapy as frontline treatment for NSCLC with *EGFR* exon 20 insertion mutations (NCT04129502).

ALK translocations

Targeting the anaplastic lymphoma kinase (ALK) in NSCLC is another successful example of genetic discoveries leading to targeted therapy. Fusions combining the ALK gene with nucleophosmin were first discovered in anaplastic lymphoma in 1994 (Morris, Kirstein, & Valentine, 1994). In 2007 it was discovered that a translocation on chromosome 2 involving the echinoderm microtubule-associated protein like 4 (*EMLA4*) and *ALK* occurred in a patient with NSCLC (Soda, Choi, & Enomoto, 2007). It is now established that *ALK* fusions occur in 3%−7% of NSCLC cases and now more than 20 fusion gene partners have been identified (Lin, Riely, & Shaw, 2017). While the function of native ALK is not clearly understood, it is an RTK and the translocation partner (*EML4*) is thought to lead to oligomerization of the fusion protein. This leads to activation of the MAPK, PI3K/AKT, and JAK/ STAT signaling pathways, leading to tumor cell survival and proliferation (Lin et al., 2017).

Crizotinib, originally developed as an MET inhibitor, was the first TKI demonstrated to have clinical activity in *ALK* fusion positive NSCLC (Kwak, Bang, & Camidge, 2010; Shaw, Kim, & Nakagawa, 2013). The PROFILE 1014 trial established the efficacy of crizotinib over frontline chemotherapy with a response rate of 74% and 3.9-month PFS benefit over chemotherapy (Solomon, Mok, & Kim, 2014). Impressively, the final analysis showed a 56.6% 4-year overall survival rate in the crizotinib arm (Solomon, Kim, & Wu, 2018). Despite initial responses, similar to EGFR inhibition, all patients will eventually develop resistance mechanisms. Initial studies discovered "gatekeeper" mutations within the tyrosine kinase domain that prevent crizotinib binding (Choi, Soda, & Yamashita, 2010) and it has been shown that these mutations occur in ∼30% of patients resistant to crizotinib (Doebele, Pilling, & Aisner, 2012; Katayama, Shaw, & Khan, 2012). Additional mechanisms of resistance include *ALK* amplification, and bypass pathways including *KIT* amplification, *EGFR* mutations, and *KRAS* mutations (McCoach, Le, & Gowan, 2018).

Several second-generation TKIs including ceritinib (Shaw, Kim, & Mehra, 2014; Shaw, Kim, & Crinò, 2017; Soria, Tan, & Chiari, 2017), alectinib (Peters, Camidge, & Shaw, 2017; Shaw, Gandhi, & Gadgeel, 2016), and brigatinib (Camidge, Kim, & Ahn, 2018) have now shown disease activity in patients refractory to crizotinib and been shown to have superior PFS when given as first-line therapies in comparison to crizotinib (Table 8.3). The second-generation TKI's are more specific for the *ALK* fusion protein and have clinical activity in patients resistant to crizotinib with known *ALK* mutations as well as those with no clear resistant mutation. This suggests many crizotinib resistant tumors may still be dependent on ALK signaling and are susceptible to more potent inhibitors (Friboulet, Li, & Katayama, 2014). The second-generation TKIs also have CNS penetration preventing a common mechanism of progression with crizotinib, which does not cross the blood−brain barrier. In the randomized Phase III ALEX trial, the group of patients receiving alectinib had a 12-month PFS rate of 68.4%, in comparison to 48.7% with crizotinib (HR 0.47, $P < .0001$) (Peters et al., 2017). In the alectinib arm, 12% of patients had an event of CNS progression in comparison to 45% of patients in the crizotinib group. Additionally, the CNS response rate of patients with measurable CNS lesions at baseline was 81% among patients treated with alectinib (Peters et al., 2017).

Resistance to second-generation TKI's show a different profile than crizotinib, with a higher portion of *ALK* mutations ($\sim 50\%$) including a ALK^{G1202R} mutation affecting the solvent front and preventing TKI binding through steric hindrance (Gainor, Dardaei, & Yoda, 2016). Lorlatinib has now been developed as a third-generation ALK TKI that can target the ALK^{G1202R} resistance mutation and has shown 39% response rate in patients who have progressed on two or more TKIs (Solomon, Besse, & Bauer, 2018). In this trial, patients who had progressed on a second-generation TKI and had a resistance mutation had higher response rates (62%) than those who progressed on a second-generation TKI but did not have a resistance mutation detected (32%) (Shaw, Solomon, & Besse, 2019). Overall, the *ALK* resistance mutations seen with *ALK* fusion positive NSCLC are much more heterogenous than *EGFR* resistance mutations seen with *EGFR* mutant NSCLC (which are commonly T790M or C979S mutations or outside of the EGFR protein as bypass pathway alterations). The ALK TKIs all have slightly different binding properties and thus their own resistance profiles. This is best exemplified by a case report of a patient who had had progressive disease on crizotinib and ceritinib and was found to have a ALK^{C1156Y} mutation so was treated with lorlatinib and had a clinical response. The patient then progressed and was found to have a new ALK^{L1198F} mutation that was shown to promote resistance to lorlatinib but this mutation was actually sensitive to crizotinib. Upon rechallenge with crizotinib the patient had a significant clinical response (Shaw, Friboulet, & Leshchiner, 2015). This highlights the ability to match precision genomic findings with specific targeted therapies.

TABLE 8.3 Pivotal trials of FDA approved TKIs for first-line treatment of locally advanced or metastatic ALK fusion positive adenocarcinoma.

Trial	Phase	Treatment	Population	N	Response	PFS	OS
Profile 1014, 2014	III	Crizotinib	ALK fusion positive	172	74%	10.9 months	NR
		Cisplatin or carboplatin + pemetrexed		171	45%	7.0 months	47.5 months
					P<.001	HR 0.45 (0.35−60)	0.76 (0.548−1.053)
ASCEND-4, 2017	III	Ceritinib	ALK fusion positive	189	72.50%	16.6 months	NR
		Cisplatin or carboplatin + pemetrexed		187	26.70%	8.1 months	26.2 months
						HR 0.55 (0.42−0.73)	HR 0.73 (0.5−1.08)
ALEX, 2017	III	Alectinib	ALK fusion positive	152	82.90%	34.8 months	NR
		Crizotinib		151	75.50%	10.9 months	57.4 months
					P=.09	0.43 (0.32−0.58)	0.67 (0.46−0.98)
ALTA-1L	III	Brigatinib	ALK fusion positive	137	71%	24 months	NR
		Crizotinib		138	60%	11 months	NR
						0.49 (0.35−0.68)	0.92 (0.57−1.47)

ROS1

The *ROS1* proto-oncogene is the third successful example of genetically targeted therapy in NSCLC. *ROS1* fusions were originally discovered in 1980s
and characterized in glioblastoma (Birchmeier, Birnbaum, Waitches, Fasano,
& Wigler, 1986). In 2007 ROS1-fusion proteins were identified in NSCLC
(Rikova, Guo, & Zeng, 2007), and it was soon found that these occur in
~1% of NSCLC (Bergethon, Shaw, & Ou, 2012). These occur more frequently in younger never smoking patients. Similar to ALK, although the
native role of ROS1 remains not well defined, *ROS*-fusion proteins function
as tyrosine kinases activating classical pathways MAPK, PI3K/AKT, and
JAK/STAT (Davies & Doebele, 2013).

ROS1 fusions are structurally very similar to ALK kinase and so crizotinib was the first TKI studied, showing a 72% objective response rate and
median duration of response of 24 months with a median overall survival of
51.4 months (Shaw, Ou, & Bang, 2014). In this study there were multiple
ROS1-fusion partners, and this did not appear to impact response to therapy
(Shaw, Riely, & Bang, 2019). Similar to patients treated for *ALK* fusions,
there is a high rate of CNS progression on crizotinib among patients with
ROS1 fusions (Patil, Smith, & Bunn, 2018). Entrectinib is an alternative TKI
that has better CNS penetration and has now shown similar frontline activity
as crizotinib in single arm Phase I and II trials (Drilon, Siena, &
Dziadziuszko, 2020). Given better CNS penetration, entrectinib has emerged
as a preferred frontline treatment option for many patients.

Resistance mutations have been identified in patients with *ROS1* fusions
treated with crizotinib. While clinical series are more limited with *ROS1*,
one study suggested up to 53% of patients had identifiable *ROS1* mutations
(Gainor, Tseng, & Yoda, 2017). A common mutation, G2032R affects the
solvent front, preventing crizotinib binding to the kinase domain—similar to
the G1202R mutation in patients with *ALK* fusions (Awad, Katayama, &
McTigue, 2013). In a Phase I–II trial the TKI lorlatinib was studied in a
subset of patients who had progressed on crizotinib and showed a 35%
response rate (Shaw, Solomon, & Chiari, 2019), including in patients with
resistance mutations. Surprisingly there were no response to lorlatinib in six
patients with G2032R mutations, despite preclinical evidence of efficacy
(Zou, Li, & Engstrom, 2015). A smaller TKI designed for less steric hindrance with the solvent front G2032R mutation, repotrectinib, is now being
studied with promising initial clinical efficacy in patients with progression
on crizotinib and the G2032R resistance mutation (Cho, Drilon, & Doebele,
2019; Drilon, Ou, & Cho, 2018; Yun, Kim, & Kim, 2020).

MET

MET is an RTK which binds to its ligand, hepatocyte growth factor to lead
to phosphorylation and downstream activation of the MAPK and PI3K/AKT

pathways. It has a significant role in embryonic development and in adults is involved in wound healing, hepatic regeneration, and angiogenesis. MET signaling has been shown to promote proliferation, migration, and survival of human cancer cells (Salgia, 2017). Similar to EGFR, several attempts at targeting MET in NSCLC have shown the benefits of precision genomics in selecting optimal patient population to achieve clinical benefit.

Initial trials focused on selecting patients with MET protein overexpression through immunohistochemical analysis (IHC). The Phase III METLung trial evaluated the monoclonal antibody onartuzumab which binds to the extracellular domain of MET. This trial selected patients who had 2 + or 3 + MET expression by IHC, but actually found a shorter median overall survival in the onartuzumab arm (Spigel, Edelman, & O'Byrne, 2017). Two other trials evaluating the MET TKI tivantinib in unselected NSCLC patient populations were stopped early due to futility and toxicity (Scagliotti, von Pawel, & Novello, 2015; Yoshioka, Azuma, & Yamamoto, 2015). Selecting patients by using MET protein overexpression (IHC) does not appear to identify a population that will benefit from MET inhibition. However, patients who have *MET* gene amplification measured by FISH or NGS testing may benefit from MET TKIs. Preliminary results from a Phase I trial of crizotinib (which also inhibits MET) show that patients with high *MET* copy number (MET/CEP7 ratio ≥ 4) have a response rate of 40% and PFS 6.7 months compared to a PFS of <2 months in patients with low copy number amplification (Camidge, Otterson, & Clark, 2018). In the Phase II GEOMETRY evaluating the MET selective TKI capmatinib, patients were enrolled in cohorts based on *MET* gene copy number. Among the cohort of patients with an *MET* copy number 6 to 9, the response rate was 12%; among those with a copy number of 4 or 5 the response rate was 9%; and in those with copy number <4 the response rate was 7%. These arms of the trial were closed early due to futility (Wolf, Seto, & Han, 2020). However, patients with a MET copy number of at least 10 did appear to have clinical benefit, with a response rate of 29% in previously treated patients (median duration of response 8.3 months) and 40% in treatment naive patients (median response duration 7.5 months) (Wolf et al., 2020). Although promising, this response rate did not meet the predefined threshold, and further data are needed before defining if and where in the treatment landscape capmatinib and other MET TKIs should be utilized in MET amplified NSCLC.

In up to 3% of patients with lung adenocarcinoma, somatic mutations affecting RNA splicing lead to skipping of *MET* exon 14 (Awad, Oxnard, & Jackman, 2016). This results in loss of a ubiquitin ligase binding site at residue Y1003, preventing physiologic degradation of MET. Preclinical modeling and early case reports suggested that these alterations may truly identify a population of patients where MET is a driver pathway and will thus be particularly sensitive to MET inhibition (Frampton, Ali, & Rosenzweig, 2015; Paik, Drilon, & Fan, 2015). The Phase 1 PROFILE 1001 trial showed a response rate of 32% and PFS of 7.3 months in patients treated with *MET* exon 14 skipping alterations treated with crizotinib (Drilon, Clark, & Weiss, 2020). Recently, next generation TKIs with selective MET inhibition have shown substantial benefit. The GEOMETRY trial discussed above also enrolled patients with *MET* exon 14 skipping mutations. In this trial, capmatinib had a 41% response rate in patients previously treated with chemotherapy (median duration of response 9.7 months) and a 68% response rate in treatment naïve patients (median duration of response 12.6 months) (Wolf et al., 2020). Capmatinib also showed CNS activity with CNS response in 7 of 13 patients with brain lesions at baseline. Tepotinib is also an MET selective TKI and has shown in the Phase II VISION trial to have a response rate of 46% and duration of response of 11.1 months (Paik, Felip, & Veillon, 2020). Tepotinib also demonstrated CNS activity with a response rate of 55% in 11 patients with brain metastasis at baseline. Savolitinib is an additional MET selective TKI with promising early data (Lu, Fang, & Cao, 2019). Early series of patients with treatment resistance suggest MET kinase domain mutation or bypass activation of the MAPK pathway can be detected as mechanisms of resistance in the majority of patients (Recondo, Bahcall, & Spurr, 2020).

RET

RET is an RTK involved in renal and nervous system development. In the early 2010s, it was discovered that gene fusions, most commonly with KIF5B lead to oncogenic signaling through Ras/MAPK and PI3K/AKT in NSCLC (Kohno, Ichikawa, & Totoki, 2012). This occurs in 1%−2% of NSCLC, most often in younger patients who are never smokers (Wang, Hu, & Pan, 2012). Early attempts at inhibiting RET with multikinase inhibitors cabozantinib and vandetanib showed modest activity with response rates 16%−53% and PFS 2.3−7.3 months (Drilon, Rekhtman, & Arcila, 2016; Lee, Lee, & Ahn, 2017; Yoh, Seto, & Satouchi, 2017). Similar mutation mechanisms to other RTKs including on target tyrosine kinase mutations and MAPK bypass pathways occur (Nakaoku, Kohno, & Araki, 2018; Nelson-Taylor, Le, & Yoo, 2017). The multikinase inhibitors are also limited by off target toxicity (related to VEGFR2) creating narrow therapeutic windows (Subbiah, Yang, Velcheti, Drilon, & Meric-Bernstam, 2020). Novel RET selective TKIs have now been developed, including selpercatinib and pralsetinib. Selpercatinib showed promising early data including response rate 64% (85% in treatment naïve patients) and median duration of response of 17.5 months in the Phase I/II LIBRETTO-001 trial (Drilon, Oxnard, & Tan, 2020). In this trial, among 11 patients with measurable CNS lesions at baseline, there was a 91% CNS response rate (Drilon et al., 2020). Additionally, in the Phase I/II ARROW trial, the RET selective TKI pralsetinib showed a 58% response rate (66% in treatment naïve setting) with median duration of response not yet reached (Gainor, Curigliano, & Kim, 2020).

BRAF

BRAF is a nonreceptor serine/threonine kinase that is part of the MAPK pathway. Classically mutated in melanoma, *BRAF* mutations are also found in 1%−3% of NSCLC (Leonetti, Facchinetti, & Rossi, 2018). Approximately 50% of the *BRAF* mutations in NSCLC are V600E, and unlike other driver mutations these can develop in heavier smokers as well as nonsmokers (Paik, Arcila, & Fara, 2011). Under physiologic conditions, BRAF acts as a dimer to activate downstream MEK; however, the BRAF V600 mutation allows BRAF to remain constitutively active as a monomer (Yaeger & Corcoran, 2019). RAF monomer inhibitors such as dabrafenib and vemurafenib have shown clinical activity in $BRAF^{V600E}$ mutant NSCLC with response rates 33% and 44.9% and duration of response 9.6 and 6.4 months, respectively (Mazieres, Cropet, & Montané, 2020; Planchard, Kim, & Mazieres, 2016). Non-V600E *BRAF* mutations act as BRAF dimers, which are not well inhibited by RAF monomer inhibitors (such as dabrafenib or vemurafenib). In a Phase II trial, patients with non-V600E *BRAF* mutations were not responsive to vemurafenib (Mazieres et al., 2020). Among $BRAF^{V600E}$ mutations treated with BRAF inhibitors, resistance often develops through reactivation of the MAPK pathway (Yaeger & Corcoran, 2019). The use of the MEK inhibitor trametinib with dabrafenib in patients with $BRAF^{V600E}$ mutations showed an improved response rate of 64% and duration of response of 10.4 months (Planchard, Smit, & Groen, 2017). Resistance can still develop though *BRAF* amplifications or splice site variations (Xue & Lito, 2020). *BRAF* mutations highlight the heterogeneity of genetic alterations in NSCLC, with personalized strategies needed to target differing *BRAF* mutations (V600 vs non-V600).

HER2

Human epidermal growth factor receptor 2 (HER2) (also known as Erbb2) is an RTK in the human epidermal growth factor family. Initial trials targeting HER2 in NSCLC selected patients based on HER2 expression by IHC. Unlike trials in breast cancer, there was no benefit to adding the anti-HER2 monoclonal antibody trastuzumab to chemotherapy-based regimens (Gatzemeier, Groth, & Butts, 2004). However, it was recognized that up to 3% of patients with NSCLC have *HER2* mutations and this population is distinct from those with *HER2* amplification or overexpression (Li, Ross, & Aisner, 2016). Clinical trials have now focused on patients with *HER2* mutations (most commonly in the exon 20 kinase domain) showing significant clinical responses with the antibody−drug conjugate ado-trastuzumab (Li, Shen, & Buonocore, 2018) or the pan HER TKI pyrotinib (Wang, Jiang, & Qin, 2019). The most promising results to date have evaluated trastuzumab deruxtecan, an antibody−drug conjugate with optimized topoisomerase payload that is membrane permeable and is linked to antibody at an 8:1 ratio.

Initial reporting from a Phase II trial shows a response rate of 61.9% and PFS of 14 months (duration of response not yet reached) (Smit, Nakagawa, & Nagasaka, 2020). Additional therapies under investigation for *HER2* mutated NSCLC include the TKIs also under investigation in *EGFR* exon 20 mutation positive NSCLC—poziotinib (NCT04172597) and mobocertinib (NCT02716116).

KRAS

KRAS is a GTPase that links RTK signaling to downstream effector pathways to promote cell proliferation and survival. Classically, KRAS not only activates the MAPK pathway but also signals though the PI3K/AKT, RAC1, and RAL pathways (Adderley, Blackhall, & Lindsay, 2019). *KRAS* is mutated in ~25% of NSCLC adenocarcinoma, and unlike *EGFR* is more likely to occur in patients with an extensive smoking history (Ferrer et al., 2018). *KRAS* is also frequently comutated with *TP53* or *STK11*. Several retrospective studies suggest inferior outcomes when *KRAS* is comutated with *STK11*, particularly due to poor response rates to ICB (El Osta, Behera, & Kim, 2019; Skoulidis, Goldberg, & Greenawalt, 2018).

KRAS mutations in NSCLC are most frequently at the G12 location and cause KRAS to remain in the activated GTP bound conformation (Ryan & Corcoran, 2018). KRAS had been termed an "undruggable" target due to the high, picomolar binding affinity of GTP for KRAS preventing GTP competitive inhibitors from being effective (Ryan & Corcoran, 2018). Initial efforts at blocking KRAS focused on post-translational modifications by blocking farnesylation, a necessary step for KRAS to bind to cell membranes for active signaling. This showed minimal clinical activity (Adjei, Mauer, & Bruzek, 2003) and it was soon discovered that alternative post-translational modifications (geranylgeranylation) could allow KRAS to bypass this inhibition. Additional therapies have targeted downstream signaling through MEK inhibition. Selumetinib reached a pivotal Phase III trial but did not improve outcomes as second-line therapy in *KRAS* mutated NSCLC when combined with docetaxel compared to docetaxel alone (Jänne, van den Heuvel, & Barlesi, 2017). This may be due to incomplete MAPK pathway inhibition and feedback pathway reactivation (Koleilat & Kwong 2020).

A novel target now under development is allosteric inhibition of $KRAS^{G12C}$ mutations through binding to the mutant cystine residue, locking KRAS in the inactive GDP bound state (Ostrem, Peters, Sos, Wells, & Shokat, 2013). A Phase I trial of sotorasib in patients with $KRAS^{G12C}$ mutant NSCLC who had progressed on prior platinum-based chemotherapy + ICB showed a response rate of 32.2% with a median duration of response of 10.9 months (Hong, Fakih, & Strickler, 2020). Further development of $KRAS^{G12C}$ inhibitors is eagerly awaited, but as in all targeted therapies resistance mechanisms limiting long-term responses are anticipated. Given significant genomic heterogeneity and coalterations in patients with *KRAS* mutant NSCLC, combination therapies with sotorasib are of particular interest and are currently under investigation (NCT04185883).

Squamous cell carcinoma and small-cell lung cancer

Unlike lung adenocarcinoma, NSCLC with squamous histology and SCLC have not seen benefit from therapies targeting genetic or alterations. Squamous cell and SCLC are universally associated with smoking and have high rates of genomic instability with high somatic mutation rates (George et al., 2015; Hammerman et al., 2012). SCLC in particular is characterized by p53 and RB loss of function alterations in virtually all cases, and no targetable therapies are available for these alterations (George et al., 2015). For squamous cell carcinoma, publication of TCGA garnered new hope of precision genomic targets. Pathway analysis suggested frequent FGFR RTK alterations (12%), PIK3CA/PTEN/AKT alterations (47%), and CDKN2A loss in 72% of cases (Hammerman et al., 2012). However, attempts at targeting these alterations both in the National Cancer Institute (NCI) Lung-MAP and independent trials have been disappointing. A Phase IB trial as well as a Phase II trial within Lung-MAP evaluated the FGFR small molecule inhibitor AD4547 among patients with FGFR alterations (predominately amplifications), but there was minimal clinical response (Aggarwal, Redman, & Lara, 2019; Paik, Shen, & Berger, 2017). Correlative analysis suggested that there was not a clear relationship between *FGFR* amplifications by RNA-based analysis and FGFR overexpression by IHC analysis, highlighting the heterogeneity of this target (Paik et al., 2017). Similarly, a Phase II trial of the pan-PI3K inhibitor buparlisib and a Phase II trial within Lung-MAP evaluating taselisib in patients with *PIK3CA* mutations were both stopped due to futility (Langer, Redman, & Wade, 2019; Vansteenkiste, Canon, & De Braud, 2015). Finally, a Phase II Lung-MAP trial evaluated CDK4/6 inhibitor palbociclib among patients with *CCND1*, *CCND2*, *CCND3*, or *CDK4* amplifications was also stopped due to futility (Edelman, Redman, & Albain, 2019). Across these trials, despite selection based on genomic alterations, there were many comutations—including *TP53*—in over 90% of patients. This highlights that in squamous cell lung cancer, patients tend to have high mutational loads and there may not be one dominate driver pathway that can be targeted.

Immunotherapy

ICB targeting program death ligand 1 (PD-L1) or the PD-1 receptor has substantially improved outcomes for a subset of patients with NSCLC. Several initial trials showed superior overall survival with PD-1 inhibitors nivolumab or pembrolizumab or PD-L1 inhibitor atezolizumab in comparison to docetaxel as second-line treatment for NSCLC (Borghaei, Paz-Ares, & Horn, 2015; Herbst, Baas, & Kim, 2016; Rittmeyer, Barlesi, & Waterkamp, 2017). However, it was noted in these trials that the subset of patients with *EGFR* mutations did not have improved median overall survival with ICB compared to docetaxel (Singhi, Horn, Sequist, Heymach, & Langer, 2019). ICB has now been established as a frontline therapy option for patients expressing PD-L1 (Mok, Wu, & Kudaba, 2019; Reck, Rodríguez-Abreu, & Robinson, 2016) and is also given in combination with chemotherapy or anti-CTLA4 for patients with lower (<50%) or no PD-L1 expression (Gandhi, Rodríguez-Abreu, & Gadgeel, 2018; Hellmann, Paz-Ares, & Bernabe Caro, 2019; Paz-Ares, Luft, & Vicente, 2018). However, these frontline therapy trials have generally excluded patients with driver mutation positive NSCLC. Multiple retrospective series have supported the finding that patients with driver RTK alterations, specifically *EGFR*, *ALK*, *ROS*, *RET*, and *HER2* have lower response rates and less durable responses with ICB compared to non-driver mutation positive NSCLC (Anagnostou, Niknafs, & Marrone, 2020; Gainor, Shaw, & Sequist, 2016; Mazieres, Drilon, & Lusque, 2019). Additionally, a Phase II trial of frontline pembrolizumab in patients with *EGFR* mutations and high PD-L1 was stopped early due to futility (Lisberg, Cummings, & Goldman, 2018). In contrast, patients with *KRAS*, *BRAF*, and possibly *MET* alterations appear to have a higher portion of patients benefiting from ICB compared to other patients with driver mutations (Jeanson, Tomasini, & Souquet-Bressand, 2019; Mazieres et al., 2019). These driver alterations (*KRAS*, *BRAF*, and *MET*) are unique in that they can be associated with smoking history, and several retrospective trials have shown that patients with smoking history or gene signatures suggesting of smoking history may be more likely to benefit from ICB (Anagnostou et al., 2020; Gainor, Rizvi, & Jimenez Aguilar, 2020). The Impower150 trial evaluated the regimen of atezolizumab + chemotherapy and bevacizumab and in exploratory analysis appeared to show survival advantage even in the *EGFR* and *ALK* alteration subgroups (Socinski, Jotte, & Cappuzzo, 2018). Thus it remains possible that combination therapies (such as with anti-VEGF) may modulate the poor response to ICB in driver mutation positive patients. It should be noted, however, that EGFR inhibition and ICB in combination are associated with high rates of pneumonitis (Schoenfeld, Arbour, & Rizvi, 2019).

In driver mutation negative NSCLC, tumor mutational burden (TMB) has emerged as an additional predictive biomarker for response to ICB independent of PD-L1 expression (Hellmann, Nathanson, & Rizvi, 2018; Rizvi, Sanchez-Vega, & La, 2018). A higher TMB may reflect an increased number of tumor-specific neoantigens that can be recognized and targeted by the immune system (Rizvi, Hellmann, & Snyder, 2015). In addition to TMB, an interferon-related gene signature suggestive of an immune activated tumor microenvironment has also predicted treatment response to ICB across multiple tumor types (Ayers, Lunceford, & Nebozhyn, 2017). Conversely, genomic features can also predict treatment resistance. *STK11* mutated NSCLC has been associated with low PD-L1 expression and an immune "desert" phenotype (Skoulidis, Byers, & Diao, 2015). Patients with *STK11* mutations treated with second-line ICB had minimal clinical benefit across multiple patient cohorts (Skoulidis et al., 2018). In the frontline treatment setting, exploratory analysis from the prospective Keynote 189 trial comparing pembrolizumab + chemotherapy to chemotherapy alone suggests that patients with *STK11* mutations still have higher response rates and survival with the addition of pembrolizumab (Gadgeel, Rodriguez-Abreu, & Felip, 2020). However, across both arms of the trial, patients with *STK11* mutations had inferior overall survival, suggesting that *STK11* mutations are also prognostic of poor outcomes regardless of treatment, which has been suggested in population level data (Papillon-Cavanagh, Doshi, Dobrin, Szustakowski, & Walsh, 2020). Mechanisms of immune therapy resistance from *STK11* mutations are not clearly defined, but preclinical data suggest reduced STING-related antitumor immunity (Kitajima, Ivanova, & Guo, 2019). Outside of *STK11* mutations, genomic testing has elucidated several mechanisms of acquired resistance including loss of antigen presentation machinery or loss of neoantigen expression (Anagnostou, Smith, & Forde, 2017; Gettinger, Choi, & Hastings, 2017). Finally, for extensive stage SCLC, the addition of atezoliumab to chemotherapy has prolonged overall survival, but to date no genomic correlates have been identified to predict patients most likely to benefit from the addition of ICB to small-cell treatment regimens (Horn, Mansfield, & Szczęsna, 2018).

Future directions

Screening and early stage disease

Low-dose CT scan screening in patients selected by smoking status has allowed for early detection of NSCLC and improved survival (de Koning et al., 2020; The National Lung Screening Trial Research Team, 2011). However, currently smoking history and age are the only criteria to select patients for screening. Attempts at personalizing care to assess lung cancer risk through GWAS have been limited in clinical application due to low penetrance of associated genes, or the low prevalence of highly penetrate genes (Musolf et al., 2017). The use of peripheral blood cell-free DNA-based platforms to detect circulating tumor DNA is one method being explored to offer precision detection of early stage lung cancer. Initial efforts have focused on common NSCLC-associated DNA mutations or tobacco signatures (Chabon, Hamilton, & Kurtz, 2020), and additional platforms have examined DNA methylation patterns that may suggest the presence of NSCLC (Liu, Ricarte Filho, & Mallisetty, 2020; Liu, Oxnard, & Klein, 2020). In the future, cfDNA may be combined with low-dose CT scanning to optimize patient selection for lung cancer screening programs (Lennon, Buchanan, & Kinde, 2020). In early stage disease, cfDNA is also being utilized to detect patients who have residual occult disease after primary surgical resections, and are expected to be at high risk of recurrence. Platforms have utilized genetic mutations from the primary tumor biopsy to generate personalized cfDNA panels that carry high sensitivity in detecting disease activity (Abbosh, Frankell, & Garnett, 2020). Finally, the treatment of early stage NSCLC may be impacted by the use of targeted therapies in the adjuvant setting. As discussed above, the Phase III ADAURA trial has shown benefit in prolonging disease-free survival with adjuvant osimertinib in EGFR mutated Stage IB through IIIA NSCLC (Wu et al., 2020). This may signal a new paradigm where other targeted agents could provide benefit (*ALK*, *ROS1*, or *RET* alterations) in the adjuvant setting. Prospective studies for these mutations are needed.

Metastatic disease

Recent advancements in targeted therapies and immunotherapy have dramatically improved outcomes for patients with metastatic lung cancer. For patients with adenocarcinoma and a genomic driver alteration, efforts are focused on exploring mechanisms of resistance to targeted therapy. In patients without targetable genomic alterations, checkpoint blockade can provide durable disease control, but only in a subset of patients (Fig. 8.2).

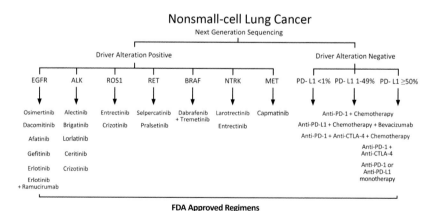

FIGURE 8.2 Frontline treatment algorithm for metastatic NSCLC. Patients initially receive genetic testing using next generation sequencing panels. There are currently seven driver alterations with FDA approved targeted therapies. If driver alteration negative, patients are considered for immune checkpoint blockade. PD-L1 immunohistochemistry is used to determine if checkpoint blockade is indicated as monotherapy or as part of combination therapy. *FDA Approvals Current as of November 2020.*

Ongoing efforts are attempting to use immune correlative biomarker studies to predict who will benefit most from ICB. Nonetheless, novel therapies are needed for patients who are refractory to genome targeted TKIs or ICB. Adoptive cellular therapy is a new class of treatment that is still early in development. Infusion of ex vivo expanded tumor infiltrating lymphocytes (TILs) has achieved initial response in patients with metastatic NSCLC, including those with *EGFR* driver mutations and low TMB who typically have low response to immune therapies (Creelan, Wang, & Teer, 2020). Other platforms attempt to produce TIL selected based off of genomic profiling and predicted neoantigens to offer truly personalized therapy (NCT04032847). Whether cellular therapy will become a standard of care in solid tumors, as it has in lymphoma with CAR-T cell therapy, will be the source of ongoing investigation.

Conclusion

Treatment of lung cancer has evolved dramatically over the past 20 years with a better understanding of underlying genomics. This is the best exemplified in metastatic lung adenocarcinoma where there are now multiple genomic alterations with effective targeted therapy. Immune therapy also has improved outcomes for many patients and ongoing efforts are focused on how genomic alterations interact with the tumor immune microenvironment to develop personalized strategies to optimize therapy. Finally, the use of cfDNA offers hope for more precision detection and prognostication in early stage NSCLC. Advancements in each of these aspects of genomic medicine are eagerly awaited as personalized medicine is expected to continue to improve outcomes for patients with lung cancer.

References

Abbosh, C., Frankell, A., Garnett, A., et al. (2020). Abstract CT023: Phylogenetic tracking and minimal residual disease detection using ctDNA in early-stage NSCLC: A lung TRACERx study. *Cancer Research, 80*, CT023.

Adderley, H., Blackhall, F. H., & Lindsay, C. R. (2019). KRAS-mutant non-small cell lung cancer: Converging small molecules and immune checkpoint inhibition. *EBioMedicine, 41*, 711–716.

Adjei, A. A., Mauer, A., Bruzek, L., Marks, R. S., Hillman, S., Geyer, S., ... Vokes, E. E. (2003). Phase II study of the farnesyl transferase inhibitor R115777 in patients with advanced non-small-cell lung cancer. *Journal of Clinical Oncology, 21*, 1760–1766.

Aggarwal, C., Redman, M. W., Lara, P. N., Jr., Borghaei, H., Hoffman, P., Bradley, J. D., ... Gandara, D. R. (2019). SWOG S1400D (NCT02965378), a Phase II study of the fibroblast growth factor receptor inhibitor AZD4547 in previously treated patients with fibroblast growth factor pathway; activated Stage IV squamous cell lung cancer (Lung-MAP Substudy). *Journal of Thoracic Oncology, 14*, 1847–1852.

Aggarwal, C., Thompson, J. C., Black, T. A., Katz, S. I., Fan, R., Yee, S. S., ... Carpenter, E. L. (2019). Clinical implications of plasma-based genotyping with the delivery of personalized therapy in metastatic non−small cell lung cancer. *JAMA Oncology, 5,* 173−180.

Amos, C. I., Pinney, S. M., Li, Y., Kupert, E., Lee, J., de Andrade, M. A., ... Anderson, MW (2010). A susceptibility locus on chromosome 6q greatly increases lung cancer risk among light and never smokers. *Cancer Research, 70,* 2359−2367.

Anagnostou, V., Niknafs, N., Marrone, K., Bruhm, D. C., White, J. R., Naidoo, J., ... Velculescu, V. E. (2020). Multimodal genomic features predict outcome of immune checkpoint blockade in non-small-cell lung cancer. *Nature Cancer, 1,* 99−111.

Anagnostou, V., Smith, K. N., Forde, P. M., Niknafs, N., Bhattacharya, R., White, J., ... Velculescu, V. E. (2017). Evolution of neoantigen landscape during immune checkpoint blockade in non−small cell lung cancer. *Cancer Discovery, 7,* 264−276.

Antonia, S. J., Borghaei, H., Ramalingam, S. S., Horn, L., De Castro Carpeño, J., Pluzanski, A., ... Brahmer, J. (2019). Four-year survival with nivolumab in patients with previously treated advanced non-small-cell lung cancer: A pooled analysis. *The Lancet Oncology, 20,* 1395−1408.

Awad, M. M., Katayama, R., McTigue, M., Liu, W., Deng, Y.-L., Brooun, A., ... Shaw, A. T. (2013). Acquired resistance to crizotinib from a mutation in CD74−ROS1. *New England Journal of Medicine, 368,* 2395−2401.

Awad, M. M., Oxnard, G. R., Jackman, D. M., Savukoski, D. O., Hall, D., Shivdasani, P., ... Shol, L. M. (2016). MET exon 14 mutations in non-small-cell lung cancer are associated with advanced age and stage-dependent MET genomic amplification and c-Met overexpression. *Journal of Clinical Oncology, 34,* 721−730.

Ayers, M., Lunceford, J., Nebozhyn, M., Murphy, E., Loboda, A., Kaufman, D. R., ... McClanahan, T. K. (2017). IFN-γ−related mRNA profile predicts clinical response to PD-1 blockade. *The Journal of Clinical Investigation, 127,* 2930−2940.

Barlesi, F., Mazieres, J., Merlio, J.-P., Debieuvre, D., Mosser, J., Lena, H., ... Biomarkers France contributors. (2016). Routine molecular profiling of patients with advanced non-small-cell lung cancer: results of a 1-year nationwide programme of the French Cooperative Thoracic Intergroup (IFCT). *The Lancet, 387,* 1415−1426.

Beer, D. G., Kardia, S. L., Huang, C. C., Giordano, T. J., Levin, A. M., Misek, D. E., ... Hanash, S. (2002). Gene-expression profiles predict survival of patients with lung adenocarcinoma. *Nature Medicine, 8,* 816−824.

Bergethon, K., Shaw, A. T., Ou, S.-H. I., Katayama, R., Lovly, C. M., McDonald, N. T., ... Iafrate, A. J. (2012). ROS1 rearrangements define a unique molecular class of lung cancers. *Journal of clinical oncology, 30,* 863−870.

Birchmeier, C., Birnbaum, D., Waitches, G., Fasano, O., & Wigler, M. (1986). Characterization of an activated human ros gene. *Molecular and Cellular Biology, 6,* 3109−3116.

Borghaei, H., Paz-Ares, L., Horn, L., Spigel, D. R., Steins, M., Ready, N. E., ... Brahmer, J. R. (2015). Nivolumab vs docetaxel in advanced nonsquamous non−small-cell lung cancer. *New England Journal of Medicine, 373,* 1627−1639.

Bossé, Y., & Amos, C. I. (2018). A decade of GWAS results in lung cancer. *Cancer Epidemiology Biomarkers and Prevention, 27,* 363−379.

Camidge, D. R., Kim, H. R., Ahn, M.-J., Yang, J. C.-H., Han, J.-Y., Lee, J.-S., ... Popat, S. (2018). Brigatinib vs crizotinib in ALK-positive non−small-cell lung cancer. *New England Journal of Medicine, 379,* 2027−2039.

Camidge, D. R., Otterson, G. A., Clark, J. W., Ignatius Ou, S.-H., Weiss, J., Ades, S., ... Villaruz, L. C. (2018). Crizotinib in patients (pts) with MET-amplified non-small cell lung cancer (NSCLC): Updated safety and efficacy findings from a phase 1 trial. *Journal of Clinical Oncology, 36,* 9062.

Chabon, J. J., Hamilton, E. G., Kurtz, D. M., Esfahani, M. S., Moding, E. J., Stehr, H., ... Diehn, M. (2020). Integrating genomic features for non-invasive early lung cancer detection. *Nature, 580,* 245−251.

Chen, H.-Y., Yu, S.-L., Chen, C.-H., Chang, G.-C., Chen, C.-Y., Yuan, A., ... Yang, P.-C. (2007). A five-gene signature and clinical outcome in non−small-cell lung cancer. *New England Journal of Medicine, 356,* 11−20.

Cho, B. C., Drilon, A. E., Doebele, R. C., Kim, D.-W., Lin, J. J., Lee, J., ... Shaw, A. T. (2019). Safety and preliminary clinical activity of repotrectinib in patients with advanced ROS1 fusion-positive non-small cell lung cancer (TRIDENT-1 study). *Journal of Clinical Oncology, 37,* 9011.

Choi, Y. L., Soda, M., Yamashita, Y., Ueno, T., Takashima, J., Nakajima, T., ... for the ALK Lung Cancer Study Group. (2010). EML4-ALK mutations in lung cancer that confer resistance to ALK inhibitors. *New England Journal of Medicine, 363,* 1734−1739.

Collisson, E. A., Campbell, J. D., Brooks, A. N., Berger, A. H., Lee, W., Chmielecki, J., ... Tsao, M.-S. (2014). Comprehensive molecular profiling of lung adenocarcinoma. *Nature, 511,* 543−550.

Coté, M. L., Kardia, S. L. R., Wenzlaff, A. S., Ruckdeschel, J. C., & Schwartz, A. G. (2005). Risk of lung cancer among white and black relatives of individuals with early-onset lung cancer. *JAMA: The Journal of the American Medical Association, 293,* 3036−3042.

Coté, M. L., Liu, M., Bonassi, S., Neri, M., Schwartz, A. G., Christiani, D. C., ... Hung, R. J. (2012). Increased risk of lung cancer in individuals with a family history of the disease: A pooled analysis from the International Lung Cancer Consortium. *European Journal of Cancer, 48,* 1957−1968.

Creelan, B., Wang, C., Teer, J., Toloza, E., Mullinax, J. E., Yao, J., ... Antonia, S. (2020). *Abstract CT056: Durable complete responses to adoptive cell transfer using tumor infiltrating lymphocytes (TIL) in non-small cell lung cancer (NSCLC): A phase I trial. Cancer Research* (80, p. CT056).

Davies, K. D., & Doebele, R. C. (2013). Molecular pathways: ROS1 fusion proteins in cancer. *Clinical Cancer Research, 19,* 4040−4045.

de Koning, H. J., van der Aalst, C. M., de Jong, P. A., Scholten, E. T., Nackaerts, K., Heuvelmans, M. A., ... Oudkerk, M. (2020). Reduced lung-cancer mortality with volume CT screening in a randomized trial. *New England Journal of Medicine, 382,* 503−513.

Doebele, R. C., Pilling, A. B., Aisner, D. L., Kutateladze, T. G., Le, A. T., Weickhardt, A. J., ... Camidge, D. R. (2012). Mechanisms of resistance to crizotinib in patients with ALK gene rearranged non-small cell lung cancer. *Clinical Cancer Research, 18,* 1472−1482.

Drilon, A., Clark, J. W., Weiss, J., Ignatius Ou, S.-H., Camidge, D. R., Solomon, B. J., ... Paik, P. K. (2020). Antitumor activity of crizotinib in lung cancers harboring a MET exon 14 alteration. *Nature Medicine, 26,* 47−51.

Drilon, A., Ou, S.-H. I., Cho, B. C., Kim, D.-W., Lee, J., Lin, J. J., ... Shaw, A. T. (2018). Repotrectinib (TPX-0005) is a next-generation ROS1/TRK/ALK inhibitor that potently inhibits ROS1/TRK/ALK solvent-front mutations. *Cancer Discovery, 8,* 1227−1236.

Drilon, A., Oxnard, G. R., Tan, D. S. W., Loong, H. H. F., Johnson, M., Gainor, J., ... Subbiah, V. (2020). Efficacy of selpercatinib in RET fusion−positive non−small-cell lung cancer. *New England Journal of Medicine, 383,* 813−824.

Drilon, A., Rekhtman, N., Arcila, M., Wang, L., Ni, A., Albano, M., ... Kris, M. G. (2016). Cabozantinib in patients with advanced RET-rearranged non-small-cell lung cancer: An open-label, single-centre, phase 2, single-arm trial. *The Lancet Oncology, 17,* 1653−1660.

Drilon, A., Siena, S., Dziadziuszko, R., Barlesi, F., Krebs, M. G., Shaw, A. T., . . . trial investigators. (2020). Entrectinib in ROS1 fusion-positive non-small-cell lung cancer: Integrated analysis of three phase 1 & 2 trials. *The Lancet Oncology*, *21*, 261−270.

Edelman, M. J., Redman, M. W., Albain, K. S., McGary, E. C., Rafique, N. M., Petro, D., . . . Herbst, R. S. (2019). SWOG S1400C (NCT02154490)-a Phase II study of palbociclib for previously treated cell cycle gene alteration-positive patients with Stage IV squamous cell lung cancer (Lung-MAP substudy). *Journal of Thoracic Oncology*, *14*, 1853−1859.

El Osta, B., Behera, M., Kim, S., Berry, L. D., Sica, G., Pillai, R. N., . . . Ramalingam, S. S. (2019). Characteristics and outcomes of patients with metastatic KRAS-mutant lung adenocarcinomas: The lung cancer mutation consortium experience. *Journal of Thoracic Oncology*, *14*, 876−889.

Ferrer, I., Zugazagoitia, J., Herbertz, S., John, W., Paz-Ares, L., & Schmid-Bindert, G. (2018). KRAS Mutant non-small cell lung cancer: From biology to therapy. *Lung Cancer (Amsterdam, Netherlands)*, *124*, 53−64.

Frampton, G. M., Ali, S. M., Rosenzweig, M., Chmielecki, J., Lu, X., Bauer, T. M., . . . Miller, V. A. (2015). Activation of MET via diverse exon 14 splicing alterations occurs in multiple tumor types and confers clinical sensitivity to MET inhibitors. *Cancer Discovery*, *5*, 850−859.

Friboulet, L., Li, N., Katayama, R., Lee, C. C., Gainor, J. F., Crystal, A. S., . . . Engelman, J. A. (2014). The ALK inhibitor ceritinib overcomes crizotinib resistance in non−small cell lung cancer. *Cancer Discovery*, *4*, 662−673.

Gadgeel, S. M., Rodriguez-Abreu, D., Felip, E., et al. (2020). Abstract LB-397: Pembrolizumab plus pemetrexed and platinum vs placebo plus pemetrexed and platinum as first-line therapy for metastatic nonsquamous NSCLC: Analysis of KEYNOTE-189 by STK11 and KEAP1 status. *Cancer Research*, *80*, LB-397-LB.

Gainor, J. F., Curigliano, G., Kim, D.-W., Lee, D. H., Besse, B., Baik, C. S., . . . Subbiah, V. (2020). Registrational dataset from the phase I/II ARROW trial of pralsetinib (BLU-667) in patients (pts) with advanced RET fusion + non-small cell lung cancer (NSCLC). *Journal of Clinical Oncology*, *38*, 9515.

Gainor, J. F., Dardaei, L., Yoda, S., Friboulet, L., Leshchiner, I., Katayama, R., . . . Shaw, A. T. (2016). Molecular mechanisms of resistance to first- and second-generation ALK inhibitors in ALK-rearranged lung cancer. *Cancer Discovery*, *6*, 1118−1133.

Gainor, J. F., Rizvi, H., Jimenez Aguilar, E., Skoulidis, F., Yeap, B. Y., Naidoo, J., . . . Hellmann, M. D. (2020). Clinical activity of programmed cell death 1 (PD-1) blockade in never, light, and heavy smokers with non-small-cell lung cancer and PD-L1 expression ≥ 50. *Annals of Oncology*, *31*, 404−411.

Gainor, J. F., Shaw, A. T., Sequist, L. V., Fu, X., Azzoli, C. G., Piotrowska, Z., . . . Mino-Kenudson, M. (2016). EGFR mutations and ALK rearrangements are associated with low response rates to PD-1 pathway blockade in non−small cell lung cancer: A retrospective analysis. *Clinical Cancer Research*, *22*, 4585−4593.

Gainor, J. F., Tseng, D., Yoda, S., Dagogo-Jack, I., Friboulet, L., Lin, J. J, . . . Shaw, A. T. (2017). Patterns of metastatic spread and mechanisms of resistance to crizotinib in ROS1-positive non−small-cell lung cancer. *JCO Precision Oncology*, 1−13.

Gandhi, L., Rodríguez-Abreu, D., Gadgeel, S., Esteban, E., Felip, E., De Angelis, F., . . . M.D. for the KEYNOTE-189 Investigators. (2018). Pembrolizumab plus chemotherapy in metastatic non−small-cell lung cancer. *New England Journal of Medicine*, *378*, 2078−2092.

Gatzemeier, U., Groth, G., Butts, C., Van Zandwijk, N., Shepherd, F., Ardizzoni, A., . . . Hirsh, V. (2004). Randomized phase II trial of gemcitabine + cisplatin with or without trastuzumab in HER2-positive non-small-cell lung cancer. *Annals of Oncology*, *15*, 19−27.

Gatzemeier, U., Pluzanska, A., Szczesna, A., Kaukel, E., Roubec, J., De Rosa, F., ... Von Pawel, J. (2007). Phase III study of erlotinib in combination with cisplatin and gemcitabine in advanced non-small-cell lung cancer: The Tarceva Lung Cancer Investigation Trial. *Journal of Clinical Oncology, 25*, 1545−1552.

George, J., Lim, J. S., Jang, S. J., Cun, Y., Ozretić, L., Kong, G., ... Thomas, R. K. (2015). Comprehensive genomic profiles of small cell lung cancer. *Nature, 524*, 47−53.

Gettinger, S., Choi, J., Hastings, K., Truini, A., Datar, I., Sowell, R., ... Politi, K. (2017). Impaired HLA class I antigen processing and presentation as a mechanism of acquired resistance to immune checkpoint inhibitors in lung cancer. *Cancer Discovery, 7*, 1420−1435.

Govindan, R., Ding, L., Griffith, M., Subramanian, J., Dees, N. D., Kanchi, K. L., ... Wilson, R. K. (2012). Genomic landscape of non-small cell lung cancer in smokers and never-smokers. *Cell, 150*, 1121−1134.

Group NM-AC. (2008). Chemotherapy in addition to supportive care improves survival in advanced non−small-cell lung cancer: A systematic review and *meta*-analysis of individual patient data from 16 randomized controlled trials. *Journal of Clinical Oncology, 26*, 4617−4625.

Guo, N. L., Wan, Y. W., Tosun, K., Lin, H., Msiska, Z., Flynn, D. C., ... Qian, Y. (2008). Confirmation of gene expression-based prediction of survival in non-small cell lung cancer. *Clinical Cancer Research, 14*, 8213−8220.

Haiman, C. A., Stram, D. O., Wilkens, L. R., Pike, M. C., Kolonel, L. N., Henderson, B. E., & Le Marchand, L. (2006). Ethnic and racial differences in the smoking-related risk of lung cancer. *New England Journal of Medicine, 354*, 333−342.

Hammerman, P. S., Lawrence, M. S., Voet, D., Jing, R., Cibulskis, K., Sivachenko, A., ... Thomson, E. (2012). Comprehensive genomic characterization of squamous cell lung cancers. *Nature, 489*, 519−525.

Hellmann, M. D., Nathanson, T., Rizvi, H., Creelan, B. C., Sanchez-Vega, F., Ahuja, A., ... Wolchok, J. D. (2018). Genomic features of response to combination immunotherapy in patients with advanced non-small-cell lung cancer. *Cancer Cell, 33*, 843−852.e4.

Hellmann, M. D., Paz-Ares, L., Bernabe Caro, R., Zurawski, B., Kim, S.-W., Costa, E. C., ... Ramalingam, S. S. (2019). Nivolumab plus ipilimumab in advanced non−small-cell lung cancer. *New England Journal of Medicine, 381*, 2020−2031.

Herbst, R. S., Baas, P., Kim, D.-W., Felip, E., Pérez-Gracia, J. L., Han, J.-Y., ... Garon, E. B. (2016). Pembrolizumab vs docetaxel for previously treated, PD-L1-positive, advanced non-small-cell lung cancer (KEYNOTE-010): A randomised controlled trial. *The Lancet, 387*, 1540−1550.

Herbst, R. S., Giaccone, G., Schiller, J. H., Natale, R. B., Miller, V., Manegold, C., ... Johnson, D. H. (2004). Gefitinib in combination with paclitaxel and carboplatin in advanced non−small-cell lung cancer: A Phase III trial—INTACT 2. *Journal of Clinical Oncology, 22*, 785−794.

Hirsch, F. R., Varella-Garcia, M., Bunn, P. A., Jr., Di Maria, M. V., Veve, R., Bremmes, R. M., ... Franklin, W. A. (2003). Epidermal growth factor receptor in non-small-cell lung carcinomas: correlation between gene copy number and protein expression and impact on prognosis. *Journal of Clinical Oncology, 21*, 3798−3807.

Hong, D. S., Fakih, M. G., Strickler, J. H., Desai, J., Durm, G. A., Shapiro, G. I., ... Li, B. T. (2020). KRASG12C inhibition with sotorasib in advanced solid tumors. *New England Journal of Medicine, 383*, 1207−1217.

Horn, L., Mansfield, A. S., Szczęsna, A., Havel, L., Krzakowski, M., Hochmair, M. J., ... for the IMpower133 Study Group. (2018). First-line atezolizumab plus chemotherapy in extensive-stage small-cell lung cancer. *New England Journal of Medicine, 379*, 2220−2229.

Hosomi, Y., Morita, S., Sugawara, S., Kato, T., Fukuhara, T., Gemma, A., . . . North-East Japan Study Group. (2020). Gefitinib alone vs gefitinib plus chemotherapy for non-small-cell lung cancer with mutated epidermal growth factor receptor: NEJ009 study. *Journal of Clinical Oncology, 38,* 115−123.

Howlader, N., Forjaz, G., Mooradian, M. J., Meza, R., Kong, C. Y., Cronin, K. A., . . . Feuer, E. J. (2020). The effect of advances in lung-cancer treatment on population mortality. *New England Journal of Medicine, 383,* 640−649.

Hsu, W. H., Yang, J. C. H., Mok, T. S., & Loong, H. H. (2018). Overview of current systemic management of EGFR-mutant NSCLC. *Annals of Oncology, 29,* i3−i9.

Imielinski, M., Berger Alice, H., Hammerman, P. S., Hernandez, B., Pugh, T. J., Hodis, E., . . . Meyerson, M. (2012). Mapping the hallmarks of lung adenocarcinoma with massively parallel sequencing. *Cell, 150,* 1107−1120.

Janne, P. A., Neal, J. W., Camidge, D. R., Spira, A. I., Piotrowska, Z., Horn, L., . . . Riely, G. J. (2019). Antitumor activity of TAK-788 in NSCLC with EGFR exon 20 insertions. *Journal of Clinical Oncology, 37,* 9007.

Jänne, P. A., van den Heuvel, M. M., Barlesi, F., Cobo, M., Mazieres, J., Crino, L., . . . Vansteenkiste, J. (2017). Selumetinib plus docetaxel compared with docetaxel alone and progression-free survival in patients with KRAS-mutant advanced non−small cell lung cancer: The SELECT-1 randomized clinical trial. *JAMA: The Journal of the American Medical Association, 317,* 1844−1853.

Jänne, P. A., Yang, J. C., Kim, D. W., Planchard, D., Ohe, Y., Ramalingam, S. S., . . . Ranson, M. (2015). AZD9291 in EGFR inhibitor-resistant non-small-cell lung cancer. *The New England Journal of Medicine, 372,* 1689−1699.

Jeanson, A., Tomasini, P., Souquet-Bressand, M., Brandone, N., Boucekine, M., Grangeon, M., . . . Mascaux, C. (2019). Efficacy of immune checkpoint inhibitors in KRAS-mutant non-small cell lung cancer (NSCLC). *Journal of Thoracic Oncology, 14,* 1095−1101.

Jonsson, S., Thorsteinsdottir, U., Gudbjartsson, D. F., Jonsson, H. H., Kristjansson, K., Arnason, S., . . . Stefansson, K. (2004). Familial risk of lung carcinoma in the Icelandic population. *JAMA: The Journal of the American Medical Association, 292,* 2977−2983.

Jordan, E. J., Kim, H. R., Arcila, M. E., Barron, D., Chakravarty, D., Gao, J., . . . Riely, G. J. (2017). Prospective comprehensive molecular characterization of lung adenocarcinomas for efficient patient matching to approved and emerging therapies. *Cancer Discovery, 7,* 596−609.

Katayama, R., Shaw, A. T., Khan, T. M., Mino-Kenudson, M., Solomon, B. J., Halmos, B., . . . Engelman, J. A. (2012). Mechanisms of acquired crizotinib resistance in ALK-rearranged lung Cancers. *Science Translational Medicine, 4,* 120ra17.

Katki, H. A., Kovalchik, S. A., Berg, C. D., Cheung, L. C., & Chaturvedi, A. K. (2016). Development and validation of risk models to select ever-smokers for CT lung cancer screening. *JAMA: The Journal of the American Medical Association, 315,* 2300−2311.

Kitajima, S., Ivanova, E., Guo, S., Yoshida, R., Campisi, M., Sundararaman, S. K., . . . Barbie, D. A. (2019). Suppression of STING associated with LKB1 loss in KRAS-driven lung cancer. *Cancer Discovery, 9,* 34−45.

Kobayashi, S., Boggon, T. J., Dayaram, T., Jänne, P. A., Kocher, O., Meyerson, M., . . . Halmos, B. (2005). EGFR mutation and resistance of non−small-cell lung cancer to gefitinib. *New England Journal of Medicine, 352,* 786−792.

Kohno, T., Ichikawa, H., Totoki, Y., Yasuda, K., Hiramoto, M., Nammo, T., . . . Shibata, T. (2012). KIF5B-RET fusions in lung adenocarcinoma. *Nature Medicine, 18,* 375−377.

Koleilat, M. K., & Kwong, L. N. (2020). Same name, different game: EGFR drives intrinsic KRAS-G12C-inhibitor resistance in colorectal cancer. *Cancer Discovery, 10,* 1094−1096.

Kovalchik, S. A., Tammemagi, M., Berg, C. D., Caporaso, N. E., Riley, T. L., Korch, M., ... Katki, H. A. (2013). Targeting of low-dose CT screening according to the risk of lung-cancer death. *New England Journal of Medicine, 369,* 245−254.

Kratz, J. R., He, J., Van Den Eeden, S. K., Zhu, Z.-H., Gao, W., Pham, P. T., ... Jablons, D. M (2012). A practical molecular assay to predict survival in resected non-squamous, non-small-cell lung cancer: Development and international validation studies. *Lancet, 379,* 823−832.

Kris, M. G., Johnson, B. E., Berry, L. D., Kwiatkowski, D. J., Iafrate, A. J., Wistuba, I. I., ... Bunn, P. A. (2014). Using multiplexed assays of oncogenic drivers in lung cancers to select targeted drugs. *JAMA: The Journal of the American Medical Association, 311,* 1998−2006.

Kwak, E. L., Bang, Y.-J., Camidge, D. R., Shaw, A. T., Solomon, B., Maki, R. G., ... Iafrate, A. J. (2010). Anaplastic lymphoma kinase inhibition in non−small-cell lung cancer. *New England Journal of Medicine, 363,* 1693−1703.

Langer, C. J., Redman, M. W., Wade, J. L., III, Aggarwal, C., Bradley, J. D., Crawford, J., ... Papadimitrakopoulou, V. A. (2019). SWOG S1400B (NCT02785913), a Phase II study of GDC-0032 (Taselisib) for previously treated PI3K-positive patients with Stage IV squamous cell lung cancer (Lung-MAP sub-study). *Journal of Thoracic Oncology, 14,* 1839−1846.

Lau, S. K., Boutros, P. C., Pintilie, M., Blackhall, F. H., Zhu, C.-Q., Strumpf, D., ... Tsao, M.-S. (2007). Three-gene prognostic classifier for early-stage non small-cell lung cancer. *Journal of Clinical Oncology, 25,* 5562−5569.

Le, X., Goldman, J. W., Clarke, J. M., Tchekmedyian, N., Piotrowska, Z., Chu, D., ... Socinski, M. A. (2020). Poziotinib shows activity and durability of responses in subgroups of previously treated EGFR exon 20 NSCLC patients. *Journal of Clinical Oncology, 38,* 9514.

Lee, S. H., Lee, J. K., Ahn, M. J., Kim, D.-W., Sun, J.-M., Keam, B., ... Park, K. (2017). Vandetanib in pretreated patients with advanced non-small cell lung cancer-harboring RET rearrangement: A phase II clinical trial. *Annals of Oncology, 28,* 292−297.

Leighl, N. B., Page, R. D., Raymond, V. M., Daniel, D. B., Divers, S. G., Reckamp, K. L., ... Papadimitrakopoulou, V. A. (2019). Clinical utility of comprehensive cell-free DNA analysis to identify genomic biomarkers in patients with newly diagnosed metastatic non−small cell lung cancer. *Clinical Cancer Research, 25,* 4691−4700.

Lennon, A. M., Buchanan, A. H., Kinde, I., Warren, A., Honushefsky, A., Cohain, A. T., ... Papadopoulos, N. (2020). Feasibility of blood testing combined with PET-CT to screen for cancer and guide intervention. *Science (New York, N.Y.), 369,* eabb9601.

Leonetti, A., Facchinetti, F., Rossi, G., Minari, R., Conti, A., Friboulet, L., ... Planchard, D. (2018). BRAF in non-small cell lung cancer (NSCLC): Pickaxing another brick in the wall. *Cancer Treatment Reviews, 66,* 82−94.

Li, B. T., Janku, F., Jung, B., Hou, C., Madwani, K., Alden, R., ... Oxnard, G. R. (2019). Ultra-deep next-generation sequencing of plasma cell-free DNA in patients with advanced lung cancers: Results from the actionable genome consortium. *Annals of Oncology, 30,* 597−603.

Li, B. T., Ross, D. S., Aisner, D. L., Chaft, J. E., Hsu, M., Kako, S. L., ... Arcila, M. E. (2016). HER2 amplification and HER2 mutation are distinct molecular targets in lung cancers. *Journal of Thoracic Oncology, 11,* 414−419.

Li, B. T., Shen, R., Buonocore, D., Olah, Z. T., Ni, A., Ginsberg, M. S., ... Kris, M. G. (2018). Ado-trastuzumab emtansine for patients with HER2-mutant lung cancers: Results from a Phase II basket trial. *Journal of Clinical Oncology, 36,* 2532−2537.

Lin, J. J., Riely, G. J., & Shaw, A. T. (2017). Targeting ALK: Precision medicine takes on drug resistance. *Cancer Discovery, 7,* 137−155.

Lisberg, A., Cummings, A., Goldman, J. W., Bornazyan, K., Reese, N., Wang, T., ... Garon, E. B. (2018). A Phase II study of pembrolizumab in EGFR-mutant, PD-L1 +, tyrosine kinase inhibitor naïve patients with advanced NSCLC. *Journal of Thoracic Oncology*, *13*, 1138−1145.

Liu, B., Ricarte Filho, J., Mallisetty, A., Villani, C., Kottorou, A., Rodgers, K., ... Hulbert, A. (2020). Detection of promoter DNA methylation in urine and plasma aids the detection of non−small cell lung cancer. *Clinical Cancer Research*, *26*, 4339−4348.

Liu, M. C., Oxnard, G. R., Klein, E. A., Swanton, C., Seiden, M. V., ... CCGA Consortium. (2020). Sensitive and specific multi-cancer detection and localization using methylation signatures in cell-free DNA. *Annals of Oncology*, *31*, 745−759.

Lu, S., Fang, J., Cao, L., Li, X., Guo, Q., Zhou, J., ... Su, W. (2019). Abstract CT031: Preliminary efficacy and safety results of savolitinib treating patients with pulmonary sarcomatoid carcinoma (PSC) and other types of non-small cell lung cancer (NSCLC) harboring MET exon 14 skipping mutations. *Cancer Research*, *79*, CT031, CT.

Lynch, T. J., Bell, D. W., Sordella, R., Gurubhagavatula, S., Okimoto, R. A., Brannigan, B. W., ... Haber, D. A. (2004). Activating mutations in the epidermal growth factor receptor underlying responsiveness of non-small-cell lung cancer to gefitinib. *The New England Journal of Medicine*, *350*, 2129−2139.

Maemondo, M., Inoue, A., Kobayashi, K., Sugawara, S., Oizumi, S., Isobe, H., ... North-East Japan Study Group. (2010). Gefitinib or chemotherapy for non−small-cell lung cancer with mutated EGFR. *New England Journal of Medicine*, *362*, 2380−2388.

Mazieres, J., Cropet, C., Montané, L., Barlesi, F., Souquet, P. J., Quantin, X., ... Blay, J. Y. (2020). Vemurafenib in non-small-cell lung cancer patients with BRAF(V600) and BRAF (nonV600) mutations. *Annals of Oncology*, *31*, 289−294.

Mazieres, J., Drilon, A., Lusque, A., Mhanna, L., Cortot, A. B., Mezquita, L., ... Gautschi, O. (2019). Immune checkpoint inhibitors for patients with advanced lung cancer and oncogenic driver alterations: Results from the IMMUNOTARGET registry. *Annals of Oncology*, *30*, 1321−1328.

Mazzone, P. J., Silvestri, G. A., Patel, S., Kanne, J. P., Kinsinger, L. S., Wiener, R. S., ... Detterbeck, F. C. (2018). Screening for lung cancer: CHEST guideline and expert panel report. *CHEST*, *153*, 954−985.

McCoach, C. E., Le, A. T., Gowan, K., Jones, K., Schubert, L., Doak, A., ... Doebele, R. C. (2018). Resistance mechanisms to targeted therapies in ROS1 + and ALK + non−small cell lung cancer. *Clinical Cancer Research*, *24*, 3334−3347.

Mitsudomi, T., Morita, S., Yatabe, Y., Negoro, S., Okamoto, I., Tsurutani, J., ... West Japan Oncology Group. (2010). Gefitinib vs cisplatin plus docetaxel in patients with non-small-cell lung cancer harbouring mutations of the epidermal growth factor receptor (WJTOG3405): An open label, randomised phase 3 trial. *The Lancet Oncology*, *11*, 121−128.

Mok, T. S., Wu, Y.-L., Ahn, M.-J., Garassino, M. C., Kim, H. R., Ramalingam, S. S., ... AURA3 Investigators. (2016). Osimertinib or platinum−pemetrexed in EGFR T790M−positive lung cancer. *New England Journal of Medicine*, *376*, 629−640.

Mok, T. S., Wu, Y.-L., Thongprasert, S., Yang, C.-H., Chu, D.-T., Saijo, N., ... Fukuoka, M. (2009). Gefitinib or carboplatin−paclitaxel in pulmonary adenocarcinoma. *New England Journal of Medicine*, *361*, 947−957.

Mok, T. S. K., Wu, Y. L., Kudaba, I., Kowalski, D. M., Cho, B. C., Turna, H. Z., ... KEYNOTE-042 Investigators. (2019). Pembrolizumab vs chemotherapy for previously untreated, PD-L1-expressing, locally advanced or metastatic non-small-cell lung cancer

(KEYNOTE-042): A randomised, open-label, controlled, phase 3 trial. *Lancet, 393,* 1819–1830.

Morris, S. W., Kirstein, M. N., Valentine, M. B., Dittmer, K. G., Shapiro, D. N., Saltman, D. L., & Look, A. T. (1994). Fusion of a kinase gene, ALK, to a nucleolar protein gene, NPM, in non-Hodgkin's lymphoma. *Science (New York, N.Y.), 263,* 1281–1284.

Moyer, V. A. (2014). Screening for lung cancer: U.S. Preventive Services Task Force recommendation statement. *Annals of Internal Medicine, 160,* 330–338.

Musolf, A. M., Simpson, C. L., de Andrade, M., Mandal, D., Gaba, C., Yang, P., ... Bailey-Wilson, J. E. (2017). Familial lung cancer: A brief history from the earliest work to the most recent studies. *Genes (Basel), 8,* 36.

Nakagawa, K., Garon, E. B., Seto, T., Nishio, M., Aix, S. P., Paz-Ares, L., ... RELAY Study Investigators. (2019). Ramucirumab plus erlotinib in patients with untreated, EGFR-mutated, advanced non-small-cell lung cancer (RELAY): A randomised, double-blind, placebo-controlled, phase 3 trial. *The Lancet Oncology, 20,* 1655–1669.

Nakaoku, T., Kohno, T., Araki, M., Niho, Seiji, Chauhan, Rakhee, Knowles, P. P., ... Goto, Koichi (2018). A secondary RET mutation in the activation loop conferring resistance to vandetanib. *Nature Communications, 9,* 625.

National Cancer Institute. (2020). *Cancer stat facts: Lung and bronchus cancer.* https://seer.cancer.gov/statfacts/html/lungb.html. Accessed 07.10.20.

Nelson-Taylor, S. K., Le, A. T., Yoo, M., Schubert, L., Mishall, K. M., Doak, A., ... Doebele, R. C. (2017). Resistance to RET-inhibition in RET-rearranged NSCLC is mediated by reactivation of RAS/MAPK signaling. *Molecular Cancer Therapeutics, 16,* 1623–1633.

Noronha, V., Patil, V. M., Joshi, A., Menon, N., Chougule, A., Mahajan, A., ... Prabhash, K. (2020). Gefitinib vs gefitinib plus pemetrexed and carboplatin chemotherapy in EGFR-mutated lung cancer. *Journal of Clinical Oncology, 38,* 124–136.

Ostrem, J. M., Peters, U., Sos, M. L., Wells, J. A., & Shokat, K. M. (2013). K-Ras(G12C) inhibitors allosterically control GTP affinity and effector interactions. *Nature, 503,* 548–551.

Oxnard, G. R., Hu, Y., Mileham, K. F., Husain, H., Costa, D. B., Tracy, P., ... Jänne, P. A. (2018). Assessment of resistance mechanisms and clinical implications in patients with EGFR T790M-positive lung cancer and acquired resistance to osimertinib. *JAMA Oncology, 4,* 1527–1534.

Oxnard, G. R., Paweletz, C. P., Kuang, Y., Mach, S. L., O'Connell, A., Messineo, M. M, ... Jänne, P. A. (2014). Noninvasive detection of response and resistance in EGFR mutant lung cancer using quantitative next-generation genotyping of cell-free plasma DNA. *Clinical Cancer Research, 20,* 1698–1705.

Oxnard, G. R., Thress, K. S., Alden, R. S., Lawrance, R., Paweletz, C. P., Cantarini, M., ... Jänne, P. A. (2016). Association between plasma genotyping and outcomes of treatment with osimertinib (AZD9291) in advanced non-small-cell lung cancer. *Journal of clinical oncology, 34,* 3375–3382.

Paez, J. G., Jänne, P. A., Lee, J. C., Tracy, S., Greulich, H., Gabriel, S., ... Meyerson, M. (2004). EGFR mutations in lung cancer: Correlation with clinical response to gefitinib therapy. *Science (New York, N.Y.), 304,* 1497–1500.

Paik, P. K., Arcila, M. E., Fara, M., Sima, C. S., Miller, V. A., Kris, M. G., ... Riely, G. J. (2011). Clinical characteristics of patients with lung adenocarcinomas harboring BRAF mutations. *Journal of Clinical Oncology, 29,* 2046–2051.

Paik, P. K., Drilon, A., Fan, P.-D., Yu, H., Rekhtman, N., Ginsberg, M. S., ... Ladanyi, M. (2015). Response to MET inhibitors in patients with stage IV lung adenocarcinomas harboring MET mutations causing exon 14 skipping. *Cancer Discovery, 5,* 842–849.

Paik, P. K., Felip, E., Veillon, R., Sakai, H., Cortot, A. B., Garassino, M. C., ... Le, X. (2020). Tepotinib in non−small-cell lung cancer with MET exon 14 skipping mutations. *New England Journal of Medicine, 383*, 931−943.

Paik, P. K., Shen, R., Berger, M. F., Ferry, D., Soria, J.-C., Mathewson, A., ... Andre, F. (2017). A phase Ib open-label multicenter study of AZD4547 in patients with advanced squamous cell lung cancers. *Clinical Cancer Research, 23*, 5366−5373.

Papillon-Cavanagh, S., Doshi, P., Dobrin, R., Szustakowski, J., & Walsh, A. M. (2020). STK11 and KEAP1 mutations as prognostic biomarkers in an observational real-world lung adeno-carcinoma cohort. *ESMO Open, 5*, e000706.

Patil, T., Smith, D. E., Bunn, P. A., Aisner, D. L., Le, A. T., Hancock, M., ... Doebele, R. C. (2018). The incidence of brain metastases in stage IV ROS1-rearranged non-small cell lung cancer and rate of central nervous system progression on crizotinib. *Journal of Thoracic Oncology, 13*, 1717−1726.

Patz, E. F., Goodman, P. C., & Bepler, G. (2000). Screening for lung cancer. *New England Journal of Medicine, 343*, 1627−1633.

Paz-Ares, L., Luft, A., Vicente, D., Tafreshi, A., Gümüş, M., Mazières, J., ... for the KEYNOTE-407 Investigators. (2018). Pembrolizumab plus chemotherapy for squamous non−small-cell lung cancer. *New England Journal of Medicine, 379*, 2040−2051.

Peters, S., Camidge, D. R., Shaw, A. T., Gadgeel, S., Ahn, J. S., Kim, D.-W., ... ALEX Trial Investigators. (2017). Alectinib vs crizotinib in untreated ALK-positive non−small-cell lung cancer. *New England Journal of Medicine, 377*, 829−838.

Piotrowska, Z., Wang, Y., Sequist, L. V., & Ramalingam, S. S. (2020). ECOG-ACRIN 5162: A phase II study of osimertinib 160mg in NSCLC with EGFR exon 20 insertions. *Journal of Clinical Oncology, 38*, 9513.

Piper-Vallillo, A. J., Sequist, L. V., & Piotrowska, Z. (2020). Emerging treatment paradigms for EGFR-mutant lung cancers progressing on osimertinib: A review. *Journal of Clinical Oncology, 38*, 2926−2936.

Planchard, D., Kim, T. M., Mazieres, J., Quoix, E., Riely, G., Barlesi, F., ... Johnson, B. E. (2016). Dabrafenib in patients with BRAF(V600E)-positive advanced non-small-cell lung cancer: A single-arm, multicentre, open-label, phase 2 trial. *The Lancet Oncology, 17*, 642−650.

Planchard, D., Smit, E. F., Groen, H. J. M., Mazieres, J., Besse, B., Helland, Å., ... Johnson, B. E. (2017). Dabrafenib plus trametinib in patients with previously untreated BRAF (V600E)-mutant metastatic non-small-cell lung cancer: An open-label, phase 2 trial. *The Lancet Oncology, 18*, 1307−1316.

Ramalingam, S. S., Vansteenkiste, J., Planchard, D., Cho, C. B., Gray, J. E., Ohe, Y., ... FLAURA Investigators. (2019). Overall survival with osimertinib in untreated, EGFR-mutated advanced NSCLC. *New England Journal of Medicine, 382*, 41−50.

Raponi, M., Zhang, Y., Yu, J., Chen, G., Lee, G., Taylor, J. M. G., ... Beer, D. G. (2006). Gene expression signatures for predicting prognosis of squamous cell and adenocarcinomas of the lung. *Cancer Research, 66*, 7466−7472.

Reck, M., Rodríguez-Abreu, D., Robinson, A. G., Hui, R., Csőszi, T., Fülöp, A., ... for the KEYNOTE-024 Investigators. (2016). Pembrolizumab vs chemotherapy for PD-L1−positive non−small-cell lung cancer. *New England Journal of Medicine, 375*, 1823−1833.

Recondo, G., Bahcall, M., Spurr, L. F., Che, J., Ricciuti, B., Leonardi, G. C., ... Awad, M. M. (2020). Molecular mechanisms of acquired resistance to MET tyrosine kinase inhibitors in patients with MET exon 14−mutant NSCLC. *Clinical Cancer Research, 26*, 2615−2625.

Reungwetwattana, T., Nakagawa, K., Cho, B. C., Cobo, M., Cho, E. K., Bertolini, A., . . . Vansteenkiste, J. (2018). CNS response to osimertinib vs standard epidermal growth factor receptor tyrosine kinase inhibitors in patients with untreated EGFR-mutated advanced non-small-cell lung cancer. *Journal of Clinical Oncology*, Jco2018783118.

Rikova, K., Guo, A., Zeng, Q., Possemato, A., Yu, J., Haack, H., . . . Comb, M. J. (2007). Global survey of phosphotyrosine signaling identifies oncogenic kinases in lung cancer. *Cell, 131,* 1190−1203.

Rittmeyer, A., Barlesi, F., Waterkamp, D., Park, K., Ciardiello, F., von Pawel, J., . . . OAK Study Group. (2017). Atezolizumab vs docetaxel in patients with previously treated non-small-cell lung cancer (OAK): A phase 3, open-label, multicentre randomised controlled trial. *Lancet, 389,* 255−265.

Rizvi, H., Sanchez-Vega, F., La, K., Chatila, W., Jonsson, P., Halpenny, D., . . . Hellmann, M. D. (2018). Molecular determinants of response to anti−programmed cell death (PD)-1 and anti−programmed death-ligand 1 (PD-L1) blockade in patients with non−small-cell lung cancer profiled with targeted next-generation sequencing. *Journal of Clinical Oncology, 36,* 633−641.

Rizvi, N. A., Hellmann, M. D., Snyder, A., Kvistborg, P., Makarov, V., Havel, J. J., . . . Chan, T. A. (2015). Mutational landscape determines sensitivity to PD-1 blockade in non−small cell lung cancer. *Science (New York, N.Y.), 348,* 124−128.

Robichaux, J. P., Elamin, Y. Y., Tan, Z., Carter, B. W., Zhang, S., Liu, S., . . . Heymach, J. V. (2018). Mechanisms and clinical activity of an EGFR and HER2 exon 20-selective kinase inhibitor in non-small cell lung cancer. *Nature Medicine, 24,* 638−646.

Rosell, R., Carcereny, E., Gervais, R., Vergnenegre, A., Massuti, B., Felip, E., . . . Spanish Lung Cancer Group in collaboration with Groupe Français de Pneumo-Cancérologie and Associazione Italiana Oncologia Toracica. (2012). Erlotinib vs standard chemotherapy as first-line treatment for European patients with advanced EGFR mutation-positive non-small-cell lung cancer (EURTAC): A multicentre, open-label, randomised phase 3 trial. *The Lancet Oncology, 13,* 239−246.

Ryan, M. B., & Corcoran, R. B. (2018). Therapeutic strategies to target RAS-mutant cancers. *Nature Reviews Clinical Oncology, 15,* 709−720.

Salgia, R. (2017). MET in lung cancer: Biomarker selection based on scientific rationale. *Molecular Cancer Therapeutics, 16,* 555−565.

Scagliotti, G., von Pawel, J., Novello, S., Ramlau, R., Favaretto, A., Barlesi, F., . . . Schwartz, B. (2015). Phase III multinational, randomized, double-blind, placebo-controlled study of tivantinib (ARQ 197) plus erlotinib vs erlotinib alone in previously treated patients with locally advanced or metastatic nonsquamous non-small-cell lung cancer. *Journal of Clinical Oncology, 33,* 2667−2674.

Schoenfeld, A. J., Arbour, K. C., Rizvi, H., Iqbal, A. N., Gadgeel, S. M., Girshman, J., . . . Hellmann, M. D. (2019). Severe immune-related adverse events are common with sequential PD-(L)1 blockade and osimertinib. *Annals of Oncology, 30,* 839−844.

Schoenfeld, A. J., Chan, J. M., Kubota, D., Sato, H., Rizvi, H., Daneshbod, Y., . . . Yu, H. A. (2020). Tumor analyses reveal squamous transformation and off-target alterations as early resistance mechanisms to first-line osimertinib in EGFR-mutant lung cancer. *Clinical Cancer Research, 26,* 2654−2663.

Sequist, L. V., Han, J.-Y., Ahn, M.-J., Cho, B. C., Yu, H., Kim, S.-W., . . . Oxnard, G. (2020). Osimertinib plus savolitinib in patients with EGFR mutation-positive, MET-amplified, non-small-cell lung cancer after progression on EGFR tyrosine kinase inhibitors: Interim results from a multicentre, open-label, phase 1b study. *The Lancet Oncology, 21,* 373−386.

Shaw, A. T., Friboulet, L., Leshchiner, I., Gainor, J. F., Bergqvist, S., Brooun, A., . . . Engelman, J. A. (2015). Resensitization to crizotinib by the lorlatinib ALK resistance mutation L1198F. *New England Journal of Medicine, 374*, 54−61.

Shaw, A. T., Gandhi, L., Gadgeel, S., Riely, G. J., Cetnar, J., West, H., . . . study investigators. (2016). Alectinib in ALK-positive, crizotinib-resistant, non-small-cell lung cancer: A single-group, multicentre, phase 2 trial. *The Lancet Oncology, 17*, 234−242.

Shaw, A. T., Kim, D.-W., Mehra, R., Tan, D. S. W., Felip, E., Chow, L. Q. M., . . . Engelman, J. A. (2014). Ceritinib in ALK-rearranged non−small-cell lung cancer. *New England Journal of Medicine, 370*, 1189−1197.

Shaw, A. T., Kim, D.-W., Nakagawa, K., Seto, T., Crinó, L., Ahn, M.-J., . . . Jänne, P. A. (2013). Crizotinib vs chemotherapy in advanced ALK-positive lung cancer. *New England Journal of Medicine, 368*, 2385−2394.

Shaw, A. T., Kim, T. M., Crinò, L., Gridelli, C., Kiura, K., Liu, G., . . . Felip, E. (2017). Ceritinib vs chemotherapy in patients with ALK-rearranged non-small-cell lung cancer previously given chemotherapy and crizotinib (ASCEND-5): A randomised, controlled, open-label, phase 3 trial. *The Lancet Oncology, 18*, 874−886.

Shaw, A. T., Ou, S. H., Bang, Y. J., Camidge, D. R., Solomon, B. J., Salgia, R., . . . Iafrate, A. J. (2014). Crizotinib in ROS1-rearranged non-small-cell lung cancer. *The New England Journal of Medicine, 371*, 1963−1971.

Shaw, A. T., Riely, G. J., Bang, Y. J., Kim, D.-W., Cambidge, D. R., Solomon, B. J., . . . Ou, S.-H. I. (2019). Crizotinib in ROS1-rearranged advanced non-small-cell lung cancer (NSCLC): Updated results, including overall survival, from PROFILE 1001. *Annals of Oncology, 30*, 1121−1126.

Shaw, A. T., Solomon, B. J., Besse, B., Bauer, T. M., Lin, C.-C., Spo, R. A., . . . Martini, J.-F. (2019). ALK resistance mutations and efficacy of lorlatinib in advanced anaplastic lymphoma kinase-positive non-small-cell lung cancer. *Journal of Clinical Oncology, 37*, 1370−1379.

Shaw, A. T., Solomon, B. J., Chiari, R., Riely, G. J., Besse, B., Soo, R. A., . . . Ou, S.-H. I. (2019). Lorlatinib in advanced ROS1-positive non-small-cell lung cancer: A multicentre, open-label, single-arm, phase 1/2 trial. *The Lancet Oncology, 20*, 1691−1701.

Shepherd, F. A., Rodrigues Pereira, J., Ciuleanu, T., Tan, E. H., Hirsh, V., Thongprasert, S., . . . National Cancer Institute of Canada Clinical Trials Group. (2005). Erlotinib in previously treated non−small-cell lung cancer. *New England Journal of Medicine, 353*, 123−132.

Siegel, R. L., Jacobs, E. J., Newton, C. C., Feskanich, D., Freedman, N. D., Prentice, R. L., . . . Jemal, A. (2015). Deaths due to cigarette smoking for 12 smoking-related cancers in the United States. *JAMA Internal Medicine, 175*, 1574−1576.

Singal, G., Miller, P. G., Agarwala, V., Li, G., Kaushik, G., Backenroth, D., . . . Miller, V. A. (2019). Association of patient characteristics and tumor genomics with clinical outcomes among patients with non−small cell lung cancer using a clinicogenomic database. *JAMA: The Journal of the American Medical Association, 321*, 1391−1399.

Singhi, E. K., Horn, L., Sequist, L. V., Heymach, J., & Langer, C. J. (2019). Advanced non-small cell lung cancer: Sequencing agents in the EGFR-mutated/ALK-rearranged populations. *American Society of Clinical Oncology Educational Book, 39*, e187−e197.

Skoulidis, F., Byers, L. A., Diao, L., Papadimitrakopoulou, V. A., Tong, P., Izzo, J., . . . Heymach, J. V. (2015). Co-occurring genomic alterations define major subsets of KRAS-mutant lung adenocarcinoma with distinct biology, immune profiles, and therapeutic vulnerabilities. *Cancer Discovery, 5*, 860−877.

Skoulidis, F., Goldberg, M. E., Greenawalt, D. M., Hellmann, M. D., Awad, M. M., Gainor, J. F., ... Heymach, J. V. (2018). STK11/LKB1 mutations and PD-1 inhibitor resistance in KRAS-mutant lung adenocarcinoma. *Cancer Discovery*, *8*, 822−835.

Smit, E. F., Nakagawa, K., Nagasaka, M., Felip, E., Goto, Y., Li, B. T., ... Janne, P. A. (2020). Trastuzumab deruxtecan (T-DXd; DS-8201) in patients with HER2-mutated metastatic non-small cell lung cancer (NSCLC): Interim results of DESTINY-Lung01. *Journal of Clinical Oncology*, *38*, 9504.

Socinski, M. A., Jotte, R. M., Cappuzzo, F., Orlandi, F., Stroyakovskiy, D., Nogami, N., ... for the IMpower150 Study Group. (2018). Atezolizumab for first-line treatment of metastatic nonsquamous NSCLC. *New England Journal of Medicine*, *378*, 2288−2301.

Soda, M., Choi, Y. L., Enomoto, M., Takada, S., Yamashita, Y., Ishikawa, S., ... Mano, H. (2007). Identification of the transforming EML4-ALK fusion gene in non-small-cell lung cancer. *Nature*, *448*, 561−566.

Solomon, B. J., Besse, B., Bauer, T. M., Felip, E., Soo, R. A., Camidge, D. R., ... Shaw, A. T. (2018). Lorlatinib in patients with ALK-positive non-small-cell lung cancer: Results from a global phase 2 study. *The Lancet Oncology*, *19*, 1654−1667.

Solomon, B. J., Kim, D. W., Wu, Y. L., Nakagawa, K., Mekhail, T., Felip, E., ... Mok, T. S. (2018). Final overall survival analysis from a study comparing first-line crizotinib vs chemotherapy in ALK-mutation-positive non-small-cell lung cancer. *Journal of Clinical Oncology*, *36*, 2251−2258.

Solomon, B. J., Mok, T., Kim, D.-W., Wu, Y.-L., Nakagawa, K., Mekhail, T., ... PROFILE 1014 Investigators. (2014). First-line crizotinib vs chemotherapy in ALK-positive lung cancer. *New England Journal of Medicine*, *371*, 2167−2177.

Soria, J. C., Tan, D. S. W., Chiari, R., Wu, Y.-L., Paz-Ares, L., Wolf, J., ... de Castro, Gilberto (2017). First-line ceritinib vs platinum-based chemotherapy in advanced ALK-rearranged non-small-cell lung cancer (ASCEND-4): A randomised, open-label, phase 3 study. *Lancet*, *389*, 917−929.

Sos, M. L., Rode, H. B., Heynck, S., Peifer, M., Fischer, F., Klüter, S., ... Rauh, D. (2010). Chemogenomic profiling provides insights into the limited activity of irreversible EGFR inhibitors in tumor cells expressing the T790M EGFR resistance mutation. *Cancer Research*, *70*, 868−874.

Spigel, D. R., Edelman, M. J., O'Byrne, K., Paz-Ares, L., Mocci, S., Phan, S., ... Mok, T. (2017). Results from the phase III randomized trial of onartuzumab plus erlotinib vs erlotinib in previously treated stage IIIB or IV non-small-cell lung cancer: METLung. *Journal of Clinical Oncology*, *35*, 412−420.

Subbiah, V., Yang, D., Velcheti, V., Drilon, A., & Meric-Bernstam, F. (2020). State-of-the-art strategies for targeting RET-dependent cancers. *Journal of Clinical Oncology*, *38*, 1209−1221.

Swanton, C., & Govindan, R. (2016). Clinical implications of genomic discoveries in lung cancer. *New England Journal of Medicine*, *374*, 1864−1873.

Tammemägi, M. C., Katki, H. A., Hocking, W. G., Church, T. R., Caporaso, N., Kvale, P. A., ... Berg, C. D. (2013). Selection criteria for lung-cancer screening. *The New England Journal of Medicine*, *368*, 728−736.

Thatcher, N., Chang, A., Parikh, P., Pereira, J. R., Ciuleanu, T., von Pawel, J., ... Carroll, K. (2005). Gefitinib plus best supportive care in previously treated patients with refractory advanced non-small-cell lung cancer: Results from a randomised, placebo-controlled, multicentre study (Iressa Survival Evaluation in Lung Cancer). *Lancet*, *366*, 1527−1537.

The National Lung Screening Trial Research Team. (2011). Reduced lung-cancer mortality with low-dose computed tomographic screening. *New England Journal of Medicine*, *365*, 395−409.

Thompson, J. C., Yee, S. S., Troxel, A. B., Savitch, S. L., Fan, R., Balli, D., . . . Carpenter, E. L. (2016). Detection of therapeutically targetable driver and resistance mutations in lung cancer patients by next-generation sequencing of cell-free circulating tumor DNA. *Clinical Cancer Research*, *22*, 5772−5782.

Timofeeva, M. N., Hung, R. J., Rafnar, T., Christiani, D. C., Field, J. K., Bickeböller, H., . . . Transdisciplinary Research in Cancer of the Lung (TRICL) Research Team. (2012). Influence of common genetic variation on lung cancer risk: *Meta*-analysis of 14 900 cases and 29 485 controls. *Human Molecular Genetics*, *21*, 4980−4995.

Vansteenkiste, J. F., Canon, J. L., De Braud, F., Grossi, F., De Pas, T., Gray, J. E., . . . Soria, J.-C. (2015). Safety and efficacy of buparlisib (BKM120) in patients with PI3K pathway-activated non-small cell lung cancer: Results from the phase II BASALT-1 study. *Journal of Thoracic Oncology*, *10*, 1319−1327.

Wang, J., Liu, Q., Yuan, S., Xie, W., Liu, Y., Xiang, Y., . . . Li, Y. (2017). Genetic predisposition to lung cancer: Comprehensive literature integration, *meta*-analysis, and multiple evidence assessment of candidate-gene association studies. *Scientific Reports*, *7*, 8371.

Wang, R., Hu, H., Pan, Y., Li, Y., Ye, T., Li, C., . . . Chen, H. (2012). RET fusions define a unique molecular and clinicopathologic subtype of non-small-cell lung cancer. *Journal of Clinical Oncology*, *30*, 4352−4359.

Wang, Y., Jiang, T., Qin, Z., Jiang, J., Wang, Q., Yang, S., . . . Hirsch, F. R. (2019). HER2 exon 20 insertions in non-small-cell lung cancer are sensitive to the irreversible pan-HER receptor tyrosine kinase inhibitor pyrotinib. *Annals of Oncology*, *30*, 447−455.

Weissfeld, J. L., Lin, Y., Lin, H.-M., Kurland, B. F., Wilson, D. O., Fuhrman, C. R., . . . Diergaarde, B. (2015). Lung cancer risk prediction using common SNPs located in GWAS-identified susceptibility regions. *Journal of Thoracic Oncology*, *10*, 1538−1545.

Wolf, J., Seto, T., Han, J.-Y., Reguart, N., Garon, E. B., Groen, H. J. M., . . . GEOMETRY mono-1 Investigators. (2020). Capmatinib in MET exon 14−mutated or MET-amplified non−small-cell lung cancer. *New England Journal of Medicine*, *383*, 944−957.

Wu, Y.-L., Tsuboi, M., He, J., John, T., Grohe, C., Majem, M., . . . for the ADAURA Investigators. (2020). Osimertinib in resected EGFR-mutated non−small-cell lung cancer. *New England Journal of Medicine*, *383*, 1711−1723.

Wu, Y.-L., Zhong, W., Wang, Q., Mao, W., Xu, S.-T., Wu, L., . . . Zhou, Q. (2020). CTONG1104: Adjuvant gefitinib vs chemotherapy for resected N1-N2 NSCLC with EGFR mutation—Final overall survival analysis of the randomized phase III trial 1 analysis of the randomized phase III trial. *Journal of Clinical Oncology*, *38*, 9005−9194.

Xu, S.-T., Xi, J.-J., Zhong, W.-Z., Mao, W.-M., Wu, L., Shen, Y., . . . Wu, Y.-L. (2019). The unique spatial-temporal treatment failure patterns of adjuvant gefitinib therapy: A post hoc analysis of the ADJUVANT trial (CTONG 1104). *Journal of Thoracic Oncology*, *14*, 503−512.

Xue, J. Y., & Lito, P. (2020). Quest for clinically effective RAF dimer inhibitors. *Journal of Clinical Oncology*, *38*, 2197−2200.

Yaeger, R., & Corcoran, R. B. (2019). Targeting alterations in the RAF−MEK pathway. *Cancer Discovery*, *9*, 329−341.

Yamamoto, H., Higasa, K., Sakaguchi, M., Shien, K., Soh, J., Ichimura, K., . . . Toyooka, S. (2014). Novel germline mutation in the transmembrane domain of HER2 in familial lung adenocarcinomas. *Journal of the National Cancer Institute*, *106*, djt338.

Yang, J. C. H., Sequist, L. V., Geater, S. L., Tsai, C.-M., Kam Mok, T. S., Schuler, M., . . . Wu, Y.-L. (2015). Clinical activity of afatinib in patients with advanced non-small-cell lung cancer harbouring uncommon EGFR mutations: A combined post-hoc analysis of LUX-Lung 2, LUX-Lung 3, and LUX-Lung 6. *The Lancet Oncology*, *16*, 830–838.

Yasuda, H., Park, E., Yun, C. H., Sng, N. J., Lucena-Araujo, A. R., Yeo, W.-L., . . . Costa, D. B. (2013). Structural, biochemical, and clinical characterization of epidermal growth factor receptor (EGFR) exon 20 insertion mutations in lung cancer. *Science Translational Medicine*, *5*, 216ra177.

Yoh, K., Seto, T., Satouchi, M., Nishio, M., Yamamoto, N., Murakami, H., . . . Goto, K. (2017). Vandetanib in patients with previously treated RET-rearranged advanced non-small-cell lung cancer (LURET): An open-label, multicentre phase 2 trial. *Lancet Respiratory Medicine*, *5*, 42–50.

Yoshioka, H., Azuma, K., Yamamoto, N., Takahashi, T., Nishio, M., Katakami, N., . . . Nakagawa, K. (2015). A randomized, double-blind, placebo-controlled, phase III trial of erlotinib with or without a c-Met inhibitor tivantinib (ARQ 197) in Asian patients with previously treated stage IIIB/IV nonsquamous nonsmall-cell lung cancer harboring wild-type epidermal growth factor receptor (ATTENTION study). *Annals of Oncology*, *26*, 2066–2072.

Yu, H. A., Arcila, M. E., Harlan Fleischut, M., Stadler, Z., Ladanyi, M., Berger, M. F., . . . Riely, G. J. (2014). Germline EGFR T790M mutation found in multiple members of a familial cohort. *Journal of Thoracic Oncology*, *9*, 554–558.

Yun, M. R., Kim, D. H., Kim, S.-Y., Joo, H.-S., Lee, Y. W., Choi, H. M., . . . Cho, B. C. (2020). Repotrectinib exhibits potent antitumor activity in treatment-naïve and solvent-front–mutant ROS1-rearranged non–small cell lung cancer. *Clinical Cancer Research*, *26*, 3287–3295.

Zou, H. Y., Li, Q., Engstrom, L. D., West, M., Appleman, V., Wong, K. A., . . . Fantin, V. R. (2015). PF-06463922 is a potent and selective next-generation ROS1/ALK inhibitor capable of blocking crizotinib-resistant ROS1 mutations. *Proceedings of the National Academy of Sciences*, 201420785.

Zugazagoitia, J., Ramos, I., Trigo, J. M., Palka, M., Gómez-Rueda, A., Jantus-Lewintre, E., . . . Paz-Ares, L. (2019). Clinical utility of plasma-based digital next-generation sequencing in patients with advance-stage lung adenocarcinomas with insufficient tumor samples for tissue genotyping. *Annals of Oncology*, *30*, 290–296.

Chapter 9

Breast cancer

Kathleen Harnden[1,2,*], Lauren Mauro[1,2,*] and Angela Pennisi[1,2,*]
[1]*Department of Medical Oncology, Inova Schar Cancer Institute, Fairfax, VA, United States,*
[2]*Department of Medicine, University of Virginia, Charlottesville, VA, United States*

Introduction

Over the past several decades, there has been an enormous research effort in the field of oncology to develop novel therapies. Great strides have been made in biomarker identification to determine which patients are the best candidates for targeted therapies. Breast cancer has been a leader in the field of precision medicine and while initial biologic subgroups were primarily treated based on the hormone receptor and human epidermal growth factor receptor 2 (HER2) status, there are now increasing options for treatment based on genomic alterations and signaling pathway aberrancies.

Estrogen receptor

The estrogen receptor (ER) is a cornerstone for treatment in breast cancer. Therapies targeting ER were some of the earliest anticancer targeted therapies and have been used as standard of care treatment for decades. In all stages of ER + breast cancer, the initial treatment strategy is to modulate estrogen's impact on breast cancer cells or to eliminate circulating estrogen. Aromatase inhibition which deprives breast cancer cells of estrogen has been more effective than selective ER modification with tamoxifen (EBCTCG, 2015). In the metastatic setting, fulvestrant is a selective ER down-regulator which can overcome resistance to tamoxifen and aromatase inhibition (Ciruelos, 2014).

Cyclin-dependent kinase, phosphatidylinositol 3-kinase, and mammalian target of rapamycin

In metastatic ER + breast cancer, treatment response rates and duration of response can be enhanced by adding novel targeted therapies to the backbone

---

* All the authors have equally contributed this chapter.

Genomic and Precision Medicine. DOI: https://doi.org/10.1016/B978-0-12-800684-9.00013-7

of endocrine therapy. Tumor sequencing is utilized to detect genomic targets that can enhance ER inhibition. Phosphatidylinositol 3-kinase (PI3K) catalytic subunit (PIK3CA) mutations occur in around 30% of breast cancers and predict response to inhibition of this pathway (Shimo, 2018). For patients whose cancers have progressed after aromatase inhibitor (AI) treatment and have a PI3KCA mutation, the SOLAR-1 trial demonstrated that the PI3K inhibitor alpelisib added to fulvestrant improved progression free survival (PFS) over fulvestrant alone (Andre, 2019). Aberrant signaling through the PI3K-Akt-mammalian target of rapamycin signaling pathway is an established mechanism of resistance to endocrine therapy (Schiff, 2004). BOLERO-2 showed improved PFS in patients treated with a combination of exemestane with everolimus compared to exemestane alone (7.8 months vs 3.2 months, respectively) (Yardley, 2013). In the first- and second-line treatment setting, the cyclin-dependent kinase (CDK) pathway has been an effective target and is upregulated in many breast cancers. CDK 4/6 inhibition causes cell cycle arrest through activation of the tumor suppressor Rb and plays a central role in the regulation of cell proliferation (Knudsen, 2010). Three CDK 4/6 inhibitors have been shown to substantially improve overall response rate (ORR), PFS, and in some cases OS when added to endocrine therapy (Gao, 2020). There is also evidence that this delays the need for chemotherapy in these patients (Turner, 2018). Of note, there are no randomized trials to date establishing the sequence of these targeted treatments in ER + metastatic breast cancer (MBC).

HER2/neu

HER2 (*ERBB2*) is a proto-oncogene located on chromosome 17 that encodes for HER2/neu (HER2), a transmembrane glycoprotein. Although HER2 cannot bind a ligand directly, it is able to couple with another ligand-bound receptor, and this heterodimerization activates the intracellular tyrosine kinase (TK) which initiates cell proliferation pathways. Although present on all breast cancer cells, about 15%−20% of breast cancers have an overexpression of HER2 leading to a constitutive activation of the TK pathway and dysregulated cell division, survival, and angiogenesis (DeSantis et al., 2019; Veeraraghavan et al., 2017; Wolff et al., 2014). It is now standard of care for all new invasive breast cancers to be tested for *HER2* gene amplification by in situ hybridization and/or HER2 protein overexpression by immunohistochemistry (Wolff, Hammond, & Allison, 2018). Before the discovery of HER2-directed therapy, HER2 + breast cancers had a more aggressive clinical course, a lower response to standard cytotoxic chemotherapy, and a worse prognosis. However, since HER2 targeted therapies have become available, HER2 positivity is now considered a good prognostic risk factor for early stage breast cancer patients who receive appropriate therapy. Patients with HER2 + MBC also now have numerous lines of therapy available to help extend their survival and quality of life (Giuliano, Connolly, & Edge, 2017).

Trastuzumab is a monoclonal antibody that binds to the extracellular domain of HER2 leading to antibody-dependent cellular cytotoxicity. As this predominantly occurs in cells that overexpress HER2, trastuzumab leads to the inhibition of tumor cell proliferation and metastases in HER2 + cancers. Trastuzumab was FDA approved for the treatment of HER2 + MBC in combination with chemotherapy in 1998 based on dramatic benefits shown in clinical trials. In a study by Slamon, Leyland-Jones, and Shak (2001), the trastuzumab/chemotherapy group had improved PFS (7.4 months vs 4.6 months), longer duration of response (9.1 months vs 6.1 months), and improved median OS (25.1 months vs 20.3 months) as compared to the chemotherapy alone group.

Trastuzumab plus a taxane was standard first-line therapy for years until the landmark CLEOPATRA trial, which demonstrated dramatic improvement in PFS from 12.4 months to 18.4 months with the addition of pertuzumab to the first-line regimen of docetaxel and trastuzumab (Baselga et al., 2012). Pertuzumab is a recombinant humanized monoclonal antibody that binds to a different HER2 epitope than trastuzumab and prevents HER2 heterodimerization. This combined blockade leads to more complete inhibition of downstream HER2 signaling. Follow-up analysis of the trial yielded a median OS of 57.1 months and a 37% 8-year OS with the triplet regimen as compared to 40.8 months and 23%, respectively, in the doublet regimen (Swain et al., 2020).

Trastuzumab emtansine (T-DM1) is an antibody-drug conjugate comprised of the trastuzumab antibody linked to the microtubule-inhibitory cytotoxic agent DM1. The antibody component of T-DM1 targets and binds to the HER2 receptor on HER2 overexpressing cells. TDM-1 enters the cells and DM1 is released, inducing cell death. As it is delivered directly into the cancer cell, it allows for high drug potency with low levels of toxicity as normal tissues are mostly spared from the cytotoxic effects (Loibl & Gianni, 2017; Verma et al., 2012). The Phase III EMILIA trial demonstrated that T-DM1 led to improved PFS (9.6 months vs 6.4 months) and OS (30.9 months and 25.1 months) as compared to lapatinib and capecitabine in second line or further therapy (Verma et al., 2012). Its efficacy was proven again in the TH3RESA trial comparing T-DM1 to physician's choice chemotherapy (Krop et al., 2017). T-DM1 monotherapy is now the second-line standard of care. Following the success of T-DM1, further research led to a second antibody-drug conjugate gaining FDA approval in 2020 for use as monotherapy in the third line and beyond setting for HER2 + MBC. Trastuzumab-deruxtecan is a humanized monoclonal antibody specifically targeting HER2 via a linker to a topoisomerase I inhibitor. It was designed with a higher drug to antibody ratio than T-DM1 and a more stable linker, thus leading to increased potency with low systemic toxicity (Cesca et al., 2020; Modi et al., 2020). The Phase II DESTINY-BREAST01 trial demonstrated a 60% response rate and long median PFS of 14.8 months in a heavily pretreated population. Trastuzumab-deruxtecan has a 9% risk of interstitial lung disease for which routine monitoring is now standard practice (Modi et al., 2020). Current studies are comparing the two antibody-drug conjugates in the second line.

Small molecule TK inhibitors (TKIs) that are able to pass through the cell membrane and interrupt intracellular protein activity have been studied for years in the HER2 + MBC setting. Lapatinib is a reversible inhibitor of the intracellular TK domains of both the HER2 and EGFR receptors. Lapatinib is approved in combination with capecitabine in HER2 + MBC. The combination demonstrated a doubling of median PFS when compared to capecitabine alone (4.4 months vs 8.4 months, respectively) (Geyer et al., 2006). In 2020, two other TKIs were approved for use in the HER2 + MBC setting. Neratinib is an irreversible inhibitor of EGFR (HER1), HER2, and HER4. The randomized Phase III NALA trial compared neratinib and capecitabine to lapatinib and capecitabine in third line and beyond setting. The neratinib and capecitabine combination demonstrated an improved PFS of 8.8 months versus 6.6 months (Saura et al., 2020). Tucatinib is another oral TKI that is highly selective for HER2 with minimal inhibition of EGFR. It was studied in HER2CLIMB, a randomized, placebo-controlled, Phase III trial that compared tucatinib with trastuzumab and capecitabine versus a trastuzumab/capecitabine doublet. Given the ability of tucatinib to cross the blood—brain barrier, patients with untreated brain metastases that did not need immediate local intervention were included in this study. The triplet regimen demonstrated an improved PFS of 7.8 months versus 5.6 months and a 1 year PFS of 33.1% versus 12.3%. The 2-year OS rate also increased from 26.6% to 44.9%. Even more significantly, in the patients with brain metastases, the 1 year PFS was 24.9% in the tucatinib combination group compared to 0% in the placebo combination group (Murthy et al., 2020). Given the propensity of HER2 + MBC to spread to the brain, this trial has led to rapid acceptance of this new regimen in clinical practice.

Although most HER2 targeted agents were first approved in the metastatic setting, many have since been studied and accepted as therapy in the (neo)adjuvant setting for early stage HER2 + breast cancer. In 2005, several clinical trials demonstrated the benefit of combining trastuzumab to cytotoxic chemotherapy in the adjuvant setting. There were improvements in both disease-free survival (with hazard ratios ranging from 0.64 to 0.76 when comparing trastuzumab-containing regimens to nontrastuzumab containing regimens) and overall survival (with hazard ratios as low as 0.63) (Cameron et al., 2017; Loibl & Gianni, 2017; Perez et al., 2014; Slamon et al., 2011; Wolff et al., 2014). Trastuzumab was FDA approved for use in the adjuvant setting in 2006. Pertuzumab was FDA approved in 2013 for use in the neoadjuvant setting in combination with trastuzumab and chemotherapy for locally advanced breast cancers based on the increase in pathologic complete response found in the NeoSphere and TRYPHAENA trials (Gianni et al., 2012; Loibl & Gianni, 2017). Pertuzumab also gained FDA approval in 2017 for use in the adjuvant setting with trastuzumab and chemotherapy based on a small improvement in disease-free survival (1.6% over 3.5 years) in the APHINITY trial (Von Minckwitz et al., 2017). One year of neratinib therapy can be used as extended anti-HER2 therapy after finishing a year of a standard trastuzumab-containing regimen based on the ExteNET trial, which demonstrated an improved 5-year disease-free survival (Martin et al., 2017). As most patients with locally advanced HER2 + disease will undergo neoadjuvant chemotherapy to allow for an in vivo assessment of the tumor's response to chemotherapy, research has evaluated the impact of changing anti-HER2-based therapy in the postoperative setting for those patients who do not achieve a pathologic complete response. In the KATHERINE trial, the use of T-DM1 to complete a year of anti-HER2 therapy following surgery as opposed to continuing trastuzumab led to an 11% reduction in invasive disease recurrence at 3 years and is now the standard approach after neoadjuvant therapy (Von Minckwitz et al., 2019).

Poly(ADP-ribose) polymerase inhibitors

Testing for *BRCA1* and *BRCA2* mutations in the metastatic setting is not only beneficial to help determine causality of the breast cancer and family cancer risks, but also to leverage therapeutic options available to those who are germline mutation carriers. Poly(ADP-ribose) polymerases (PARPs) are a family of enzymes that help repair single-strand DNA breaks. *BRCA1/2* mutations lead to an impairment in the ability to repair double-strand DNA breaks. Cells with these mutations are therefore particularly sensitive to PARP inhibitors as they will acquire DNA errors that they are unable to repair, thus leading to cell cycle arrest and apoptosis. Both olaparib and talazoparib are PARP inhibitors that were shown to improve PFS (by roughly 3 months each) in MBC as compared to standard single-agent chemotherapy in *BRCA1/2* germline carriers (Litton et al., 2018; Robson et al., 2017). Given the ease of use and the tolerability of PARP inhibitors as compared to chemotherapy, it is imperative to test all patients with MBC to determine if they are eligible to receive either of these agents. More recently, the OlympiA trial also demonstrated the role of PARP inhibitors in the early stage setting and thus expands the number of patients who should be tested for BRCA1 and 2 mutations. For patients with high risk HER2 negative early breast cancer and germline BRCA1 or BRCA mutations, adjuvant olaparib given for 1 year improved invasive disease free survival by 8.8% (HR 0.58) and improved 3 year DFS by 7.1% compared to placebo (87.5% vs 80.4% respectively) (Andrew et al., 2021).

Immunotherapy

Immune checkpoint antagonists specific for immunosuppressive receptors such as CTLA-4 (i.e., ipilimumab), PD-1 (i.e., pembrolizumab, nivolumab), and PD-L1 (i.e., atezolizumab, durvalumab, avelumab) improve the cytotoxic and proliferative capacity of tumor infiltrating lymphocytes (TILs). Immune checkpoints inhibitors (ICIs) have revolutionized cancer therapy, inducing durable objective responses that sometimes translate into an OS benefit in multiple cancer types (El-Khoueiry et al., 2017; Garon et al., 2015; Kim et al., 2018; Motzer et al., 2015; Ribas & Wolchok, 2018; Rosenberg et al., 2016; Topalian et al., 2012; Wolchok et al., 2017) and represent a promising treatment for breast cancer, particularly in triple negative breast cancer (TNBC). Several key characteristics make TNBC more likely to respond to immunotherapy than other breast cancer subtypes. First, TNBC have more TILs (Denkert et al., 2018), which correlate with better response to ICIs (Topalian et al., 2012). In early stage TNBC, high levels of TILs are associated with improved survival, reduced risk of recurrence, and increased likelihood of response to neoadjuvant chemotherapy (Denkert et al., 2018; Loi et al., 2014; Salgado et al., 2015). Second, TNBC has higher levels of PD-L1 expression on both tumor and immune cells (Gatalica et al., 2014; Mittendorf et al., 2014), which also correlate with response to anti-PD-1 therapies (Topalian et al., 2012). Finally, TNBC has a higher frequency of immunogenic mutations, which is associated with improved survival following immunotherapy.

Although response rates to ICIs are higher in TNBC than other subtypes of breast cancers, the single-agent efficacy is still low, with response rates ranging from ∼5% in unselected patients to ∼23% in treatment naïve PDL-1-positive patients (Adams, Loi, et al., 2019; Adams, Schmid, et al., 2019; Dirix et al., 2018; Emens et al., 2019; Nanda et al., 2016). Notably, in the 84 treatment naïve PDL-1-positive patients enrolled in the cohort B of the Phase II KEYNOTE-086 study, the ORR to pembrolizumab was 21.4%, suggesting that ICIs have greater efficacy in the first-line metastatic setting (Adams, Diamond, et al., 2019). Combination regimens of PD-1/PD-L1 inhibitors plus chemotherapy were more successful than single agents in mTNBC. These findings are not surprising as chemotherapy is known to have various immunomodulatory effects and restoration of host immune surveillance may be part of the benefit seen with conventional therapies (Wang, Fletcher, Yu, & Zhang, 2018; Zitvogel, Galluzzi, & Smyth, Kroemer, 2013). In particular, taxanes are considered beneficial therapeutic partners for ICIs as they promote Toll-like receptor activity and dendritic activation (Emens & Middleton, 2015). In the Phase 1b/2 ENHANCE-1 trial, the combination of pembrolizumab with eribulin yielded an ORR of 26.4% (Tolaney et al., 2018). Similarly, the smaller Phase Ib trial of atezolizumab and nab-paclitaxel (NCT01375842) showed a promising ORR of 39.4% in 33 patients also treated with 0−2 prior lines (Adams, Diamond, et al., 2019). Finally, the randomized IMpassion 130 Phase III trial evaluated efficacy of nab-paclitaxel plus atezolizumab versus placebo in first-line mTNBC (Schmid et al., 2018). In the intention-to-treat analysis, the median PFS was 7.2 months with atezolizumab plus nab-paclitaxel, as compared with 5.5 months with placebo plus nab-paclitaxel [HR for progression or death, 0.80; 95% confidence interval (CI), 0.69−0.92; $P = .002$]; among patients with PD-L1−positive tumors the median PFS was 7.5 and 5.0 months, respectively (HR, 0.62; 95% CI, 0.49−0.78; $P < .001$). This led to the approval of atezolizumab and nab-paclitaxel only for PDL-1-positive tumors. The subsequent KEYNOTE-355 trial (Cortes et al., 2020) showed that first-line chemotherapy with pembrolizumab significantly improved PFS (9.7 months vs 5.6 months; HR, 0.65; 95% CI 0.49−0.0.86; *P* = .0012) compared to chemotherapy alone in patients with mTNBC expressing PD-L1. In 2021, pembrolizumab was approved for use in high risk TNBC in the neo/adjuvant setting. The addition of pembrolizumab to standard chemotherapy improved pathologic complete response by 13.6% to 64.8% and after a median follow up of 15.5 months there was an improvement of event free survival by 4.4% over the placebo arm (Schmid et al., 2020).

Conclusion

Currently, there are several ongoing clinical trials with novel targeted therapies for breast cancer. Promising therapeutic strategies include immunotherapy combinations, signal transduction pathway inhibition, and inhibition of resistance pathways. Further work is needed to identify patients who may benefit from precision therapies based on genomic alterations.

References

Adams, S., Diamond, J. R., Hamilton, E., et al. (2019). Atezolizumab plus nab-paclitaxel in the treatment of metastatic triple-negative breast cancer with 2-year survival follow-up: A Phase 1b clinical trial. *JAMA Oncology, 5*(3), 334−342.

Adams, S., Loi, S., Toppmeyer, D., et al. (2019). Pembrolizumab monotherapy for previously untreated, PD-L1-positive, metastatic triple-negative breast cancer: Cohort B of the phase II KEYNOTE-086 study. *Annals of Oncology, 30*(3), 405−411.

Adams, S., Schmid, P., Rugo, H. S., et al. (2019). Pembrolizumab monotherapy for previously treated metastatic triple-negative breast cancer: Cohort A of the phase II KEYNOTE-086 study. *Annals of Oncology, 30*(3), 397−404.

Baselga, J., Cortes, J., Kim, S.-B., et al. (2012). Pertuzumab plus trastuzumab plus docetaxel for metastatic breast cancer. *The New England Journal of Medicine, 366*(2), 109−119.

Cameron, D., Piccart-Gebhart, M. J., Gelber, R. D., et al. (2017). 11 years' follow-up of trastuzumab after adjuvant chemotherapy in HER2-positive early breast cancer: Final analysis of the HERceptin Adjuvant (HERA) trial. *Lancet, 389*, 1195−1205.

Cesca, M. G., Vian, L., Cristovao-Ferreira, S., et al. (2020). HER2-positive advanced breast cancer treatment in 2020. *Cancer Treatment Reviews, 88*. Available from https://doi.org/10.1016/j.ctrv.2020.102033.

Ciruelos, E., Pascual, T., Vozmediano, M. L. A., et al. (2014). The therapeutic role of fulvestrant in the management of patients with hormone receptor-positive breast cancer. *Breast,* 201−208.

Cortes, J., Cescon, D. W., Rugo, H. S., et al. (2020). KEYNOTE-355: Randomized, double-blind, phase III study of pembrolizumab + chemotherapy vs placebo + chemotherapy for previously untreated locally recurrent inoperable or metastatic triple-negative breast cancer. *Journal of Clinical Oncology, 38*(15), 1000.

Dean, J. L., Thangavel, C., McClendon, A. K., et al. (2010). Therapeutic CDK4/6 inhibition in breast cancer: key mechanisms of response and failure. *Oncogene, 29*, 4018−4032.

Denkert, C., von Minckwitz, G., Darb-Esfahani, S., et al. (2018). Tumour-infiltrating lymphocytes and prognosis in different subtypes of breast cancer: A pooled analysis of 3771 patients treated with neoadjuvant therapy. *The Lancet Oncology, 19*(1), 40−50.

DeSantis, C. E., Ma, J., Gaudet, M. M., et al. (2019). Breast cancer statistics, 2019. *CA: A Cancer Journal for Clinicians, 69*(6), 438−451.

Dirix, L. Y., Takacs, I., Jerusalem, G., et al. (2018). Avelumab, an anti-PD-L1 antibody, in patients with locally advanced or metastatic breast cancer: A phase 1b JAVELIN solid tumor study. *Breast Cancer Research and Treatment, 167*(3), 671−686.

El-Khoueiry, A. B., Sangro, B., Yau, T., et al. (2017). Nivolumab in patients with advanced hepatocellular carcinoma (CheckMate 040): An open-label, non-comparative, phase 1/2 dose escalation and expansion trial. *Lancet, 389*(10088), 2492−2502.

Emens, L. A., Cruz, C., Eder, J. P., et al. (2019). Long-term clinical outcomes and biomarker analyses of atezolizumab therapy for patients with metastatic triple-negative breast cancer: A Phase 1 study. *JAMA Oncology, 5*(1), 74−82.

Emens, L. A., & Middleton, G. (2015). The interplay of immunotherapy and chemotherapy: Harnessing potential synergies. *Cancer Immunology Research, 3*(5), 436−443.

Fabrice André, M. D., Eva Ciruelos, M. D., Gabor Rubovszky, M. D., et al. (2019). lpelisib for PIK3CA-mutated, hormone receptor−positive advanced breast cancer. *New England Journal of Medicine, 380*, 1929−1940.

Gao, J. J., Cheng, J., Bloomquist, E., et al. (2020). CDK4/6 inhibitor treatment for patients with hormone receptor-positive, HER2-negative, advanced or metastatic breast cancer: a US Food and Drug Administration pooled analysis. *Oncology, 21*(2), 250−260.

Garon, E. B., Rizvi, N. A., Hui, R., et al. (2015). Pembrolizumab for the treatment of non-small-cell lung cancer. *The New England Journal of Medicine, 372*(21), 2018−2028.

Gatalica, Z., Snyder, C., Maney, T., et al. (2014). Programmed cell death 1 (PD-1) and its ligand (PD-L1) in common cancers and their correlation with molecular cancer type. *Cancer Epidemiology, Biomarkers & Prevention, 23*(12), 2965−2970.

Geyer, C. E., Forster, J., Lindquist, D., et al. (2006). Lapatinib plus capecitabine for HER2-positive advanced breast cancer. *The New England Journal of Medicine, 355*, 2733−2743.

Gianni, L., Pienkowski, T., Im, Y.-H., et al. (2012). Efficacy and safety of neoadjuvant pertuzumab and trastuzumab in women with locally advanced, inflammatory or early HER2-positive breast cancer (NeoSphere): A randomized multicentre, open-label, phase 2 trial. *The Lancet Oncology, 13*, 25−32.

Giuliano, A. E., Connolly, J. L., Edge, S. B., et al. (2017). Breast cancer − Major changes in the American Joint Committee on cancer eighth edition cancer staging manual. *CA: A Cancer Journal for Clinicians, 67*(4), 290-303.

Kim, S. T., Cristescu, R., Bass, A. J., et al. (2018). Comprehensive molecular characterization of clinical responses to PD-1 inhibition in metastatic gastric cancer. *Nature Medicine, 24*(9), 1449−1458.

Krop, I. E., Kim, S.-B., Martin, A. G., et al. (2017). Trastuzumab emtansine vs treatment of physician's choice in patients with previously treated HER2-positive metastatic breast cancer (TH3RESA): Final overall survival results from a randomized open-label phase 3 trial. *The Lancet Oncology, 18*(6), 743−754.

Litton, J. K., Rugo, H. S., Ettl, J., et al. (2018). Talazoparib in patients with advanced breast cancer and a germline BRCA mutation. *The New England Journal of Medicine, 379*(8), 753−763.

Loi, S., Michiels, S., Salgado, R., et al. (2014). Tumor infiltrating lymphocytes are prognostic in triple negative breast cancer and predictive for trastuzumab benefit in early breast cancer: Results from the FinHER trial. *Annals of Oncology, 25*(8), 1544−1550.

Loibl, S., & Gianni, L. (2017). Her2-positive breast cancer. *Lancet, 389*, 2415−2429.

Martin, M., Holmes, F. A., Ejlertsen, B., et al. (2017). Neratinib after trastuzumab-based adjuvant therapy in HER2-positive breast cancer (ExteNET): 5-year analysis of a randomized, double-blind, placebo-controlled, phase 3 trial. *The Lancet Oncology, 18*, 1688−1700.

Mittendorf, E. A., Philips, A. V., Meric-Bernstam, F., et al. (2014). PD-L1 expression in triple-negative breast cancer. *Cancer Immunology Research, 2*(4), 361−370.

Modi, S., Saura, C., Yamashita, T., et al. (2020). Trastuzumab deruxtecan in previously treated HER2-positive breast cancer. *The New England Journal of Medicine, 382*(7), 610−621.

Motzer, R. J., Rini, B. I., McDermott, D. F., et al. (2015). Nivolumab for metastatic renal cell carcinoma: Results of a randomized Phase II trial. *Journal of Clinical Oncology, 33*(13), 1430−1437.

Murthy, R. K., Loi, S., Okines, A., et al. (2020). Tucatinib, trastuzumab, and capecitabine for HER-2 positive metastatic breast cancer. *The New England Journal of Medicine, 382*(7), 597−609.

Nanda, R., Chow, L. Q., Dees, E. C., et al. (2016). Pembrolizumab in patients with advanced triple-negative breast cancer: Phase Ib KEYNOTE-012 study. *Journal of Clinical Oncology, 34*(21), 2460−2467.

Perez, E. A., Romond, E. H., Suman, V. J., et al. (2014). Trastuzumab plus adjuvant chemotherapy for human epidural growth factor receptor 2-positive breast cancer: Planned joint analysis of overall survival from NSABP B-31 and NCCTG N9831. *Journal of Clinical Oncology, 32*, 3744−3753.

Ribas, A., & Wolchok, J. D. (2018). Cancer immunotherapy using checkpoint blockade. *Science (New York, N.Y.), 359*(6382), 1350−1355.

Robson, M., Im, S.-A., Senkus, E., et al. (2017). Olaparib for metastatic breast cancer in patients with a germline BRCA mutation. *The New England Journal of Medicine, 377*(6), 523−533.

Rosenberg, J. E., Hoffman-Censits, J., Powles, T., et al. (2016). Atezolizumab in patients with locally advanced and metastatic urothelial carcinoma who have progressed following treatment with platinum-based chemotherapy: A single-arm, multicentre, phase 2 trial. *Lancet, 387*(10031), 1909−1920.

Salgado, R., Denkert, C., Demaria, S., et al. (2015). The evaluation of tumor-infiltrating lymphocytes (TILs) in breast cancer: Recommendations by an International TILs Working Group 2014. *Annals of Oncology, 26*(2), 259−271.

Saura, C., Oliveira, M., Feng, Y.-H., et al. (2020). Neratinib plus capecitabine vs lapatinib plus capecitabine in HER2-positive metastatic breast cancer previously treated with ≥ 2 HER2-directed regimens: Phase III NALA trial. *Journal of Clinical Oncology, 38*, 3138−3149.

Schiff, R., Massarweh, S. A., Shoum, J., et al. (2004). Cross-talk between estrogen receptor and growth factor pathways as a molecular target for overcoming endocrine resistance. *Clinical Cancer Research, 10*(1 Pt 2), 331S−336S.

Schmid, P., Adams, S., Rugo, H. S., et al. (2018). Atezolizumab and Nab-paclitaxel in advanced triple-negative breast cancer. *The New England Journal of Medicine, 379*(22), 2108−2121.

Shimoi, T., Hamada, A., Marifu Yamagishi, M., et al. (2018). PIK3CA mutation profiling in patients with breast cancer, using a highly sensitive detection system. *Cancer Science, 109* (8), 2558−2566.

Slamon, D., Eiermann, W., Robert, N., et al. (2011). Adjuvant trastuzumab in HER2-positive breast cancer. *The New England Journal of Medicine, 365*(14), 1273−1283.

Slamon, D. J., Leyland-Jones, B., Shak, S., et al. (2001). Use of chemotherapy plus a monoclonal antibody against HER2 for metastatic breast cancer that overexpresses HER2. *The New England Journal of Medicine*, *344*(11), 783−792.

Swain, S. M., Miles, D., Kim, S.-B., et al. (2020). Pertuzumab, trastuzumab, and docetaxel for HER2-positive metastatic breast cancer (CLEOPATRA): End-of-study results from a double-blind, randomized, placebo-controlled, phase 3 study. *The Lancet Oncology*, *21*, 519−530.

Tolaney, S. M., Kalinsky, K., Kaklamani, V., et al. (2018). Phase 1b/2 study to evaluate eribulin mesylate in combination with pembrolizumab in patients with metastatic triple-negative breast cancer [abstract]. In: *Proceedings of the 2017 San Antonio breast cancer symposium*, December 5−9, 2017, San Antonio, TX. Philadelphia, PA: AACR; *Cancer Research* 2018;*78*(4 Suppl.):Abstract nr PD6-13.

Topalian, S. L., Hodi, F. S., Brahmer, J. R., et al. (2012). Safety, activity, and immune correlates of anti-PD-1 antibody in cancer. *The New England Journal of Medicine*, *366*(26), 2443−2454.

Turner, N. C., Dennis, J., Slamon, D. J., Ro, J., et al. (2018). Overall Survival with Palbociclib and Fulvestrant in Advanced Breast Cancer. *New England Journal of Medicine*, *379*, 1926−1936.

Veeraraghavan, J., De Angelis, C., Reis-Filho, J. S., et al. (2017). De-escalation of treatment in HER2 positive breast cancer: Determinants of response and mechanisms of resistance. *The Breast*, *34*, S19−S26.

Verma, S., Miles, D., Gianni, L., et al. (2012). Trastuzumab emtansine for HER2 positive advanced breast cancer. *The New England Journal of Medicine*, *367*(19), 1783−1791.

Von Minckwitz, G., Huang, C.-S., Mano, M. S., et al. (2019). Trastuzumab emtansine for residual invasive HER2-positive breast cancer. *The New England Journal of Medicine*, *380*(7), 617−628.

Von Minckwitz, G., Procter, M., de Azambuja, E., et al. (2017). Adjuvant pertuzumab and trastuzumab in early HER2 positive breast cancer. *The New England Journal of Medicine*, *377* (2), 122−131.

Wang, Y. J., Fletcher, R., Yu, J., & Zhang, L. (2018). Immunogenic effects of chemotherapy-induced tumor cell death. *Genes & Diseases.*, *5*(3), 194−203.

Wolchok, J. D., Chiarion-Sileni, V., Gonzalez, R., et al. (2017). Overall survival with combined nivolumab and ipilimumab in advanced melanoma [published correction appears in N Engl J Med. 2018 Nov 29;379(22):2185]. *The New England Journal of Medicine*, *377*(14), 1345−1356.

Wolff, A. C., Domchek, S. M., Davidson, N. E., et al. (2014). Cancer of the Breast. In J. E. Neiderhuber, et al. (Eds.), *Abeloff's clinical oncology* (pp. 1630−1692). Philadelphia, PA: Elsevier Saunders.

Wolff, A. C., Hammond, M. E., Allison, K. H., et al. (2018). Human epidermal growth factor receptor 2 testing in breast cancer: American Society of Clinical Oncology/College of American Pathologists clinical practice guideline focused update. *Archives of Pathology & Laboratory Medicine*, *142*(11), 1364−1382.

Yardley, D. A., Noguchi, S., Pritchard, K. I., et al. (2013). Everolimus plus exemestane in postmenopausal patients with HR(+) breast cancer: BOLERO-2 final progression-free survival analysis. *Advances in Therapy*, *30*(10), 870−874.

Zitvogel, L., Galluzzi, L., Smyth, M. J., & Kroemer, G. (2013). Mechanism of action of conventional and targeted anticancer therapies: Reinstating immunosurveillance. *Immunity*, *39*(1), 74−88.

Chapter 10

Epithelial ovarian cancer

Tanja Pejovic[1], Kunle Odunsi[2] and Matthew L. Anderson[3]

[1]Department of Obstetrics and Gynecology, Oregon Health Sciences University, Portland, OR, United States, [2]Department of Gynecologic Oncology, Roswell Park Cancer Center, Buffalo, NY, United States, [3]Department of Obstetrics and Gynecology, University of South Florida Morsani College of Medicine and H. Lee Moffitt Cancer Center and Research Institute, Tampa, FL, United States

Inherited syndromes

Some of the earliest insights into the genetic basis of epithelial ovarian cancer (EOC) were generated by studying familial syndromes of breast and ovarian cancer (Black & Solomon, 1993). These efforts initially identified two genes, *BRCA1* and *BRCA2*, associated with significantly increased risk of developing a high-grade serous ovarian cancer (HGSOC). The earliest aspects of this work pinpointed specific *BRCA1* mutations, 185delAG and 5382insC, which are found in 1% and 0.1% of Ashkenazi Jewish women, respectively.

Since this initial work, hundreds of other inactivating loss-of-function, nonsense, or frameshift *BRCA1* mutations have been identified that are associated with an enhanced susceptibility to developing high-grade serous carcinoma (HGSC). More recently, large genomic rearrangements have also been found to contribute to the spectrum of genomic events leading to *BRCA1* loss, accounting for as many as 8% of ovarian cancers where BRCA1 expression has been lost (Puget et al., 1999). This latter discovery is noteworthy, as these rearrangements were not detected and/or reported by the earliest versions of clinical tests designed to identify *BRCA1/2* mutation carriers (Mazoyer, 2005).

Functionally, *BRCA1* plays an important role in regulating the activity of the key tumor suppressor TP53. In *BRCA* wt cells, loss of *BRCA1* function allows DNA damage to accumulate by impairing the activity of the key tumor suppressor *TP53*. Regulation of the cell cycle at its G1-S checkpoint plays a key role in ensuring genomic integrity. Mutations and/or other genomic events that lead to *TP53* inactivation or dysfunction can be found in more than 95% of EOCs (The Cancer Genome Atlas Research Network,

Genomic and Precision Medicine. DOI: https://doi.org/10.1016/B978-0-12-800684-9.00011-3
173

2011). Thus the accumulation of genomic damage subsequent to loss of *BRCA1/2* likely plays a key role in the early genetic events predisposing women to HGSC. Evidence also continues to build that mutations in *BRCA1* likely contribute to the development of an EOC independent of its ability to interact with *TP53*. These include its ability to specifically regulate X chromosome gene expression mediated by an association of *Xist* with the inactive X chromosome (Ganesan et al., 2004). Consistent with this observation, site-specific dysregulation of X-linked gene expression in *BRCA1*-associated epithelial ovarian malignancies has been described.

Clinically, a greater proportion of familial ovarian cancers (76%−92%) are associated with mutations at the *BRCA1* locus, located on 17q21 (The Cancer Genome Atlas Research Network, 2011). It is important to note that disease penetrance and timing of ovarian cancer susceptibility differ significantly between women who carry high-risk *BRCA1* and *BRCA2* mutations. The cumulative lifetime risk of developing HGSC in carriers of a high-risk mutation in *BRCA1* has been found to be as high as 59%, while significantly fewer women who carry a high-risk, inactivating *BRCA2* mutation will develop EOC. In addition, multiple population-based studies have shown that *BRCA2* mutation carriers develop disease 10 or more years later than women who carry a high-risk *BRCA1* mutation. These latter observations have important implications for the timing of prophylactic, risk-reducing salpingo-oophorectomy, potentially allowing many women to achieve the maximal cardio-protective benefit associated with preserving ovarian function until age 50. Recent work has also focused on more precisely timing risk-reducing salpingo-oophorectomy based on detailed analyses of risk associated with distinct patterns of mutations (Solsky, Chen, & Rebbeck, 2020).

A number of other familial syndromes are also known to be associated with increased heritable risks of both epithelial and non-EOC. For example, germline mutations leading to the lost expression of specific enzymes involved in DNA mismatch repair (MMR), such as *MSH2*, *MLH1*, *MSH6*, *PMS2*, and *EPCAM*, lead to Lynch syndrome, which not only predisposes women to developing HGSC but also carcinomas of the uterus, colon, and pancreas. In general, the cumulative lifetime risk of developing an ovarian cancer for women known to carry a pathogenic mutation in one of the Lynch syndrome genes ranges from 6% to 8% (Daniels & Lu, 2016). Pathogenic variants in the noncoding RNA-processing enzyme *DICER1* have been recently shown to predispose young women not only to a type of non-EOC known as Sertoli-Leydig cell tumors, but also pleuropulmonary blastoma, pulmonary cysts, multinodular goiter, and thyroid adenomas/cancers (Schultz et al., 1993). The majority of these tumors occur in women less than 40 years of age. Other syndromes that confer an increased risk of developing ovarian cancer include Li-Fraumeni syndrome (pathogenic *TP53* mutations), Cowden syndrome (*PTEN*), Peutz-Jeghers syndrome (*STK11*), and neurofibromatosis type 1 syndrome (*NF1*) (Eng & Ponder, 1993; Piombino et al., 2020).

Over the past 5 years, a number of other important advances in our understanding of the genetic basis of EOC have occurred (Pietragalla, Arcieri, Marchetti, Scambia, & Fagotti, 2020). First, it is now clear that somatic events associated with a loss of BRCA1/2 function can be found in as many as 30% of EOCs, a proportion significantly greater than the incidence at which these mutations can be attributed to a heritable germline mutation. Of note, these mutations are often associated with long disease-free intervals and favorable survival (Norquist et al., 2018). As discussed below, this finding has important implications for the clinical management of EOC and has led to increasing use of clinical platforms for somatic mutational profiling in patients. Second, it is also now well understood that diverse genetic events can also predispose women to developing breast and ovarian cancer. These genes include *BRIP1*, *PALB2*, *RAD51C*, and *RAD51D* (Dansonka-Mieszkowska et al., 2010; Suszynska, Ratajska & Kozlowski, 2020). The association between mutations and other gene products known to increase risk of breast cancer, such as *ATM*, *BARD1*, or *CDH*, is less clear or, as in the case of *CHEK2*, has been assessed and found to be nonexistent (Alenezi, Fierheller, Recio & Tonin, 2020; Baysal et al., 2004; Thorstenson et al., 2003). The reasons for these distinct susceptibilities are presently unclear.

One of the notable themes that emerges from these data is that many of the gene products that predispose women to developing ovarian cancer play a key role in the repair of double-stranded DNA damage by homologous recombination (HR) (Walsh et al., 2011). Mutations in more than one of these gene products have been shown to be heritable, although not all heritable mutations contributing to homologous recombination deficiency (HRD) equally predispose women to develop breast and ovarian cancer. Lastly, it remains important to emphasize that any understanding of the role of *BRCA1* in ovarian cancer is further complicated by reports of women with high-risk mutations in *BRCA1* who fail to develop ovarian cancer. These observations speak clearly to the role of genetic modifiers in determining whether *BRCA1/2* mutations ultimately lead to malignancy.

Histologic origins of ovarian cancer

The past decade has witnessed an important transformation of our insight into the histologic origins of EOC. Historically, nearly all EOCs were thought to arise from the relatively nondescript, cuboidal epithelium covering the ovarian surface (Levanon, Crum & Drapkin, 2008). Developmentally, the ovarian surface epithelium (OSE) derives from the celomic epithelium that also gives rise to the peritoneal mesothelium and oviductal epithelium (Auersperg, Wong, Choi, Kang & Leung, 2001).

At baseline, the OSE appears generally stable, uniform, and quiescent. Significant effort has been invested in an attempt to understand how this epithelium might undergo malignant transformation and reconcile the diverse histologic patterns exhibited by ovarian cancers with its putative origins in the relatively simple monolayer structure of the OSE. It is known, for example, that OSE is competent to undergo proliferation in vivo (Wright, Pejovic, Fanton, & Stouffer, 2008). Ovulation and/or possibly other types of trauma are believed to disrupt the ovarian surface in a manner that triggers a repair response and leads to long-lasting alterations in the molecular phenotype of regions within the EOC as evidenced by altered expression of cadherins. In turn, these alterations have been hypothesized to subsequently contribute to an enhanced proliferative capacity of these cells. However, in contrast to other organs, such as the colon, it has not been possible to identify precursor lesions in the OSE that accumulate genetic defects that ultimately result in malignancy. In large part, this failure has been attributed to the fact that healthy ovaries are only rarely biopsied or examined clinically.

More recently, however, a number of investigators have described histologic findings consistent with a preinvasive lesion in ovaries removed from women who eventually developed peritoneal carcinomas, fallopian tubes removed from women at high risk for developing ovarian cancer, and in areas of ovarian epithelium adjacent to early-stage ovarian cancers (Schlosshauer et al., 2003). Notably, regions of epithelial irregularity have been identified in the distal tubal epithelia, which express findings suggestive of both TP53 dysfunction and increased cellular proliferation. These lesions, known as serous tubal intraepithelial carcinomas, were first reported in 8%— 12% of fallopian tubes removed from *BRCA1/2* mutation carriers undergoing risk-reducing salpingo-oophorectomy (Crum et al., 2007). Support for the hypothesis that these lesions are premalignant is strengthened by observations that regions of epithelial irregularity have also been found in the tubal epithelia adjacent to clearly invasive serous ovarian cancers and that a clear histologic progression from preinvasive to invasive lesions has been observed (Carlson et al., 2008). These observations have profound implications not only for understanding molecular drivers but also for developing specific interventions designed to prevent these diseases. For example, the need to carefully surveil women for these lesions has led to the use of histologic protocols that require serial evaluation of the entire fallopian tube when women undergo risk-reducing salpingectomy (the SEE-FIM protocol) (Conner et al., 2014).

Genetic hallmarks of ovarian cancer

Genomic instability

Genomic instability—manifested as a cell's ability to tolerate DNA damage—is a hallmark of EOCs. Tolerance to DNA damage can be achieved by alterations in any of the six major DNA repair pathways: base excision repair (BER), MMR, nucleotide excision repair (NER), HR, non-HR, and translesion DNA synthesis (Fig. 10.1). HGSOC, the most prevalent histologic subtype of EOC, has defects in the HR pathway in 50% of cases (The Cancer Genome Atlas Research Network, 2011). Mutations in either *BRCA1/2* genes or in other genes in the HR repair pathway causes HRD phenotype, also known as BRCAness (Tan et al., 2008). HRD carcinomas exhibit synthetic lethality with PARP inhibitors. PARP inhibitors act by exploiting a tumor cell's defect in HR, a type of DNA repair. This is because of the following PARP inhibition; cells require HR to repair common types of DNA damage. While normal cells can use HR for repair of this damage and survive, certain types of tumors (e.g., those with *BRCA1* or *BRCA2* defects) have lost the ability to repair by HR and will die. Consequently, the United States Food and Drug Administration (FDA) has approved four PARP inhibitor drugs as active or maintenance therapy for women with HGSOC-exhibiting HRD (Mateo et al., 2019). These agents are highly effective at not only inducing a regression of platinum-sensitive disease both as monotherapy and in combination with other agents (Ray-Coquard et al., 2019), they have also been shown to be highly effective at preventing ovarian cancer recurrence following first and second line therapy (Moore et al., 2018). However, accumulated evidence suggests that not all HRD carcinomas respond equally to the treatment and some non-HRD tumors also respond to PARP inhibitors (Gourley et al., 2019). The main problem in our ability to identify all cases with HRD

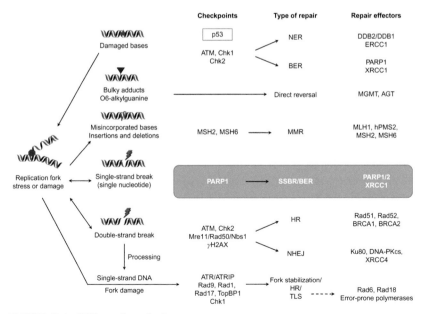

FIGURE 10.1   DNA repair mechanisms.

is the lack of a functional test of HRD. The hallmark of HRD is the lack of RAD51 foci formation; however, commercial testing for RAD51 foci formation is not available. Therefore surrogate markers for HRD are being utilized via several panel tests relying on NGS analysis, LOH, and in some cases on analysis epigenetic silencing of HRD genes. Clinically, sensitivity to platinum drugs has been used as a surrogate for response to PARP inhibitor. However, resistance to PARP inhibitors is thought to develop via inactivation of DNA repair proteins leading to restoration of HR, demethylation of BRCA1 promoter, loss of PARP1 protein itself or secondary mutations in *BRCA* genes that then restore HR (D'Andrea, 2018). Despite these shortcomings, treatment with PARP inhibitors has significantly improved survival in HGSOC and shifted the treatment paradigm from palliation to cure.

Defects in DNA repair pathways

Similar to the BRCA pathway, disruptions of other DNA repair pathways are frequently observed in EOC. These disruptions account, at least in part, for the specific drug sensitivity of EOC. Recent studies indicate that defects in translesion DNA synthesis in ovarian cancer are a consequence of elevation in activity of POLB, an error-prone polymerase. Inhibition of POLB in these cells results in resensitization to cisplatin (Boudsocq et al., 2005). Although inactivation of one DNA repair pathway may confer survival advantage to tumors, cancer cells may rely more on other repair pathways. Therefore inactivation of the second pathway would be deleterious for the cells, causing synthetic lethality. An RNA interference screen identified the ATM pathway to be synthetically lethal with analogous mutations leading to Fanconi anemia (Kennedy et al., 2007).

Identification of common ovarian cancer susceptibility variants may have clinical implications in the future for identifying patients at greatest risk of the disease. In this regard, several genome-wide association studies have been performed in ovarian cancer. The most striking of these was a recent study by the Ovarian Cancer Association Consortium designed to identify common ovarian cancer susceptibility alleles (Song et al., 2009). A total of 507,094 single nucleotide polymorphisms (SNPs) were genotyped in 1817 cases and 2353 controls from the United Kingdom; 22,790 top-ranked SNPs were also genotyped in 4274 cases and 4809 controls of European ancestry from Europe, the United States, and Australia. Twelve SNPs associated with disease risk were identified at 9p22. The most significant SNP (rs3814113; $P = 2.5 \times 10^{-17}$) was genotyped in a further 2670 ovarian cancer cases and 4668 controls, confirming its association [combined data odds ratio (OR) = 0.82, 95% confidence interval (CI) 0.79−0.86, Pi, end = 5.1×10^{-19}]. The association was strongest for serous ovarian cancers (OR = 0.77, 95% CI: 0.73−0.81; Pi, end = 4.1×10^{-21}).

Transcriptional profiling of ovarian cancer histologic subtypes

Patterns of gene expression in EOC have now been extensively profiled under diverse clinical contexts. Several gene expression studies using cDNA microarrays have been performed in ovarian cancer. In addition, several studies have focused on alterations in DNA copy number (Mayr et al., 2006; Meinhold-Heerlein et al., 2005; Nakayama et al., 2007). Use of both next-generation sequencing and array-based technologies has repeatedly demonstrated not only that the different histological subtypes of ovarian carcinoma are distinguishable based on their overall genetic expression profiles but also that distinct molecularly defined subtypes of EOC can be defined among histologically similar cancers (The Cancer Genome Atlas Research Network, 2011). A common finding among several studies is the ability to distinguish between low-grade serous ovarian carcinoma and high-grade carcinoma based on gene expression profiles (Bonome et al., 2005; Gilks, Vanderhyden, Zhu, Van De Rijn & Longacre, 2005; Hough, Cho, Zonderman, Schwartz & Morin, 2001; Meinhold-Heerlein et al., 2005; Schwartz et al., 2002). A number of genes shown to be differentially expressed in EOC are known to be involved in many important cellular mechanisms, including cell cycle regulation, apoptosis, tumor invasion, and control of local immunity (Berchuck et al., 2004; Gilks et al., 2005; Landen, Birrer, & Sood, 2008).

Increased mutagenic signaling by receptor tyrosine kinases plays a major role in ovarian carcinogenesis. Overexpression of EGFR (ERBB1), ERBB2/ HER-2/neu, and c-FMS has been reported. One of the major downstream mediators of signaling initiated by these receptors is the phosphatidylinositol 3-kinase (PIK3K)-AKT pathway. Aberrations in this pathway, including increased AKT1 kinase activity, AKT2 and PIK3 amplification, and PIK3R1 mutations, may provide opportunities for therapeutic intervention. It has been reported that more than 75% of ovarian carcinomas are resistant to transforming growth factor-β (TGF-β), and the loss of TGF-β responsiveness may play an important role in the pathogenesis and/or progression of ovarian cancer (Hu et al., 2000). It has also been shown that expression of TGF-β the TGF-β receptors (TBR-11 and TBR-I), and the TGFβ-signaling component Smad2 are altered in ovarian cancer. Alterations in TBR-11 have been identified in 25% of ovarian carcinomas, whereas mutations in TBR-I were reported in 33% of such cancers (Chen et al., 2001). Proto-oncogene transformation might lead either to an overexpression of mitogenic molecules or an inactivation of those with inhibitory action, thus contributing to neoplastic transformation and development. The most important proto-oncogenes of the first group are c-FMS and HER-2/neu. The first of these encodes a transmembrane tyrosine kinase receptor that binds macrophage colony stimulating factor (MCSF). It is possible that c-FMS stimulates epithelial cell proliferation and induces a chemical attraction for macrophages, which, in turn, can produce mitogenic-stimulating factor. Elevated plasma concentrations of MCSF are present in the sera of 70% of patients with ovarian cancer (Van Haaften-Day et al., 2001). The second proto-oncogene, HER-2/neu, encodes another tyrosine kinase that is similar to the epidermal growth factor receptor (EGFR). Its action may consist of amplification of mitogenic action in target cells. This oncogene is overexpressed in 30%−35% of ovarian cancer and is associated with a poor prognosis (Coronado Martín, Fasero Laiz, García Santos, Ramírez Mena & Vidart Aragón, 2007).

Genetic events contributing to ovarian cancer metastasis

Metastasis involves the invasion of transformed epithelial cells across their basement membrane, through underlying stroma, and, lastly, into blood vessels and lymphatic channels, where they are able to subsequently disseminate to distant sites. Only a tiny fraction of cells released into the circulation by a tumor ever result in metastasis. A wide variety of genes and gene products implicated in the metastasis of other cancers have also been implicated in the metastasis of ovarian cancer. These include growth factor receptors such as EGFR, insulin-like growth factor receptors, and kinases such as *jak/stat*, focal adhesion kinase, PI3 kinase, and c-Met. Comparisons of primary and metastatic ovarian cancers by transcriptional profiling have failed to reveal significant differences in the expression of gene products likely related to the metastatic process.

Particular attention has recently focused on the role of lysophosphatidic acid (LPA) in promoting the metastasis of ovarian cancers. LPA is constitutively produced by mesothelial cells lining the peritoneal cavity. Its levels are increased in the ascites of women with both early- and late-stage ovarian cancers (Ren et al., 2006). In vitro, LPA promotes both the migration of these cells and their invasion across artificial barriers analogous to a basement membrane. At the molecular level, exogenous LPA enhances ovarian cancer invasiveness both by activating matrix metalloproteinase-2 via membrane-type-1-matrix metalloproteinase (MTI-MMP) and by down-regulating the expression of specific tissue inhibitors of metalloproteinases (TIMP-2 and -3) (Sengupta et al., 2007). Other observations are consistent with the idea that LPA promotes dissemination of ovarian cancer by loss of cell adhesion (Do et al., 2007). However, LPA has been shown to promote the invasiveness of ovarian cancers by additional mechanisms dependent on interleukin-8. The G12/13-RhoA and cyclooxygenase pathways have also been implicated in the LPA-induced migration of ovarian cancers. These mechanisms appear to be independent of the ability of LPA to induce changes in MMP2 expression.

Until recently, ovarian cancer metastasis has been almost exclusively studied as a process involving individual cells. However, multicellular clusters of self-adherent cells known as spheroids can be isolated from the ascitic fluid of women with ovarian cancer. Spheroids readily adhere to both extra-cellular matrix proteins such collagen IV and mesothelial cells in monolayer culture using beta-1 integrins. Once adherent, the cells contained in spher-oids disaggregate, allowing them to invade underlying mesothelial cells and create invasive foci (Burleson, Hansen, & Skubitz, 2004). These observations are consistent with the hypothesis that ovarian cancer spheroids play an important role in the metastatic potential of ovarian cancer. Recent evidence has shown that a loss of circulating gonadotropins results in a dose-dependent decrease in the expression of vascular endothelial growth factor (VEGF) in the outer, proliferating cells of ovarian cancer spheroids, indicat-ing that these cell clusters remain responsive to signals in their microenvi-ronment that may promote metastasis (Schiffenbauer et al., 1997).

The presence of spheroids in ascites may also help to explain the frequent persistence and frequent recurrence of ovarian cancer after treatment. Spheroids express high levels of cyclin-dependent kinase inhibitor 1B and P-glycoprotein that contribute, at least in part, to their relative resistance to the cytotoxic effects of paclitaxel when compared with ovarian cancer cells in monolayer culture. Ovarian cancer spheroids have also been shown to be relatively resistant to the cytotoxic effects of radiation (Griffon et al., 1995). These observations are consistent with in vitro studies that demonstrate that the signals generated by adhesion to specific components of the extracellular matrix, such as collagen IV, can modify the sensitivity of ovarian cancers to chemotherapy. However, the mechanisms by which the aggregation of malig-nant cells promotes or enhances cell survival remain unclear. It is also unclear how the aggregation of these malignant cells might promote or enhance the migration, attachment, or invasion of ovarian cancer cells.

A number of recent discoveries help shed insight into how interactions of spheroids with their local environment within the peritoneal cavity promote not only metastasis but also chemotherapy resistance (Mukherjee et al., 2020; Nieman et al., 2011). For example, investigators have found that direct transfer of fatty acids into ovarian cancer cells from adipocytes within the omentum plays a key role in promoting EOC metastasis. More recently, investigators have begun to understand pathways by which the omental microenvironment promotes intraperitoneal spread of ovarian cancer metasta-ses. The mechanisms by which ovarian cancer spheroids promote ovarian cancer metastasis and generate chemotherapy resistance include the overex-pression of gene products involved in epigenetic regulation. These include EZH2 and other members of the PRC2 complex. Data indicate that increased expression of EZH2 can be associated with chemotherapy resistance. Notably, the PRC2 complex has been implicated in regulating the pluripo-tency of stem cells, which has been shown to a key hallmark of HGSC. Thus formation of spheroids in ovarian cancer ascites may enable the survival of ovarian cancer cells in women undergoing front line treatment by creating a unique niche, in which expression of gene products typically associated with pluripotency not only enables metastasis but also the ability of these cells to persist through treatment. The molecular nature of this niche is not currently known. It should also be understood that the role of an "ovarian cancer stem cell" remains controversial (Ottevanger, 2017).

Epigenetic alterations in patterns of gene expression may play an important role in contributing to both ovarian cancer metastasis and/or the dedifferentiation of a tissue-specific stem cell precursor. There is now little question that epigenetic events can predispose individuals to cancer as frequently as loss of gene function due to mutations or loss of heterozygosity. The overall level of genomic methylation is reduced in cancer (global hypomethylation), but hypermethylation of promoter regions of specific genes is a common event often associated with transcriptional inactivation of specific genes (Baylin & Ohm, 2006). This is critical because the silenced genes are often tumor suppressor genes. Epigenetic gene silencing is a complex series of events that includes DNA hypermethylation of CpG islands within gene-promoter regions, histone deacetylation, methylation or phosphorylation, or histone demethylation. Global hypermethylation of CpG islands appears to be prevalent but highly variable in ovarian cancer tissue (Wei et al., 2002). Multiple genes are abnormally methylated in ovarian cancer compared with normal ovarian tissue, including the genes for p16, RAR-β, H-cadherin, leukotriene B4 receptor, progesterone receptor, and estrogen receptor-a, GSTP1, MGMT, RASSF1A, MTHFR, CDH1, GSF4, BRCA1, TMS1, the putative tumor suppressor km23 (TGF-β component), and others (Balch, Huang, Brown & Nephew, 2004). The degree of DNA methylation at these gene loci is now widely accepted as important for ensuring the success of ovarian cancer treatment and determining its long-term outcomes.

Role of immune responses

The novel observation by William Coley in the 1890s that severe bacterial infections could induce an antitumor response in patients with partially resected tumors has evolved into an understanding that the immune system can recognize tumor-associated antigens and direct a targeted response. The concept of "cancer immunoediting" suggests that the immune system not only protects the host against the development of primary cancers but also sculpts tumor immunogenicity (Zhang et al., 2003). In EOC, support for the role of immune surveillance of tumors comes from observations that the presence of tumor infiltrating lymphocytes (TILs) is associated with improved survival of patients with the disease (Smyth, Dunn & Schreiber, 2006; Zhang et al., 2003). A recent *meta*-analysis of 10 studies with 1815 ovarian cancer patients confirmed the observation that a lack of TILs is significantly associated with poor survival in ovarian cancer patients (pooled HR: 2.24) (Tone et al., 2008). This effect was evident regardless of tumor grade, stage, or histologic subtype.

Finally, advanced-stage ovarian cancer patients can have detectable tumor-specific cytotoxic T cells and antibody immunity as shown in a study implicating immunity to p53 as a predictor of improved survival in patients with advanced-stage disease (Goodell et al., 2006). All of these observations support clinical trials of immunotherapy for EOC in an effort to elicit effective antitumor responses. Major obstacles include the identification of tumor-restricted immunogenic targets, generation of a sufficient immune response to cause tumor rejection, and approaches to overcome tumor evasion of immune attack.

Ovarian cancer-specific antigens

Human tumor antigens can be classified into one of several categories: (1) differentiation antigens, such as tyrosinase, Melan-A/MART-1, and gp 100; (2) mutational antigens, such as CDK4, β-catenin, caspase-8, and P53; (3) amplification antigens, such as Her2/neu and p53; (4) splice variant antigens, such as NY-CO-37/PDZ-45 and Ing1; (5) viral antigens, such as human papillomavirus and Epstein–Barr virus; and (6) cancer-testis antigens, such as MAGE, NY-ESO-1, and LAGE-1. Some antigens may play a crucial role in progression of tumor cells (e.g., Her2/neu) and could be useful as biomarkers of disease progression and targets of therapy. On the other hand, in considering an antigenic target for ovarian cancer immunotherapy, an ideal candidate antigen should demonstrate high-frequency expression in the tumor tissues and restricted expression in normal tissues, and also provide evidence for inherent immunogenicity. In this regard, the cancer-testis antigens are a distinct and unique class of differentiation antigens with high levels of expression in adult male germ cells, but generally not in other normal adult tissues, and aberrant expression in a variable proportion of a wide range of different cancer types. Among cancer-testis antigens, NY-ESO-1 is particularly immunogenic, eliciting both cellular and humoral immune responses in a high proportion of patients with advanced ovarian cancer (Odunsi et al., 2003). The reasons for the aberrant expression of cancer-testis antigens in cancer are currently unknown. Nevertheless, the fact that the expression of these antigens is restricted to cancers, gametes, and trophoblast suggests a link between cancer and gametogenesis.

A recently described mechanism in ovarian cancer is the expression of inhibitory molecules such as programmed death-1 (PD-1) and lymphocyte activation gene-3 (LAG-3) (Matsuzaki et al., 2010). Together, these molecules render ovarian tumor infiltrating CD8 + T cells "hyporesponsive," wherein effector function is most impaired in antigen-specific LAG-3 + PD-1 + CD8 + TILs.

Angiogenesis

Angiogenesis is tightly regulated by a balance of pro- and antiangiogenic factors. These include growth factors such as TGF-β, VEGF, and platelet-derived growth factor; prostaglandins such as prostaglandin E2; cytokines such as interleukin-8; and other factors, such as the angiopoietins (Ang-1, Ang-2), and hypoxia-inducible factor1 alpha (HIF-lα). Many of these angiogenic factors have been implicated in ovarian cancer. For example, VEGF is a family of secreted polypeptides with critical roles in both normal development and human disease. Many cancers, including ovarian carcinomas, release VEGF in response to the hypoxic or acidic conditions typical in solid tumors. Variable levels of VEGF expression have been reported in ovarian cancers, in which higher levels correlate with advanced disease and poor clinical prognosis (Kassim et al., 2004). Circulating levels of VEGF have also been reported to be higher in the serum of women with ovarian cancers when compared with those with benign tumors. Expression of HIF-1 correlates well with microvessel density in ovarian cancers and has been proposed to upregulate VEGF expression (Jiang & Feng, 2006). Culturing ovarian cancer cell lines under hypoxic conditions stimulates the expression of both HIF-lα and VEGF in ovarian cancer cell lines; the addition of prostaglandin E2 potentiates the ability of hypoxia to induce the expression of both proangiogenic factors (Zhu, Saed, Deppe, Diamond & Munkarah, 2004).

Ironically, many of the molecules implicated in regulating angiogenesis in cancer also regulate other processes critical for cancer metastasis, such as cell migration and invasiveness. Inhibition of PI3 kinase decreases transcription of VEGF in ovarian cancer cells, an effect that is reversed by the expression of AKT. Such observations are consistent with reports that hypoxia not only induces angiogenesis, but also increases the invasiveness of ovarian cancer cells (Imai et al., 2003). Likewise, an acidic environment induces increased interleukin-8 expression in ovarian cancer in a manner dependent on transcription factors AP-1 and nuclear products associated with predominant features of that group. For example, immunoreactive ovarian cancers are particularly noted for their expression of T-cell chemokine ligands CXCL11 and CXCL10, as well as the receptor CXCR3. Analyses have identified a set of 193 genes whose patterns of transcription predicted ovarian survival (108 correlated with poor survival and 85 associated with good survival).

Summary

Recent years have witnessed tremendous advancements in identifying molecular determinants of ovarian cancer predisposition, as well as biochemical determinants of its outcomes. These successes have been facilitated by development of high-throughput technologies, as well as large-scale genomic association studies in clinically well-described populations. These analyses, when combined with advancements in our ability to develop small molecule inhibitors and other novel targeted therapies, have resulted in significant new options not only for ovarian cancer risk reduction but also improving clinical outcomes.

In response to stress/DNA damage, cells activate one of the six major DNA repair mechanisms—NER, BER, MMR, nonhomologous end joining, HR, and translesion synthesis. Defects in HR are a dominant molecular phenotype of EOCs. Robust clinical evidence supports the use of small molecule inhibitors for PARP1 as a synthetic lethal strategy for preventing disease recurrences (highlighted).

References

Alenezi, W. M., Fierheller, C. T., Recio, N., & Tonin, P. N. (2020). Literature review of BARD1 as a cancer predisposing gene with a focus on breast and ovarian cancers. *Genes, 11*(8), 856. Available from https://doi.org/10.3390/genes11080856.

Auersperg, N., Wong, A. S., Choi, K. C., Kang, S. K., & Leung, P. C. (2001). Ovarian surface epithelium: Biology, endocrinology, and pathology. *Endocrine Reviews, 22*(2), 255−288. Available from https://doi.org/10.1210/edrv.22.2.0422.

Balch, C., Huang, T. H., Brown, R., & Nephew, K. P. (2004). The epigenetics of ovarian cancer drug resistance and resensitization. *American Journal of Obstetrics and Gynecology, 191*(5), 1552−1572. Available from https://doi.org/10.1016/j.ajog.2004.05.025.

Baylin, S. B., & Ohm, J. E. (2006). Epigenetic gene silencing in cancer - A mechanism for early oncogenic pathway addiction? *Nature Reviews Cancer, 6*(2), 107−116. Available from https://doi.org/10.1038/nrc1799.

Baysal, B. E., DeLoia, J. A., Willett-Brozick, J. E., Goodman, M. T., Brady, M. F., Modugno, F., & Gallion, H. H. (2004). Analysis of CHEK2 gene for ovarian cancer susceptibility. *Gynecologic Oncology, 95*(1), 62−69. Available from https://doi.org/10.1016/j.ygyno.2004.07.015.

Berchuck, A., Iversen, E. S., Lancaster, J. M., Dressman, H. K., West, M., Nevins, J. R., & Marks, J. R. (2004). Prediction of optimal vs suboptimal cytoreduction of advanced-stage serous ovarian cancer with the use of microarrays. *American Journal of Obstetrics and Gynecology, 190*(4), 910−925. Available from https://doi.org/10.1016/j.ajog.2004.02.005.

Black, D. M., & Solomon, E. (1993). The search for the familial breast/ovarian cancer gene. *Trends in Genetics, 9*(1), 22−26. Available from https://doi.org/10.1016/0168-9525(93)90068-s.

Bonome, T., Lee, J. Y., Park, D. C., Radonovich, M., Pise-Masison, C., Brady, J., & Birrer, M. J. (2005). Expression profiling of serous low malignant potential, low-grade, and high-grade tumors of the ovary. *Cancer Research, 65*(22), 10602−10612. Available from https://doi.org/10.1158/0008-5472.Can-05-2240.

Boudsocq, F., Benaim, P., Canitrot, Y., Knibiehler, M., Ausseil, F., Capp, J. P., & Cazaux, C. (2005). Modulation of cellular response to cisplatin by a novel inhibitor of DNA polymerase beta. *Molecular Pharmacology, 67*(5), 1485−1492. Available from https://doi.org/10.1124/mol.104.001776.

Burleson, K. M., Hansen, L. K., & Skubitz, A. P. (2004). Ovarian carcinoma spheroids disaggregate on type I collagen and invade live human mesothelial cell monolayers. *Clinical & Experimental Metastasis, 21*(8), 685−697. Available from https://doi.org/10.1007/s10585-004-5768-5.

Carlson, J. W., Miron, A., Jarboe, E. A., Parast, M. M., Hirsch, M. S., Lee, Y., & Crum, C. P. (2008). Serous tubal intraepithelial carcinoma: Its potential role in primary peritoneal serous carcinoma and serous cancer prevention. *Journal of Clinical Oncology: Official Journal of the American Society of Clinical Oncology, 26*(25), 4160−4165. Available from https://doi.org/10.1200/jco.2008.16.4814.

Chen, T., Triplett, J., Dehner, B., Hurst, B., Colligan, B., Pemberton, J., & Carter, J. H. (2001). Transforming growth factor-beta receptor type I gene is frequently mutated in ovarian carcinomas. *Cancer Research, 61*(12), 4679−4682.

Conner, J. R., Meserve, E., Pizer, E., Garber, J., Roh, M., Urban, N., & Feltmate, C. (2014). Outcome of unexpected adnexal neoplasia discovered during risk reduction salpingo-oophorectomy in women with germ-line BRCA1 or BRCA2 mutations. *Gynecologic Oncology, 132*(2), 280−286. Available from https://doi.org/10.1016/j.ygyno.2013.12.009.

Coronado Martín, P. J., Fasero Laiz, M., García Santos, J., Ramírez Mena, M., & Vidart Aragón, J. A. (2007). Overexpression and prognostic value of p53 and HER2/neu proteins in benign ovarian tissue and in ovarian cancer. *Medicina Clínica (Barc), 128*(1), 1−6. Available from https://doi.org/10.1157/13096935.

Crum, C. P., Drapkin, R., Kindelberger, D., Medeiros, F., Miron, A., & Lee, Y. (2007). Lessons from BRCA: The tubal fimbria emerges as an origin for pelvic serous cancer. *Clinical Medicine & Research, 5*(1), 35−44. Available from https://doi.org/10.3121/cmr.2007.702.

D'Andrea, A. D. (2018). Mechanisms of PARP inhibitor sensitivity and resistance. *DNA Repair (Amst), 71*, 172−176. Available from https://doi.org/10.1016/j.dnarep.2018.08.021.

Daniels, M. S., & Lu, K. H. (2016). Genetic predisposition in gynecologic cancers. *Seminars in Oncology, 43*(5), 543−547. Available from https://doi.org/10.1053/j.seminoncol.2016.08.005.

Dansonka-Mieszkowska, A., Kluska, A., Moes, J., Dabrowska, M., Nowakowska, D., Niwinska, A., & Kupryjanczyk, J. (2010). A novel germline PALB2 deletion in Polish breast and ovarian cancer patients. *BMC Medical Genetics, 11*, 20. Available from https://doi.org/10.1186/1471-2350-11-20.

Do, T. V., Symowicz, J. C., Berman, D. M., Liotta, L. A., Petricoin, E. F., III, Stack, M. S., & Fishman, D. A. (2007). Lysophosphatidic acid down-regulates stress fibers and up-regulates pro-matrix metalloproteinase-2 activation in ovarian cancer cells. *Molecular Cancer Research, 5*(2), 121−131. Available from https://doi.org/10.1158/1541-7786.Mcr-06-0319.

Eng, C., & Ponder, B. A. (1993). The role of gene mutations in the genesis of familial cancers. *The FASEB Journal, 7*(10), 910−919. Available from https://doi.org/10.1096/fasebj.7.10.8102106.

Ganesan, S., Silver, D. P., Drapkin, R., Greenberg, R., Feunteun, J., & Livingston, D. M. (2004). Association of BRCA1 with the inactive X chromosome and XIST RNA. *Philosophical Transactions of the Royal Society of London. Series B, Biological Sciences, 359*(1441), 123−128. Available from https://doi.org/10.1098/rstb.2003.1371.

Gilks, C. B., Vanderhyden, B. C., Zhu, S., Van De Rijn, M., & Longacre, T. A. (2005). Distinction between serous tumors of low malignant potential and serous carcinomas based on global mRNA expression profiling. *Gynecologic Oncology, 96*(3), 684−694. Available from https://doi.org/10.1016/j.ygyno.2004.11.039.

Goodell, V., Salazar, L. G., Urban, N., Drescher, C. W., Gray, H., Swensen, R. E., & Disis, M. L. (2006). Antibody immunity to the p53 oncogenic protein is a prognostic indicator in ovarian cancer. *Journal of Clinical Oncology: Official Journal of the American Society of Clinical Oncology, 24*(5), 762−768. Available from https://doi.org/10.1200/JCO.2005.03.2813.

Gourley, C., Balmaña, J., Ledermann, J. A., Serra, V., Dent, R., Loibl, S., & Boulton, S. J. (2019). Moving from poly (ADP-ribose) polymerase inhibition to targeting DNA repair and DNA damage response in cancer therapy. *Journal of Clinical Oncology: Official Journal of the American Society of Clinical Oncology, 37*(25), 2257−2269. Available from https://doi.org/10.1200/jco.18.02050.

Griffon, G., Marchal, C., Merlin, J. L., Marchal, S., Parache, R. M., & Bey, P. (1995). Radiosensitivity of multicellular tumour spheroids obtained from human ovarian cancers.

European Journal of Cancer, 31a(1), 85−91. Available from https://doi.org/10.1016/0959-8049(94)00377-h.

Hough, C. D., Cho, K. R., Zonderman, A. B., Schwartz, D. R., & Morin, P. J. (2001). Coordinately up-regulated genes in ovarian cancer. *Cancer Research, 61*(10), 3869−3876.

Hu, W., Wu, W., Nash, M. A., Freedman, R. S., Kavanagh, J. J., & Verschraegen, C. F. (2000). Anomalies of the TGF-beta postreceptor signaling pathway in ovarian cancer cell lines. *Anticancer Research, 20*(2a), 729−733.

Imai, T., Horiuchi, A., Wang, C., Oka, K., Ohira, S., Nikaido, T., & Konishi, I. (2003). Hypoxia attenuates the expression of E-cadherin via up-regulation of SNAIL in ovarian carcinoma cells. *The American Journal of Pathology, 163*(4), 1437−1447. Available from https://doi.org/10.1016/s0002-9440(10)63501-8.

Jiang, H., & Feng, Y. (2006). Hypoxia-inducible factor 1alpha (HIF-1alpha) correlated with tumor growth and apoptosis in ovarian cancer. *International Journal of Gynecological Cancer: Official Journal of the International Gynecological Cancer Society, 16*(1), 405−412. Available from https://doi.org/10.1111/j.1525-1438.2006.00310.x.

Kassim, S. K., El-Salahy, E. M., Fayed, S. T., Helal, S. A., Helal, T., Azzam Eel, D., & Khalifa, A. (2004). Vascular endothelial growth factor and interleukin-8 are associated with poor prognosis in epithelial ovarian cancer patients. *Clinical Biochemistry, 37*(5), 363−369. Available from https://doi.org/10.1016/j.clinbiochem.2004.01.014.

Kennedy, R. D., Chen, C. C., Stuckert, P., Archila, E. M., De la Vega, M. A., Moreau, L. A., & D'Andrea, A. D. (2007). Fanconi anemia pathway-deficient tumor cells are hypersensitive to inhibition of ataxia telangiectasia mutated. *The Journal of Clinical Investigation, 117*(5), 1440−1449. Available from https://doi.org/10.1172/jci31245.

Landen, C. N., Jr., Birrer, M. J., & Sood, A. K. (2008). Early events in the pathogenesis of epithelial ovarian cancer. *Journal of Clinical Oncology: Official Journal of the American Society of Clinical Oncology, 26*(6), 995−1005. Available from https://doi.org/10.1200/jco.2006.07.9970.

Levanon, K., Crum, C., & Drapkin, R. (2008). New insights into the pathogenesis of serous ovarian cancer and its clinical impact. *Journal of Clinical Oncology: Official Journal of the American Society of Clinical Oncology, 26*(32), 5284−5293. Available from https://doi.org/10.1200/jco.2008.18.1107.

Mateo, J., Lord, C. J., Serra, V., Tutt, A., Balmaña, J., Castroviejo-Bermejo, M., & De Bono, J. S. (2019). A decade of clinical development of PARP inhibitors in perspective. *Annals of Oncology: Official Journal of the European Society for Medical Oncology, 30*(9), 1437−1447. Available from https://doi.org/10.1093/annonc/mdz192.

Matsuzaki, J., Gnjatic, S., Mhawech-Fauceglia, P., Beck, A., Miller, A., Tsuji, T., & Odunsi, K. (2010). Tumor-infiltrating NY-ESO-1-specific CD8 + T cells are negatively regulated by LAG-3 and PD-1 in human ovarian cancer. *Proceedings of the National Academy of Sciences of the United States of America, 107*(17), 7875−7880. Available from https://doi.org/10.1073/pnas.1003345107.

Mayr, D., Kanitz, V., Anderegg, B., Luthardt, B., Engel, J., Löhrs, U., & Diebold, J. (2006). Analysis of gene amplification and prognostic markers in ovarian cancer using comparative genomic hybridization for microarrays and immunohistochemical analysis for tissue microarrays. *American Journal of Clinical Pathology, 126*(1), 101−109. Available from https://doi.org/10.1309/n6x5mb24bp42kp20.

Mazoyer, S. (2005). Genomic rearrangements in the BRCA1 and BRCA2 genes. *Human Mutation, 25*(5), 415−422. Available from https://doi.org/10.1002/humu.20169.

Meinhold-Heerlein, I., Bauerschlag, D., Hilpert, F., Dimitrov, P., Sapinoso, L. M., Orlowska-Volk, M., & Hampton, G. M. (2005). Molecular and prognostic distinction between serous ovarian carcinomas of varying grade and malignant potential. *Oncogene*, 24(6), 1053−1065. Available from https://doi.org/10.1038/sj.onc.1208298.

Moore, K., Colombo, N., Scambia, G., Kim, B. G., Oaknin, A., Friedlander, M., & DiSilvestro, P. (2018). Maintenance olaparib in patients with newly diagnosed advanced ovarian cancer. *The New England Journal of Medicine*, 379(26), 2495−2505. Available from https://doi.org/ 10.1056/NEJMoa1810858.

Mukherjee, A., Chiang, C. Y., Daifotis, H. A., Nieman, K. M., Fahrmann, J. F., Lastra, R. R., & Lengyel, E. (2020). Adipocyte-induced FABP4 expression in ovarian cancer cells promotes metastasis and mediates carboplatin resistance. *Cancer Research*, 80(8), 1748−1761. Available from https://doi.org/10.1158/0008-5472.Can-19-1999.

Nakayama, K., Nakayama, N., Jinawath, N., Salani, R., Kurman, R. J., Shih Ie, M., & Wang, T. L. (2007). Amplicon profiles in ovarian serous carcinomas. *International Journal of Cancer*, 120 (12), 2613−2617. Available from https://doi.org/10.1002/ijc.22609.

Nieman, K. M., Kenny, H. A., Penicka, C. V., Ladanyi, A., Buell-Gutbrod, R., Zillhardt, M. R., & Lengyel, E. (2011). Adipocytes promote ovarian cancer metastasis and provide energy for rapid tumor growth. *Nature Medicine*, 17(11), 1498−1503. Available from https://doi.org/ 10.1038/nm.2492.

Norquist, B. M., Brady, M. F., Harrell, M. I., Walsh, T., Lee, M. K., Gulsuner, S., & Swisher, E. M. (2018). Mutations in Homologous recombination genes and outcomes in ovarian carcinoma patients in GOG 218: An NRG oncology/gynecologic oncology group study. *Clinical Cancer Research: An Official Journal of the American Association for Cancer Research*, 24(4), 777−783. Available from https://doi.org/10.1158/1078-0432.Ccr-17-1327.

Odunsi, K., Jungbluth, A. A., Stockert, E., Qian, F., Gnjatic, S., Tammela, J., & Old, L. J. (2003). NY-ESO-1 and LAGE-1 cancer-testis antigens are potential targets for immunotherapy in epithelial ovarian cancer. *Cancer Research*, 63(18), 6076−6083. Available from http://www.ncbi.nlm. nih.gov/entrez/query.fcgi?cmd = Retrieve&db = PubMed&dopt = Citation&list_ uids = 14522938.

Ottevanger, P. B. (2017). Ovarian cancer stem cells more questions than answers. *Seminars in Cancer Biology*, 44, 67−71. Available from https://doi.org/10.1016/j.semcancer.2017.04.009.

Pietragalla, A., Arcieri, M., Marchetti, C., Scambia, G., & Fagotti, A. (2020). Ovarian cancer predisposition beyond BRCA1 and BRCA2 genes. *International Journal of Gynecological Cancer: Official Journal of the International Gynecological Cancer Society*. Available from https://doi.org/10.1136/ijgc-2020-001556.

Piombino, C., Cortesi, L., Lambertini, M., Punie, K., Grandi, G., & Toss, A. (2020). Secondary prevention in hereditary breast and/or ovarian cancer syndromes other than BRCA. *Journal of Oncology*, 2020, 6384190. Available from https://doi.org/10.1155/2020/6384190.

Puget, N., Stoppa-Lyonnet, D., Sinilnikova, O. M., Pagès, S., Lynch, H. T., Lenoir, G. M., & Mazoyer, S. (1999). Screening for germ-line rearrangements and regulatory mutations in BRCA1 led to the identification of four new deletions. *Cancer Research*, 59(2), 455−461.

Ray-Coquard, I., Pautier, P., Pignata, S., Pérol, D., González-Martín, A., Berger, R., & Harter, P. (2019). Olaparib plus bevacizumab as first-line maintenance in ovarian cancer. *The New England Journal of Medicine*, 381(25), 2416−2428. Available from https://doi.org/10.1056/ NEJMoa1911361.

Ren, J., Xiao, Y. J., Singh, L. S., Zhao, X., Zhao, Z., Feng, L., & Xu, Y. (2006). Lysophosphatidic acid is constitutively produced by human peritoneal mesothelial cells and enhances adhesion, migration, and invasion of ovarian cancer cells. *Cancer Research*, 66(6), 3006−3014. Available from https://doi.org/10.1158/0008-5472.Can-05-1292.

Schiffenbauer, Y. S., Abramovitch, R., Meir, G., Nevo, N., Holzinger, M., Itin, A., & Neeman, M. (1997). Loss of ovarian function promotes angiogenesis in human ovarian carcinoma. *Proceedings of the National Academy of Sciences of the United States of America*, *94*(24), 13203−13208. Available from https://doi.org/10.1073/pnas.94.24.13203.

Schlosshauer, P. W., Cohen, C. J., Penault-Llorca, F., Miranda, C. R., Bignon, Y. J., Dauplat, J., & Deligdisch, L. (2003). Prophylactic oophorectomy: A morphologic and immunohistochemical study. *Cancer*, *98*(12), 2599−2606. Available from https://doi.org/10.1002/cncr.11848.

Schultz, K. A. P., Stewart, D. R., Kamihara, J., Bauer, A. J., Merideth, M. A., Stratton, P., & Hill, D. A. (1993). DICER1 tumor predisposition. In M. P. Adam, H. H. Ardinger, R. A. Pagon, S. E. Wallace, L. J. H. Bean, K. Stephens, & A. Amemiya (Eds.), *GeneReviews(®)*. Seattle, WA: University of Washington, Seattle.

Schwartz, D. R., Kardia, S. L., Shedden, K. A., Kuick, R., Michailidis, G., Taylor, J. M., & Cho, K. R. (2002). Gene expression in ovarian cancer reflects both morphology and biological behavior, distinguishing clear cell from other poor-prognosis ovarian carcinomas. *Cancer Research*, *62*(16), 4722−4729.

Sengupta, S., Kim, K. S., Berk, M. P., Oates, R., Escobar, P., Belinson, J., & Xu, Y. (2007). Lysophosphatidic acid downregulates tissue inhibitor of metalloproteinases, which are negatively involved in lysophosphatidic acid-induced cell invasion. *Oncogene*, *26*(20), 2894−2901. Available from https://doi.org/10.1038/sj.onc.1210093.

Smyth, M. J., Dunn, G. P., & Schreiber, R. D. (2006). Cancer immunosurveillance and immunoediting: The roles of immunity in suppressing tumor development and shaping tumor immunogenicity. *Advances in Immunology*, *90*, 1−50. Available from https://doi.org/10.1016/S0065-2776(06)90001-7.

Solsky, I., Chen, J., & Rebbeck, T. R. (2020). Precision prophylaxis: Identifying the optimal timing for risk-reducing salpingo-oophorectomy based on type of BRCA1 and BRCA2 cluster region mutations. *Gynecologic Oncology*, *156*(2), 363−376. Available from https://doi.org/10.1016/j.ygyno.2019.11.036.

Song, H., Ramus, S. J., Tyrer, J., Bolton, K. L., Gentry-Maharaj, A., Wozniak, E., & Gayther, S. A. (2009). A genome-wide association study identifies a new ovarian cancer susceptibility locus on 9p22.2. *Nature Genetics*, *41*(9), 996−1000. Available from https://doi.org/10.1038/ng.424.

Suszynska, M., Ratajska, M., & Kozlowski, P. (2020). BRIP1, RAD51C, and RAD51D mutations are associated with high susceptibility to ovarian cancer: Mutation prevalence and precise risk estimates based on a pooled analysis of ∼30,000 cases. *Journal of Ovarian Research*, *13*(1), 50. Available from https://doi.org/10.1186/s13048-020-00654-3.

Tan, D. S., Rothermundt, C., Thomas, K., Bancroft, E., Eeles, R., Shanley, S., & Gore, M. E. (2008). "BRCAness" syndrome in ovarian cancer: A case-control study describing the clinical features and outcome of patients with epithelial ovarian cancer associated with BRCA1 and BRCA2 mutations. *Journal of Clinical Oncology: Official Journal of the American Society of Clinical Oncology*, *26*(34), 5530−5536. Available from https://doi.org/10.1200/jco.2008.16.1703.

The Cancer Genome Atlas Research Network. (2011). Integrated genomic analyses of ovarian carcinoma. *Nature*, *474*(7353), 609−615. Available from https://doi.org/10.1038/nature10166.

Thorstenson, Y. R., Roxas, A., Kroiss, R., Jenkins, M. A., Yu, K. M., Bachrich, T., & Oefner, P. J. (2003). Contributions of ATM mutations to familial breast and ovarian cancer. *Cancer Research*, *63*(12), 3325−3333.

Tone, A. A., Begley, H., Sharma, M., Murphy, J., Rosen, B., Brown, T. J., & Shaw, P. A. (2008). Gene expression profiles of luteal phase fallopian tube epithelium from BRCA mutation carriers resemble high-grade serous carcinoma. *Clinical Cancer Research: An Official*

Journal of the American Association for Cancer Research, 14(13), 4067–4078. Available from https://doi.org/10.1158/1078-0432.Ccr-07-4959.

Van Haaften-Day, C., Shen, Y., Xu, F., Yu, Y., Berchuck, A., Havrilesky, L. J., & Hacker, N. F. (2001). OVX1, macrophage-colony stimulating factor, and CA-125-II as tumor markers for epithelial ovarian carcinoma: A critical appraisal. *Cancer, 92*(11), 2837–2844. Available from https://doi.org/10.1002/1097-0142(20011201)92:112837:aid-cncr100933.0.co;2-5.

Walsh, T., Casadei, S., Lee, M. K., Pennil, C. C., Nord, A. S., Thornton, A. M., & Swisher, E. M. (2011). Mutations in 12 genes for inherited ovarian, fallopian tube, and peritoneal carcinoma identified by massively parallel sequencing. *Proceedings of the National Academy of Sciences of the United States of America, 108*(44), 18032–18037. Available from https://doi.org/10.1073/pnas.1115052108.

Wei, S. H., Chen, C. M., Strathdee, G., Harnsomburana, J., Shyu, C. R., Rahmatpanah, F., & Huang, T. H. (2002). Methylation microarray analysis of late-stage ovarian carcinomas distinguishes progression-free survival in patients and identifies candidate epigenetic markers. *Clinical Cancer Research: An Official Journal of the American Association for Cancer Research, 8*(7), 2246–2252.

Wright, J. W., Pejovic, T., Fanton, J., & Stouffer, R. L. (2008). Induction of proliferation in the primate ovarian surface epithelium in vivo. *Human Reproduction (Oxford, England), 23*(1), 129–138. Available from https://doi.org/10.1093/humrep/dem347.

Zhang, L., Conejo-Garcia, J. R., Katsaros, D., Gimotty, P. A., Massobrio, M., Regnani, G., & Coukos, G. (2003). Intratumoral T cells, recurrence, and survival in epithelial ovarian cancer. *The New England Journal of Medicine, 348*(3), 203–213. Available from https://doi.org/10.1056/NEJMoa020177.

Zhu, G., Saed, G. M., Deppe, G., Diamond, M. P., & Munkarah, A. R. (2004). Hypoxia upregulates the effects of prostaglandin E2 on tumor angiogenesis in ovarian cancer cells. *Gynecologic Oncology, 94*(2), 422–426. Available from https://doi.org/10.1016/j.ygyno.2004.05.010.

Chapter 11

Endometrial cancer

Daniel Spinosa[1], Mary Katherine Montes de Oca[1], Catherine H. Watson[3], Andrew Berchuck[1,2] and Rebecca Ann Previs[1,2]
[1]Department of Obstetrics and Gynecology, Duke University, Durham, NC, United States, [2]Duke Cancer Institute, Durham, NC, United States, [3]Vanderbilt University Medical Center, Nashville, TN, United States

The Bokhman classification of endometrial cancers

"Historically, endometrial cancer has been divided into two pathogenic types based on endocrine and metabolic features." Type I tumors were more common (65%) and described as highly differentiated, low-grade tumors with a favorable prognosis. These tumors were thought to arise from atypical hyperplasia as a result of hyperestrogenism, often in women with obesity or metabolic syndrome. Type II tumors were less common (35%) and described as poorly differentiated, high-grade tumors with deep invasion and poor prognosis (Bokhman et al., 1983).

This dogma provided a useful framework for decades of scientific progress. Nevertheless, studies have expanded the complexity and heterogeneity of endometrial cancer by examining histologic subtypes and molecular features. On the basis of histologic features, type I tumors are primarily well-differentiated endometrioid adenocarcinomas, while serous, clear cell, and poorly differentiated endometrioid tumors fall into the type II category. Genomic testing in gynecologic cancers through next-generation sequencing (NGS) has provided a foundation for understanding the molecular alterations that are characteristic of these types of tumors. The *PTEN/PI3K/AKT/mTOR* pathway is frequently mutated in Type I endometrial cancers, and the *PTEN* tumor suppressor gene is the most commonly mutated gene, with alterations occurring in 50%−70% of Type 1 tumors (Mittica et al., 2017; Risinger et al., 1998; Tashiro et al., 1997). *PTEN* mutations often coincide with phosphatidylinositol-3-OH kinase (PI3K) pathway mutations (Cheung et al., 2011). Other common gene mutations in type I cancers include *FGFR2*, *ARID1A*, *CTNNB1*, *PIK3CA*, *PIK3R1*, and *KRAS* (Byron et al., 2012; McConechy et al., 2012; Urick et al., 2011). Type I tumors may also demonstrate high microsatellite instability (MSI-H) or loss of mismatch repair (MMR) proteins, including MLH1, MSH2, MSH6, and PMS2, which play an

Genomic and Precision Medicine. DOI: https://doi.org/10.1016/B978-0-12-800684-9.00014-9

191

important role in recognizing and repairing DNA replication errors. Loss of MMR leads to accumulation of genetic mutations, particularly in short repetitive DNA sequences called microsatellites (Baretti et al., 2018). Germline mutations in any of the MMR genes, commonly referred to as Lynch Syndrome, account for 2%−3% of uterine cancers. Patients with Lynch syndrome have a 40%−60% lifetime risk of endometrial and colon cancer (Aarnio et al., 1999; Stoffel et al., 2009).

Type II tumors frequently have mutations in *TP53*, *PIK3CA*, and *PPP2R1A*. *TP53* mutations are found in about 90% of type II tumors, particularly in the serous subtype. Increased expression of the HER-2/neu receptor tyrosine kinase was initially noted in 9% of endometrial cancers and was associated with advanced stage disease and increased mortality from cancer (Berchuck et al., 1991). *HER2* (*ERBB2*) is amplified in about 1% of endometrioid cancers and 25% of uterine serous carcinomas (Cancer Genome Atlas Research Network et al., 2013).

The Cancer Genome Atlas

The most significant step beyond Bokhman's schema for classification of endometrial cancers came with The Cancer Genome Atlas' (TCGA) systematic molecular characterization of endometrial cancers in 2013 (Cancer Genome Atlas Research Network et al., 2013). Whereas prior studies molecularly characterizing endometrial cancers had focused on molecular phenotypes within specific histologies or grades, the TCGA study examined endometrial cancers across all grades and histologies.

TCGA collected tumor samples and germline DNA from 373 patients with endometrial cancer, including 307 endometrioid, 53 serous, and 13 mixed histology cases. Specimens were analyzed across multiple genomic and proteomic platforms including exome sequencing, whole genome sequencing, RNA sequencing, miRNA sequencing, DNA methylation, DNA copy number, and reverse phase protein arrays (293 samples).

One of the defining contributions of the TCGA study was the clustering of tumors into four molecularly and clinically distinct groups based on somatic nucleotide substitutions, MSI, and somatic copy number alterations (SCNAs).

POLE (ultramutated): *POLE* is a catalytic subunit of DNA polymerase epsilon, and all tumors in this group had a mutation in the exonuclease domain of POLE. These tumors were additionally characterized by an unusually high mutation rates (232×10^{-6} mutations/megabase [Mb]), with characteristic nucleotide changes including increased C → A transversion. While histology cannot predict *POLE* status, roughly 50% of these tumors demonstrate Grade 3 histology. Seventeen (7%) tumors fell into this category, and these cases are associated with improved progression-free survival (PFS), 100% at 5 years after first-line therapy, despite the significant portion of high-grade histology. This may be due to enhanced immune recognition of these cancers due to their expression of many new recognizable antigens.

MSI-H (Hypermutated): MSI-H is a phenotype to describe tumors with deficient DNA MMR that results in frequent mutations in repetitive DNA sequences. This manifests in whole exome sequencing as a hypermutated phenotype (18×10^{-6} mutations/Mb). Most MSI-H disease is due to MLH1 promoter methylation that inactivates this MMR gene. MSI-H endometrioid tumors had a mutation frequency approximately 10-fold greater than microsatellite stable (MSS) endometrioid tumors, few SCNAs, frameshift deletions in *RPL22*, frequent nonsynonymous *KRAS* mutations, and few mutations in *FBXW7, CTNNB1, PPP2R1A*, or *TP53*. Sixty-five (28%) tumors fell into this category, and these tumors are associated with roughly 75% PFS at 5 years.

Copy number low (Endometrioid): This group is predominantly composed of MSS endometrioid tumors, with a substantially lower mutation frequency (2.9×10^{-6} mutations/Mb) than those with POLE mutations or MSI. This group is also notable for increased expression of progesterone receptor (PR). Pathway analysis revealed mutually exclusive mutations between *CTNNB1, KRAS*, and *SOX17* in this cohort, suggesting alternate mechanisms for activating WNT signaling. Ninety (39%) tumors were in this category, and these tumors are associated with roughly 75% PFS at 5 years after first-line therapy.

Copy number high (Serous-like): This group is notable for a significant burden of SCNAs and a low mutation rate (2.3×10^{-6} mutations/Mb). Histologically, this group contains serous tumors and ∼25% of Grade 3 endometrioid tumors. In addition to SCNAs, this group has a high rate of *TP53, FBXW7*, and *PPP2R1A* mutations, all previously reported as common in serous, but not endometrioid tumors. RNA analysis of this cohort revealed upregulation of proteins associated with cell cycle deregulation, including CCNE1, PIK3CA, MYC, and CDK2NA. Reverse phase protein array demonstrated increased protein expression of TP53, suggesting loss of function missense mutations, as opposed to protein truncating mutations. *ERBB2* is focally amplified with protein overexpression in roughly 25% of these tumors, suggesting a role for treatment with HER2 targeted inhibitors. Roughly 40% of these tumors have *PIK3CA* mutations which may also have therapeutic implications. Sixty (26%) tumors fell into this category, and these tumors tend to recur lending to worse survival outcomes when compared to the other molecular subtypes, with roughly 50% PFS at 5 years.

The results of TCGA suggest that molecular classification may offer some unique insights and advantages over histologic categorization. For instance, a single serous tumor was identified without a *TP53* mutation or SCNA. The sample was reassessed in light of these molecular attributes and was reclassified as Grade 3 endometrioid. Additionally, a subset (∼25%) of Grade 3 endometrioid tumors demonstrate frequent *TP53* mutations and extensive SCNA. In view of this, there may be benefit in treating this subset of Grade 3 endometrioid tumors as though they were serous tumors. Finally, molecular characterization is less prone to interobserver discordance than traditional classification techniques (Gilks, Oliva, & Soslow, 2013; Han et al., 2013; Hoang et al., 2013).

The proactive molecular risk classifier for endometrial cancer (*ProMisE*) algorithm

Molecular categorization of all endometrial cancers with the comprehensive approach used in the TCGA study is expensive and difficult to justify, since about 75% of cases are cured with surgery alone. The Proactive Molecular Risk Classifier for Endometrial Cancer (ProMisE), first described in 2015 (Talhouk et al., 2015) offers an alternative approach for molecularly classifying tumors that relies on immunohistochemistry (IHC) and single gene sequencing for *POLE*. The authors analyzed 142 previously collected endometrial cancer specimens and were able to closely reapproximate the survival curves published by TCGA using a simplified and more affordable testing paradigm. The ProMisE authors reanalyzed the TCGA dataset to identify *POLE* hotspot mutations and surrogates for copy number status that recapitulate the findings from the initial database. Specifically, they determined that *TP53*/p53 was a good surrogate for copy number status. They ultimately tested 16 different algorithms that varied across both type of assay used [e.g., p53 IHC vs *TP53* sequencing vs florescence in situ hybridization (FISH) analysis], as well as order of testing (e.g., MSI testing prior to *POLE* testing). The favored model had the highest level of discrimination and also has suitable clinical feasibility. The favored model first tests for MMR status using IHC, and if this is normal *POLE* mutational status is tested using single gene sequencing, and finally p53 IHC is done if a *POLE* mutation is not found (Fig. 11.1).

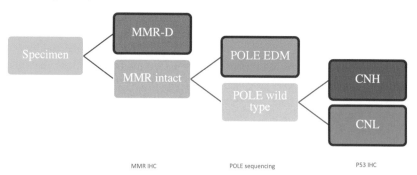

FIGURE 11.1   Clinically applicable steps of molecular classification from the ProMisE algorithm. Tumors are assessed for mismatch repair (MMR) deficiency (MMR-D), polymerase epsilon (*POLE*) exonuclease domain mutations (EDMs), and protein 53 (p53) abnormalities. The four resulting subgroups are MMR-D, *POLE* EDM, copy number high (CNH, p53 wild type), and copy number low (CNL, p53 null or mutated).

Since initially publishing the ProMisE algorithm, the same group has published a number of follow-up studies, including a validation of the testing paradigm on a nonoverlapping cohort of 319 specimens (Talhouk et al., 2017). All specimens were successfully categorized, and the ProMisE categorization was an independent predictor of survival (overall, disease specific, and progression free). The prognostic performance of ProMisE was compared to the prognostic performance of the European Society of Medical Oncology (ESMO) risk-stratification system, a widely used system based on traditional histologic and clinical features. When comparing predictive performance of ProMisE as compared to ESMO criteria, ProMisE was superior. There were four distinct survival curves based on ProMisE categorization, whereas low and intermediate risk survival curves were overlapping in ESMO categorization. Of particular interest is the *POLE* ultramutated cohort, which demonstrated favorable outcomes despite frequent high-risk features [70% were Grade 3 tumors, 35% had deep myometrial invasion, and 40% had positive lymphovascular space involvement (LVSI)]. Due to this histologic variability, *POLE* ultramutated tumors were classified across all risk levels according to ESMO categorization, and thus would have received different treatments despite similar molecular phenotypes. The predictive utility of ProMisE categorization improved when it was combined with known clinic-pathologic data including stage, LVSI, myometrial invasion, and lymph node status. The ProMisE algorithm also has utility when applied to endometrial biopsy and curettage specimens(Talhouk et al., 2016).

Though not initially part of TCGA categorization or the ProMisE algorithm, L1-cell adhesion molecule (*L1CAM*) has been demonstrated to be prognostically significant in endometrial cancer when overexpressed. Recent reanalysis of the TCGA dataset demonstrated that high L1CAM expression was associated with advanced stage, high grade histology, serous histology, and nodal involvement (Dellinger et al., 2016). Prior studies have demonstrated poorer progression free and overall survival in patients with increased *L1CAM* expression (Zeimet et al., 2011). In the post hoc reanalysis of the seminal PORTEC trials below, increased *L1CAM* was associated with distant metastasis and poor survival (Fogel et al., 2003). Assessing for L1CAM overexpression is relatively easy, and the prognostic significance seems high; therefore, it is another marker that is being incorporated into prospective studies of the clinical utility of molecular characterization.

Significant evidence regarding tailored therapy for endometrial cancer comes from the PORTEC studies (Post-Operative Radiation Therapy for Endometrial Carcinoma). PORTEC 1−3 each initially recruited patients prior to TCGA molecular characterization, but post hoc analysis has used molecular characterization on specimens from those studies. Post hoc analysis of PORTEC 1 and 2 revealed that 96% of specimens could be categorized according to TCGA subtypes, and the previously described prognostic significance of molecular characterization was confirmed (Stelloo et al., 2016).

Post hoc analysis of PORTEC 3 revealed feasibility of molecularly characterizing endometrial tumors, with 97% of specimens successfully categorized. Survival curves again demonstrated the worst survival in the group with p53 alterations, best survival with POLE mutant cases, and intermediate survival with MSI and "no specific molecular profile" (analogous to copy number low) (León-Castillo et al., 2020). Of note, nonendometrioid histologies were found in all four molecular categories, as well as patients with Stage III disease. Taken together, these findings suggest that traditional histologic and stage-based risk stratification does not capture important determinants of outcomes, including molecular phenotype. There was a trend toward improved progression free and overall survival in patients with MSI-H tumors with adjuvant radiation (RT) as compared to combined chemotherapy and radiation (CTRT), though the effect was not statistically significant (León-Castillo et al., 2020). Conversely, the group with abnormal p53 had a statistically improved 5-year recurrence-free survival and overall survival with CTRT.

Though preliminary and retrospective, these findings suggest a role for molecular characterization prior to initiation of first-line adjuvant therapy, a hypothesis that is being prospectively studied in PORTEC 4a. PORTEC 4a is a multicenter randomized Phase 3 trial that began recruiting in 2016. Eligible patients with Stage I or II endometrial cancer are randomized to either adjuvant vaginal brachytherapy, or adjuvant chemotherapy based on molecular characterization of the tumor. Treatment arms include no treatment for the favorable prognostic group (*POLE* mutated or MSS with *CTNNB1* wild type), vaginal brachytherapy for the intermediate prognostic group (MSI and/or *CTNNB1* mutant), or whole pelvis external beam radiation for the unfavorable cohort (*TP53* mutant or >10% *L1CAM* expression) (Wortman et al., 2018).

Targetable alterations

Despite promising advances in molecular categorization and prognostication, molecular classification of endometrial cancers has not become the standard of care. At present there are few molecular alterations that have treatment implications. The most actionable alterations include MSI-H, *HER2* amplification in serous cancers, and estrogen (ER) and PR positivity.

The Society of Gynecologic Oncology recommends screening all endometrial cancer cases for MSI-H to identify the subset of these patients who have inherited MMR mutations due to Lynch Syndrome. This can facilitate prevention of a second cancer in the patient and cancers in other family members through increased surveillance and risk-reducing strategies. In addition, MSI testing can be used to guide the use of immune checkpoint inhibitor drugs. Across all cancer types, including endometrial cancer, treatment with pembrolizumab has yielded a 53% overall response rate and 21%

complete response rate in MSI-H cancers, even in the setting of prior treatments (U.S. Food & Drug Administration, 2017). In Keynote-158, an open-label Phase 2 trial of 49 patients with MSI-H recurrent endometrial cancers, there was a 57% ORR and median PFS of 26 months in patients treated with pembrolizumab (Omalley et al., 2019).

Patients with advanced stage or recurrent serous uterine cancer with amplification of *HER2* benefit from the addition of trastuzumab, a monoclonal antibody targeting HER2/neu. A Phase 2 trial randomized patients with newly diagnosed Stage III or IV, or recurrent *HER2* amplified serous cancer to carboplatin and paclitaxel for six cycles with or without trastuzumab. In the recurrent setting, median PFS was 8.0 months versus 12.6 months in favor of the trastuzumab arm. In the primary, advanced stage setting, PFS was 9.3 months versus 17.9 months, also in favor of the trastuzumab arm (Fader et al., 2018).

ERs increase proliferation of endometrial epithelial cells and progestins cause differentiation, and these effects are mediated by nuclear receptors for these hormones. Prior to the era of molecular profiling, expression of these receptors was shown to be predictive of response to hormonal therapies. Grade 1 and 2 endometrioid tumors are more likely to express ER and PRs and to respond to hormonal therapy (Lentz, Brady, Major, Reid, & Soper, 1996). The progestin megestrol acetate and the "antiestrogen" tamoxifen are most frequently used in hormonal treatment of endometrial cancer. A regimen that alternates the use of these two drugs was evaluated in a Phase 2 clinic trial by the Gynecologic Oncology Group. In this study, women with advanced or recurrent endometrial cancer, regardless of ER/PR status, had a 27% response rate to this hormonal regimen (Fiorica et al., 2004). Mutations in estrogen receptor-alpha (*ESR1*) have previously been shown to be a marker of potential resistance to hormonal therapy in breast cancer. *ESR1* mutations also have been noted in about 2% of endometrial cancers subjected to molecular profiling and may indicate decreased potency of these therapies (Gaillard et al., 2019).

Biomarker-directed therapies such as pembrolizumab, trastuzumab, and hormonal therapy have an established role in treatment of endometrial cancer. Other potential targets with potential treatment implications that can be identified on NGS include homologous recombination deficiency pathway genes, tumor mutational burden (TMB), *PIK3/AKT/mTOR* pathway alterations, and *CCNE1* amplification.

Mutations in homologous recombination genes identify ovarian cancer patients as candidates for poly(ADP-ribose) polymerase inhibitors (PARPIs). In a study that characterized somatic *BRCA* mutations in endometrial cancers, 15% had a somatic *BRCA1* or *BRCA2* mutation. In this group, 38% had Grade 1 or 2 endometrioid histology, and 31% were Grade 3 endometrioid (Fehninger et al., 2020). The presence of *BRCA* or other homologous recombination deficiency alterations could identify certain endometrial cancers as targets for PARPIs. Preclinical data also exists that suggest *PTEN* deficient endometrial cancer cells are more sensitive to PARPIs than *PTEN* wild type. While there are currently no approved indications for PARPIs in uterine cancer, there are multiple ongoing trials regarding the potential targeting of homologous recombination deficient tumors. For example, UTOLO (NCT03745950) is a randomized Phase II trial investigating the use of maintenance PARPI after primary platinum-based chemotherapy for both advanced and recurrent endometrial cancer. Several studies also combine PARPIs with other targeted agents or immunotherapies: NCT03572478 is a Phase I/II study that is assessing the safety of combining a checkpoint inhibitor with a PARPI in patients with advanced or recurrent endometrial cancer (Clinicaltrials.gov, n.d.-a, n.d.-b).

Only about 25% of all endometrial cancers are MSI-H and the incidence is lower in cases with metastatic or recurrent disease, as these are more often in the *TP53* mutated copy number high group. This has limited the use of immune check point inhibitor therapy. However, NGS platforms also report TMB. In 2020, the U.S. Food & Drug Administration (FDA) approved the use of pembrolizumab for all patients with metastatic or recurrent solid tumors having a high TMB (defined as more than 10 mutations/Mb as per FoundationOne CDx). This was based on findings from a retrospective study that demonstrated an overall response rate of 29% for pembrolizumab in 102 patients with high TMB (TMB-H), regardless of PD-L1 status (Goodman et al., 2017). Approximately 40% of endometrial cancers are TMB-H; Grade 3 tumors are significantly more likely to have a high TMB than Grade 1 tumors (49% vs 13%) (Jones et al., 2020). Thus, even for endometrial cancers that are MSS, immunotherapy is an option based on the results of NGS.

Another indication for the use of pembrolizumab has emerged in MSS endometrial cancers. The KEYNOTE-146/Study 111 trial is a recent single-arm, multicenter Phase Ib/II study that evaluated the use of lenvatinib and pembrolizumab in 108 patients with previously treated advanced endometrial cancer. Lenvatinib is a multikinase inhibitor of VEGF receptors, FGF receptors, PDGF receptors, RET, and KIT. The combination of lenvatinib and pembrolizumab resulted in a response rate of 37% in patients with MSS tumors, and the study concluded that this combination showed promising activity regardless of tumor MSI status. The FDA recently approved lenvatinib plus pembrolizumab for the treatment of patients with advanced endometrial cancer (Makker et al., 2019).

In 2018 the FDA approved the use of tropomyosin receptor kinase (TRK) inhibitors for patients with solid tumors that have neurotrophic tyrosine receptor kinase (*NTRK*) gene fusions. This approval was based on three separate protocols that showed an overall response rate of 75% of tumors with *NTRK* fusions treated with larotrectinib, a TRK inhibitor (Drilon et al., 2018). About 71% of these responses were persistent after 12 months of treatment. *NTRK* fusions, while uncommon in gynecologic malignancies, have been associated with a unique subset of uterine sarcomas (Chiang et al., 2018).

The PI3K/AKT/mTOR pathway is the most commonly altered pathway in type I endometrial cancer (Fiorica et al., 2004). Activation of PI3K leads to AKT activation and increased activity of mTOR which promotes cell growth and proliferation, reduces apoptosis, and increases angiogenesis (Chiang et al., 2018). This process is regulated by *PTEN*, a tumor suppressor gene that is mutated in about 30%−60% of early stage endometrial carcinomas (Mutter et al., 2000). Loss of PTEN protein expression or increased PIK3CA leads to activation of AKT and increased mTOR activity which deregulates PI3K/AKT/mTOR signaling. Genomic alterations in the PI3K pathway in endometrial cancer make this an attractive target for PI3K/AKT/mTOR pathway inhibitors. Everolimus is a rapamycin derivative that acts by selectively inhibiting mTOR (Husseinzadeh & Husseinzadeh, 2014). Although it is not FDA approved for this indication, recent Phase II studies have investigated the role of everolimus in patients with recurrent endometrial cancer who were treated with up to two prior cytotoxic agents. Everolimus showed encouraging single agent clinical benefit rate (CBR), defined as confirmed complete or partial response or prolonged stable disease, in 22% of patients at 20 weeks of therapy (Slomovitz et al., 2010). A study on everolimus plus letrozole resulted in a CBR of 40% at 16 weeks of therapy (Slomovitz et al., 2015).

The TCGA study concluded that copy number high tumors, which include serous and serous-like endometrioid tumors, exhibit upregulation of the *CCNE1* gene (Aarnio et al., 1999). *CCNE1* amplification is prevalent in many gynecologic malignancies, including 40% of uterine carcinosarcomas and 7.5% of other uterine histologic subtypes (Cerami et al., 2012; Gao et al., 2013). *CCNE1* encodes the protein cyclin E1, and amplification of *CCNE1* leads to disruption in the cell cycle. Wee1 kinase acts as a "gatekeeper" by inactivating cyclin E1 to control DNA replication, and inhibition of Wee1 can further interfere with cell cycle regulation and increase stress on DNA replication (Brown, O'Carrigan, Jackson, & Yap, 2017). A recent Phase II study examined the use of an oral Wee1 inhibitor, adavosertib, in patients with recurrent uterine serous cancer. Six of 21 patients demonstrated a response for an overall response rate of 30%. The most common adverse effects were anemia, diarrhea, nausea, and fatigue (Liu et al., 2020). FDA approval for Wee1 inhibitors for this indication has not yet been obtained.

Molecular findings may also influence prognosis, which could subsequently impact treatment strategy. β-Catenin, a protein encoded by the *CTNNB1* gene that is involved with the Wnt signaling pathway, may predict poorer outcomes (Gao et al., 2018; Kurnit et al., 2017). In one recent study looking at low grade endometrioid (Grades 1 and 2) and early stage (Stages 1 and 2) endometrial cancer, *CTNNB1* mutation was found to have the highest hazard ratio for recurrence in multivariate analysis that controlled for standard clinicopathologic features and molecular alterations including *TP53* mutations (Kurnit et al., 2017). *CTNNB1* mutations are associated with younger age at time of diagnosis, higher rates of low grade tumors, lower rates of LVSI, and lower rates of deep myometrial invasion. These characteristics make routine assessment of *CTNNB1* status all the more compelling, as patients with the mutation may otherwise be incorrectly deemed to have a very favorable prognosis (Liu et al., 2014). Used in conjunction with TCGA molecular characterization, *CTNNB1* has been shown to be prognostically significant in the copy number low cohort that is otherwise intermediate risk (Stelloo et al., 2016). In PORTEC 4a, the experimental arm provides vaginal brachytherapy to these intermediate risk patients based on their *CTNNB1* status. Depending on the results of this trial, *CTNNB1* characterization standard could directly impact clinical practice. In GOG-86P, a study on the incorporation of bevacizumab or temsirolimus to standard initial therapy for advanced endometrial cancer, patients with CTNNB-1 mutated tumors had longer PFS when treated with bevacizumab compared to patients without mutations who received bevacizumab (Aghajanian et al., 2018).

References

Aarnio., et al. (1999). Cancer risk in mutation carriers of DNA-mismatch-repair genes. *International Journal of Cancer, 81*(2), 214−218.

Aghajanian, C., Filiaci, V., Dizon, D. S., et al. (2018). A phase II study of frontline paclitaxel/carboplatin/bevacizumab, paclitaxel/carboplatin/temsirolimus, or ixabepilone/carboplatin/bevacizumab in advanced/recurrent endometrial cancer. *Gynecologic Oncology, 150*(2), 274−281. Available from https://doi.org/10.1016/j.ygyno.2018.05.018.

Baretti., et al. (2018). DNA mismatch repair in cancer. *Parmacology & Therapeutics, 189*, 45−62.

Berchuck, A., et al. (1991). Overexpression of HER-2/neu in endometrial cancer is associated with advanced stage disease. *American Journal of Obstetrics and Gynecology, 164*(1 Pt 1), 15−21.

Bokhman, J. V., et al. (1983). Two pathogenetic types of endometrial carcinoma. *Gynecologic Oncology, 15*(1), p. 10-7.2.

Brown, J. S., O'Carrigan, B., Jackson, S. P., & Yap, T. A. (2017). Targeting DNA repair in cancer: Beyond PARP inhibitors. *Cancer Discovery., 7*(1), 20−37.

Byron, S. A., et al. (2012). FGFR2 point mutations in 466 endometrioid endometrial tumors: Relationship with MSI, KRAS, PIK3CA, CTNNB1 mutations and clinicopathological features. *PLoS One, 7*(2), e30801.

Cancer Genome Atlas Research Network, et al. (2013). Integrated genomic characterization of endometrial carcinoma. *Nature, 497*(7447), 67−73.

Cerami, E., Gao, J., Dogrusoz, U., et al. (2012). The cBio cancer genomics portal: An open platform for exploring multidimensional cancer genomics data. *Cancer Discovery., 2*(5), 401−404.

Cheung, L. W., et al. (2011). High frequency of PIK3R1 and PIK3R2 mutations in endometrial cancer elucidates a novel mechanism for regulation of PTEN protein stability. *Cancer Discovery, 1*(2), 170−185.

Chiang., et al. (2018). NTRK fusions define a novel uterine sarcoma subtype with features of fibrosarcoma. *The American Journal of Surgical Pathology, 42*(6), 791−798.

Clinicaltrials.gov(n.d.-a). https://clinicaltrials.gov/ct2/show/NCT03745950

Clinicaltrials.gov(n.d.-b). https://clinicaltrials.gov/ct2/show/NCT03572478

Dellinger, T. H., et al. (2016). L1CAM is an independent predictor of poor survival in endometrial cancer—An analysis of The Cancer Genome Atlas (TCGA). *Gynecologic Oncology, 141*(2), 336−340.

Drilon., et al. (2018). Efficacy of larotrectinib in TRK fusion-positive cancer in adults and children. *The New England Journal of Medicine, 378,* 731−739.

Fader, A. N., et al. (2018). Randomized phase II trial of carboplatin-paclitaxel vs carboplatin-paclitaxel-trastuzumab in uterine serous carcinomas that overexpress human epidermal growth factor receptor 2/neu. *Journal of Clinical Oncology, 36*(20), 2044−2051.

Fehninger, et al. (2020). BRCA1/2 somatic mutations in patients with advanced or recurrent endometrial cancer. In *Presented at SGO 25th annual winter meeting,* February 6−8, Snowmass, CO.

Fiorica., et al. (2004). Phase II trial of alternative courses of megestrol acetate and tamoxifen in advanced endometrial carcinoma: A Gynecologic Oncology Group study. *Gynecologic Oncology, 92*(1), 10−14.

Fogel, M., et al. (2003). L1 expression as a predictor of progression and survival in patients with uterine and ovarian carcinomas. *The Lancet, 362*(9387), 869−875.

Gaillard, S. L., Andreano, K. J., Gay, L. M., Steiner, M., Jorgensen, M. S., Davidson, B. A., & Elvin, J. A. (2019). Constitutively active ESR1 mutations in gynecologic malignancies and clinical response to estrogen-recept directed therapy. *Gynecologic Oncology, 154*(1), 199−206.

Gao, J., Aksoy, B. A., Dogrusoz, U., et al. (2013). Integrative analysis of complex cancer genomics and clinical profiles using the cBioPortal. *Science Signaling, 6*(269), pl1.

Gao, C., et al. (2018). Exon 3 mutations of CTNNB1 drive tumorigenesis: A review. *Oncotarget, 9*(4), 5492.

Gilks, C. B., Oliva, E., & Soslow, R. A. (2013). Poor interobserver reproducibility in the diagnosis of high-grade endometrial carcinoma. *The American Journal of Surgical Pathology, 37* (6), 874−881.

Goodman, A. M., Kato, S., Bazhenova, L., Patel, S. P., Frampton, G. M., Miller, V., & Kurzrock, R. (2017). Tumor mutational burden as an independent predictor of response to immunotherapy in diverse cancers. *Molecular Cancer Therapeutics, 16*(11), 2598−2608.

Han, G., et al. (2013). Reproducibility of histological cell type in high-grade endometrial carcinoma. *Modern Pathology, 26*(12), 1594−1604.

Hoang, L. N., et al. (2013). Histotype-genotype correlation in 36 high-grade endometrial carcinomas. *The American Journal of Surgical Pathology, 37*(9), 1421−1432.

Husseinzadeh, N., & Husseinzadeh, H. D. (2014). mTOR inhibitors and their clinical application in cervical, endometrial and ovarian cancers: A critical review. *Gynecologic Oncology, 133* (2), 375−381.

Jones., et al. (2020). Immune checkpoint expression, microsatellite instability, and mutational burden: Identifying immune biomarker phenotypes in uterine cancer. *Gynecologic Oncology, 156,* 393−399.

Kurnit, K. C., et al. (2017). CTNNB1 (beta-catenin) mutation identifies low grade, early stage endometrial cancer patients at increased risk of recurrence. *Modern Pathology, 30*(7), 1032−1041.

Lentz, S. S., Brady, M. F., Major, F. J., Reid, G. C., & Soper, J. T. (1996). High-dose megestrol acetate in advanced or recurrent endometrial carcinoma: A Gynecologic Oncology Group Study. *Journal of Clinical Oncology, 14*(2), 357–361. Available from https://doi.org/10.1200/JCO.1996.14.2.357.

León-Castillo, A., et al. (2020). Molecular classification of the PORTEC-3 trial for high-risk endometrial cancer: Impact on prognosis and benefit from adjuvant therapy. *Journal of Clinical Oncology, 38*(29), 3388–3397.

Liu, J., Tayob, N., Campos, S., et al. (2020). A phase II trial of the Wee1 inhibitor adavosertib (AZD1775) in recurrent uterine serous carcinoma. *2020 SGO annual meeting on women's cancer*. Abstract 7.

Liu, Y., et al. (2014). Clinical significance of CTNNB1 mutation and Wnt pathway activation in endometrioid endometrial carcinoma. *Journal of the National Cancer Institute, 106*(9), dju245.

Makker, V., Rasco, D., Vogelzang, N. J., et al. (2019). Lenvatinib plus pembrolizumab in patients with advanced endometrial cancer: An interim analysis of a multicentre, open-label, single-arm, phase 2 trial. *The lancet oncology, 20*(5), 711–718. Available from https://doi.org/10.1016/S1470-2045(19)30020-8.

McConechy, M. K., et al. (2012). Use of mutation profiles to refine the classification of endometrial carcinomas. *The Journal of Pathology, 228*(1), 20–30.

Mittica, G., Ghisoni, E., Giannone, G., Aglietta, M., Genta, S., & Valabrega, G. (2017). Checkpoint inhibitors in endometrial cancer: Preclinical rationale and clinical activity. *Oncotarget, 8*(52), 90532–90544.

Mutter, G. L., Lin, M. C., Fitzgerald, J. T., et al. (2000). Altered PTEN expression as a diagnostic marker for the earliest endometrial precancers. *Journal of the National Cancer Institute, 92*(11), 924–930.

Omalley, et al. (2019). Pembrolizumab in patients with MSI-H advanced endometrial cancer from the KEYNOTE-158 Study. *Presented at ESMO*.

Risinger, J. I., et al. (1998). PTEN mutation in endometrial cancers is associated with favorable clinical and pathologic characteristics. *Clinical Cancer Research, 4*(12), 3005–3010.

Slomovitz, B. M., Jiang, Y., Yates, M. S., et al. (2015). Phase II study of everolimus and letrozole in patients with recurrent endometrial carcinoma. *Journal of Clinical Oncology, 33*(8), 930–936.

Slomovitz, B. M., Lu, K. H., Johnston, T., et al. (2010). A phase 2 study of the oral mammalian target of rapamycin inhibitor, everolimus, in patients with recurrent endometrial carcinoma. *Cancer., 116*(23), 5415–5419.

Stelloo, E., et al. (2016). Improved risk assessment by integrating molecular and clinicopathological factors in early-stage endometrial cancer—Combined analysis of the PORTEC cohorts. *Clinical Cancer Research, 22*(16), 4215–4224.

Stoffel., et al. (2009). Calculation of risk of colorectal and endometrial cancer among patients with Lynch syndrome. *Gastroenterology, 137*(5), 1621–1627.

Talhouk, A., et al. (2015). A clinically applicable molecular-based classification for endometrial cancers. *British Journal of Cancer, 113*(2), 299–310.

Talhouk, A., et al. (2017). Confirmation of ProMisE: A simple, genomics-based clinical classifier for endometrial cancer. *Cancer, 123*(5), 802–813.

Talhouk, A., et al. (2016). Molecular classification of endometrial carcinoma on diagnostic specimens is highly concordant with final hysterectomy: Earlier prognostic information to guide treatment. *Gynecologic Oncology, 143*(1), 46–53.

Tashiro, H., et al. (1997). Mutations in PTEN are frequent in endometrial carcinoma but rare in other common gynecological malignancies. *Cancer Research, 57*(18), 3935−3940.

U.S. Food & Drug Administration. (2017, May 23). *Drug Administration press release. FDA approves first cancer treatment for any solid tumor with a specific genetic feature.* August 10, 2020. https://www.fda.gov/news-events/press-announcements/fda-approves-first-cancer-treatment-any-solid-tumor-specific-genetic-feature.

Urick, M. E., et al. (2011). PIK3R1 (p85alpha) is somatically mutated at high frequency in primary endometrial cancer. *Cancer Research, 71*(12), 4061−4067.

Wortman, B., et al. (2018). Molecular-integrated risk profile to determine adjuvant radiotherapy in endometrial cancer: Evaluation of the pilot phase of the PORTEC-4a trial. *Gynecologic Oncology, 151*(1), 69−75.

Zeimet, A. G., et al. (2011). Large international multicenter evaluation of the clinical significance of L1-CAM expression in FIGO stage I, type 1 endometrial cancer. *Journal of Clinical Oncology, 29*(15_suppl), 5091.

Chapter 12

Colorectal cancer

Marc T. Roth[1] and Kristen K. Ciombor[2]

[1]*Medical Oncology, Sarah Cannon Cancer Institute, Kansas City, MO, United States,*
[2]*Division of Hematology/Oncology, Department of Internal Medicine, Vanderbilt University Medical Center, Nashville, TN, United States*

Introduction

Colorectal cancer (CRC) is the second leading cause of cancer-related death in the United States, and it is estimated that there will be nearly 148,000 new diagnoses in 2020 (Siegel et al., 2020). Over the past few decades, as the human genome has become the focus of intense study, we have begun to understand the etiology of CRC as well as appreciate its genomic diversity. Many factors are now thought to contribute to carcinogenesis, including genetic events, environmental exposures and the gut microbiome, among others. This chapter will focus on the genomic component of colorectal tumorigenesis and how molecular alterations have been leveraged to facilitate development of targeted treatment options.

Colorectal cancer genetics

Early carcinogenesis studies suggested that most colorectal carcinomas arise from early lesions called adenomas, as mutations in individual genes accumulate in a stepwise fashion (Fearon & Vogelstein, 1990; Vogelstein et al., 1988). The most common gene implicated was adenomatous polyposis coli (*APC*), a tumor suppressor gene, which was discovered and named by its association with an inherited cancer syndrome—familial adenomatous polyposis (FAP) (Powell et al., 1993). FAP is responsible for 1% of CRC cases annually (Jasperson et al., 2010). The pathognomonic germline mutation in *APC* causes innumerable polyps to develop throughout the colon, which incurs a lifetime risk of cancer close to 90%.

    The most common heritable CRC syndrome, however, is hereditary non-polyposis CRC, or Lynch syndrome, which accounts for roughly 3% of annual CRC cases (Hampel et al., 2008). Germline mutations in various mismatch repair (MMR) genes (*MLH1, MSH2, MSH6, PMS2, EPCAM*) lead to

Genomic and Precision Medicine. DOI: https://doi.org/10.1016/B978-0-12-800684-9.00020-4

the development of multiple tumor types with microsatellite instability (MSI-H) or deficient mismatch repair (dMMR). Rates of CRC development in Lynch syndrome ranged from 43% to 66% in one analysis (Stoffel et al., 2009).

Additional less well-known familial CRC syndromes exist (Table 12.1). *MUTYH*-associated polyposis (MAP) is associated with the base excision repair gene *MUTYH*, as the name implies (Stoffel, 2016). Although patients with MAP can have a similar phenotype to FAP without the germline *APC* mutation, most have fewer adenomas. Peutz-Jeghers syndrome carries an increased risk for a variety of cancers, including breast, genitourinary, lung, and multiple sites in the gastrointestinal tract, although GI polyps are hamartomatous in nature. Germline mutations in *STK11* characterize this syndrome. Juvenile polyposis syndrome is seen in those with *SMAD4* and *BMPR1A* germline mutations and also characterized by hamartomatous polyps in the stomach and colon. Cowden syndrome is associated with a germline *PTEN* mutation, characteristic physical exam findings, and gastrointestinal lesions of various pathologies, which may increase the risk of CRC.

Even within familial cancer syndromes, many variables factor into the risk of developing CRC. Therefore it is understandable why predicting sporadic

TABLE 12.1 Genes of interest in familial colorectal cancers.

Abbreviation	Gene name	Familial syndrome
APC	Adenomatous polyposis coli	Familial adenomatous polyposis
BMPR1A	Bone morphogenic protein receptor 1A	Juvenile polyposis syndrome
EPCAM	Epithelial cell adhesion molecule	Lynch syndrome
MLH1	MutL homolog 1	Lynch syndrome
MSH2	MutS homolog 2	Lynch syndrome
MSH6	MutS homolog 6	Lynch syndrome
MUTYH	MutY homolog	MUTYH-associated polyposis
PMS2	PMS1 homolog 2	Lynch syndrome
PTEN	Phosphatase and tensin homolog	Cowden syndrome
SMAD4	Mothers against decapentaplegic 4	Juvenile polyposis syndrome
STK11	Serine/threonine kinase 11	Peutz-Jeghers syndrome

CRC among the general population is difficult. Thanks to genome-wide association studies comparing thousands of CRC patients, pertinent single-nucleotide polymorphisms (SNPs) in tumor development are being identified (Lu et al., 2019). *MLH1* is known to be associated with familial CRC, but variants likely also have a role in sporadic cases (Pardini et al., 2020).

We are now seeing a rise in the incidence of CRC in patients under the age of 50, which accounts for 10% of new diagnoses (Siegel et al., 2017). Although the rate of germline mutations has nearly doubled in the population younger than 50 years old (Mork et al., 2015), up to 80% of these individuals have no identifiable germline predisposition. Retrospective reviews have revealed that early-onset CRC is more likely to be MSI-H and have *ATM* and *CTNNB1* mutations, with fewer *BRAF V600E* and *APC* mutations (Willauer et al., 2019). Hypomethylation appears to be a pertinent epigenetic feature of some early onset cancer, which could lead to chromosomal instability and the aforementioned genomic profile (Cavestro et al., 2018).

Diagnosis and prognosis of colorectal cancer

The rate of diagnosis of CRC has been influenced by the incorporation of routine screening practices as recommended by most national guidelines (Anon., 2000; Rex et al., 2017), allowing for earlier detection of disease. In addition, the study of familial cancer syndromes and description of their course has allowed for individualized screening recommendations in these populations (Syngal et al., 2015). For example, patients with classic FAP are recommended to begin screening colonoscopies between age of 10 and 12 years old at a frequency of every $1-2$ years, while Lynch syndrome screening is routinely initiated at age 20 with a similar frequency.

Models have been developed to account for family history and genetic risk based on the presence of CRC-associated SNPs, but this has yet to be incorporated into routine screening practices, nor they have been validated prospectively in a widespread manner (Hsu et al., 2015; Jeon et al., 2018; Weigl et al., 2018).

Although genome-specific prognostic models do not exist for CRC and in the same way they do for hematologic malignancies, associations with specific genes do. *BRAF V600E*-mutated CRC is generally associated with more aggressive disease and poorer survival (Blaker et al., 2019). Conversely, although patients with early-stage MSI-H CRC do not usually benefit from adjuvant 5-fluorouracil chemotherapy, it is generally associated with more favorable survival (Popat et al., 2005). Mutant *TP53* in a variety of cancers is considered a poor prognostic feature, but significance in CRC is not well understood (Munro et al., 2005).

National guidelines recommend testing mCRC for MSI-H, *KRAS/NRAS* mutations, *BRAF* mutations, and *HER2* amplification, either individually or as part of a next-generation sequencing (NGS) panel (National Comprehensive Cancer Network, 2020). If the latter is chosen, depending on the platform utilized, rare targetable alterations such as neurotrophic receptor tyrosine kinase (*NTRK*) fusions and others may be detected, as well as variants of unknown significance (National Comprehensive Cancer Network, 2020).

Precision therapeutics for colorectal cancer

Cytotoxic chemotherapies have long been the cornerstone of CRC treatment, and while this is still true, targeted therapies and other precision therapeutics are of increasing interest. Checkpoint inhibition for MSI-H/dMMR mCRC represents one of the more recent examples. As mentioned previously, Lynch syndrome is defined by germline mutations in specific MMR genes, causing accumulation of mutations and MSI-H/dMMR. While the familial syndrome represents 3% of CRC cases, additional 10%−17% have sporadic mutations in MMR genes, most commonly MLH1 (Aaltonen et al., 1998; Hampel et al., 2005). The accumulation of neoantigens due to deficient MMR and intratumoral infiltration of immune cells in these patients was hypothesized to prime the tumor for immunotherapy (Le et al., 2015). Pembrolizumab (Keytruda, 2020), which inhibits Programmed Death-1 (PD-1), thereby activating the immune system, was shown to have an effect in some patients with dMMR/MSI-H mCRC by Le et al. (2015). This trial noted an immune-related objective response rate (ORR) of 40% in dMMR/MSI-H patients versus 0% in microsatellite stable mCRC. KEYNOTE-164 confirmed this observation in a larger number of patients with treatment-refractory dMMR/MSI-H mCRC, finding an ORR of 33% and a duration of response (DOR) that was not reached at a median follow up of 31.3 and 24.2 months between two cohorts (Le et al., 2020). Checkmate-142 utilized a different PD-1 inhibitor, nivolumab (Opdivo, 2020), in a similar population. At a median follow up of 12 months, 23% of patients achieved an objective response and the median DOR was not reached (Overman et al., 2017). In a separate arm, the study evaluated the combination of nivolumab and ipilimumab (Yervoy, 2019), a cytotoxic T-lymphocyte-associated protein 4 inhibitor, based on their synergistic response seen in other malignancies. The ORR was 55% with a DOR that was not reached at a median follow up of 19.9 months (Morse et al., 2019). This improved response came with an increase in immune-related side effects, however, with 20% of patients experiencing grade 3 or 4 treatment-related adverse events. Following these trials, the checkpoint inhibitors as monotherapy, as well as the nivolumab/ipilimumab combination, have been approved by the United States Food and Drug Administration (FDA) for dMMR/MSI-H mCRC after progression on a fluoropyrimidine, oxaliplatin, and irinotecan.

The epidermal growth factor receptor (EGFR) is overexpressed in nearly 50% of CRC cases and is thought to play a role in its carcinogenesis through its activation of multiple cellular proliferation pathways (Antonacopoulou et al., 2008; McKay et al., 2002). This receptor upregulation is now an established target for inhibition. Two anti-EGFR monoclonal antibodies, cetuximab (Erbitux Cetuximab, 2018) and panitumumab (Vectibix, 2014), are approved for use in mCRC as monotherapy and in combination with cytotoxic chemotherapy (National Comprehensive Cancer Network, 2020). Interestingly, EGFR expression levels are not associated with treatment efficacy when using these agents (Hecht et al., 2010). Mutations in *KRAS*, an established oncogene in CRC, have been shown to negatively impact response to EGFR inhibition (Amado et al., 2008; Benvenuti et al., 2007). This is problematic given that *KRAS* is thought to become mutated early in CRC carcinogenesis (Fearon & Vogelstein, 1990), and the most common aberrations in codons 12 and 13 of exon 2 have been seen in up to 40% of sporadic CRC (Amado et al., 2008). Given this finding, wild-type results of extended *RAS* testing in exons 2−4 of *KRAS*, as well as exons 2−4 of *NRAS*, have been correlated with response to EGFR inhibition across multiple randomized clinical trials in mCRC (Sorich et al., 2015). Findings supported the hypothesis that mCRC patients with *RAS* mutations do not experience benefit, and potentially have a detrimental effect, to anti-EGFR treatment (Douillard et al., 2013; Heinemann et al., 2014). Current guidelines recommend use of these agents only if extended *RAS* testing yields no mutations (Allegra et al., 2016). Furthermore, pooled analyses show that patients with *RAS* wild-type right-sided primary tumors fare worse with chemotherapy plus EGFR inhibition compared to chemotherapy plus VEGF inhibition in the first-line setting; therefore, first-line use of EGFR inhibitors in right-sided *RAS* wild-type tumors is not recommended (Arnold et al., 2017; National Comprehensive Cancer Network, 2020). This may be due to distribution of varied molecular subtypes along the colon, a hypothesis still under investigation (Loree et al., 2018).

BRAF mutations are found in approximately 8%−19% of mCRC, most commonly *BRAF V600E* (Nunes et al., 2020; Souglakos et al., 2009). This mutation is housed within the kinase domain, promoting constitutive downstream signaling (through RAS, MEK, ERK, MAP) and proliferation (Davies et al., 2002). BRAF inhibitors were studied as monotherapy for mCRC, and tumors were found to develop rapid resistance through upregulation of EGFR (Kopetz et al., 2015; Prahallad et al., 2012). This formed the basis for subsequent clinical trials targeting both pathways concomitantly. The BEACON study is the largest trial for BRAF *V600E*-mutated mCRC to date investigating BRAF, MEK, and EGFR inhibition with encorafenib (Braftovi, 2019), binimetinib (Mektovi, 2019), and cetuximab (Kopetz et al., 2019), respectively. The trial included three arms—triplet therapy with the aforementioned drugs, doublet therapy with encorafenib and cetuximab, and standard treatment with cetuximab plus fluorouracil, leucovorin, and irinotecan or irinotecan with cetuximab. The study yielded favorable survival statistics in a cohort of poor prognosis patients with a median overall survival of 9.0 months in the triplet arm [hazard ratio (HR) = 0.52, $P < 0.0001$) and 8.4 months in the doublet arm (HR = 0.60, $P = 0.003$) compared to 5.4 months in the control arm. Due to similar outcomes between the doublet and triplet arms and increased toxicity with the triplet arm, an FDA application for use of the doublet (encorafenib and cetuximab) as a second-line treatment option in *BRAF V600E*-mutated mCRC was submitted and recently approved (Kopetz et al., 2020). Of note, BRAF non-*V600E* mutations have not been associated with the same poor prognosis (Jones et al., 2017).

HER2 gene amplification occurs in up to 6% of CRC (Seo et al., 2014). Much of the therapeutic groundwork targeting HER2 overexpression had been established in breast cancer previously, but not tested in CRC until recently. Resistance with anti-HER2 monotherapy in mCRC is common; therefore, the combination of trastuzumab (Herceptin, 2017) and lapatinib (Tykerb, 2018) was investigated in heavily pretreated mCRC in the HERACLES phase II clinical trial, finding an ORR of 30%, including one complete response (Sartore-Bianchi et al., 2016). Additional agents under investigation include pertuzumab (Perjeta, 2017), tucatinib, and T-DM1 (KADCYLA, 2019), among others (Sartore-Bianchi et al., 2016; Strickler et al., 2017). Although not yet approved by FDA, HER2 testing is recommended by the National Comprehensive Cancer Network (NCCN) guidelines, with anti-HER2 treatment recommended for those exhibiting HER2 overexpression after failure of other standard of care lines of chemotherapy (National Comprehensive Cancer Network, 2020).

The *NTRK* genes can create oncogenic fusion products that serve as drivers in various cancer and are seen in 0.3% of mCRC, more commonly in MSI-H tumors (Deihimi et al., 2017; Solomon et al., 2020). Despite the low prevalence of NTRK fusions, a number of basket studies using two small molecular NTRK inhibitors—larotrectinib (VITRAKVI, 2019) and entrectinib (Entrectinib, 2019)—were developed for patients with NTRK fusions, and favorable outcomes led to their FDA approval in 2018 and 2019, respectively (Doebele et al., 2020; Siena et al., 2019). Integrated analyses show that 1/4 (25%) of mCRC patients achieved a partial response (PR) with entrectinib, while 2/4 (50%) of mCRC patients on larotrectinib demonstrated a PR (Drilon et al., 2018; Siena et al., 2019). The approval of larotrectinib represents the second tissue-agnostic treatment for cancer, with pembrolizumab for MSI-H tumors being the first.

Novel and emerging therapeutics and future opportunities

Although NGS for mCRC is standard of care and targeted treatment options are expanding, getting the right drug to the right patient at the right time can be a challenge. Various efforts are underway to identify actionable genomic alterations in patients with mCRC and treat these patients accordingly. CRC and Liquid Biopsy Screening Protocol for Molecularly Assigned Therapy, or

TABLE 12.2 Selected studies of targeted agents in metastatic colorectal cancer.

Gene	Agents	Studies	Phase
BRAF	Encorafenib + binimetinib + cetuximab (first-line)	NCT03693170 (ANCHOR-CRC)	II
FGFR	Pemigatinib	NCT04096417	II
	AZD4547	NCT02465060 (MATCH)	II
	Erdafitinib	NCT02465060 (MATCH)	II
HER2	Tucatinib + trastuzumab	NCT03043313 (MOUNTAINEER)	II
	Neratinib trastuzumab vs. neratinib + cetuximab	NCT03457896	II
	Afatinib	NCT02465060 (MATCH)	II
c-MET	Crizotinib	NCT02465060 (MATCH)	II
	Savolitinib	NCT03592641	II
dMMR/ MSI-H	mFOLFOX6/ bevacizumab ± atezolizumab vs. atezolizumab monotherapy (first-line)	NCT02997228 (COMMIT)	III
	Avelumab (or *POLE mutations*)	NCT03150706	II
	Durvalumab (or *POLE mutations*)	NCT03435107	II
NTRK	Repotrectinib	NCT03093116	I/II
	Selitrectinib	NCT03215511	I/II
PIK3CA	CB-839 + capecitabine	NCT02861300	I/II
	Taselisib	NCT02465060 (MATCH)	II
	Copanlisib	NCT02465060 (MATCH)	II
KRAS^{G12C}	JNJ-74699157	NCT04006301	I
	MRTX849	NCT03785249	I/II
	LY3499446	NCT04165031	I/II

COLOMATE, is an example of a phase II clinical trial (NCT03765736), which utilizes an umbrella protocol to match patients with mCRC to various targeted therapy companion studies based on their genomic profile (Ciombor et al., 2019). The study aims to use cell-free DNA from peripheral blood to identify actionable genomic alterations in mCRC patients, assess the impact of molecularly assigned therapy accordingly under its individual companion studies, correlate patient outcomes with mutational burden, and explore mechanisms of resistance to targeted therapy. Platforms such as COLOMATE and their associated companion trials represent innovative methods to investigate novel genomically selected treatments for patients with mCRC.

The number of targeted agents under investigation in mCRC has grown tremendously in recent years. As mentioned previously, *RAS* is an important oncogene in CRC carcinogenesis, but it has been notoriously difficult to directly target *RAS* for a number of reasons (Gysin et al., 2011). Targeting approaches, therefore, have focused on inhibiting targets either up- or downstream of the protein (Ryan & Corcoran, 2018). There has been recent success in developing $KRAS^{G12C}$ inhibitors, however, which are currently being studied in mCRC and nonsmall cell lung cancer (NCT04006301, NCT03785249, NCT04165031), but unfortunately this specific mutation only represents a small percentage of *KRAS*-mutant CRC (Cox et al., 2014). Other genomic targets under investigation in mCRC include FGFR, *PIK3CA*, RET, and c-MET (Ahn et al., 2016), among others, in addition to potentially expanding the use of immunotherapy to other genetic variants such as *POLE* and *POLD1* (Palles et al., 2013; Wang et al., 2019) (Table 12.2).

Conclusions

CRC is a complex genomic disease. Due to massive sequencing and analysis efforts, we are learning more about the underlying pathogenesis at a rapid pace. New technologies in drug development have allowed for an increase in precision therapeutics, and collaborative clinical trial efforts have expanded access to these treatments. Although we have much more to understand, it is likely that targeted treatment options will continue to play an important role in improving outcomes for metastatic CRC patients in the years to come.

Disclosures

No funding was received for the publication of this chapter.

Dr. Roth has no disclosures to report.

Dr. Ciombor has received research grants to her institution from Incyte, BMS, Merck, Array, Daiichi Sankyo, Nucana, Abbvie; she is a consultant for Bayer, Foundation Medicine, Taiho, Natera and Array.

References

Aaltonen, L. A., et al. (1998). Incidence of hereditary nonpolyposis colorectal cancer and the feasibility of molecular screening for the disease. *The New England Journal of Medicine, 338*(21), 1481−1487.

Ahn, D. H., et al. (2016). Genomic diversity of colorectal cancer: Changing landscape and emerging targets. *World Journal of Gastroenterology, 22*(25), 5668−5677.

Allegra, C. J., et al. (2016). Extended RAS gene mutation testing in metastatic colorectal carcinoma to predict response to anti-epidermal growth factor receptor monoclonal antibody therapy: American Society of Clinical Oncology Provisional Clinical Opinion Update 2015. *Journal of Clinical Oncology: Official Journal of the American Society of Clinical Oncology, 34*(2), 179−185.

Amado, R. G., et al. (2008). Wild-type KRAS is required for panitumumab efficacy in patients with metastatic colorectal cancer. *Journal of Clinical Oncology: Official Journal of the American Society of Clinical Oncology, 26*(10), 1626−1634.

Anon. (2000). Colon cancer screening (USPSTF recommendation). U.S. Preventive Services Task Force. *Journal of the American Geriatrics Society. 48*(3), 333−335.

Antonacopoulou, A. G., et al. (2008). EGFR, HER-2 and COX-2 levels in colorectal cancer. *Histopathology, 53*(6), 698−706.

Arnold, D., et al. (2017). Prognostic and predictive value of primary tumour side in patients with RAS wild-type metastatic colorectal cancer treated with chemotherapy and EGFR directed antibodies in six randomized trials. *Annals of Oncology: Official Journal of the European Society for Medical Oncology, 28*(8), 1713−1729.

Benvenuti, S., et al. (2007). Oncogenic activation of the RAS/RAF signaling pathway impairs the response of metastatic colorectal cancers to anti-epidermal growth factor receptor antibody therapies. *Cancer Research, 67*(6), 2643−2648.

Blaker, H., et al. (2019). The association between mutations in BRAF and colorectal cancer-specific survival depends on microsatellite status and tumor stage. *Clinical Gastroenterology and Hepatology: The Official Clinical Practice Journal of the American Gastroenterological Association, 17*(3), 455−462, e6.

Braftovi® (encorafenib). (2019). *Prescribing information.* Boulder, CO: Array BioPharma, Inc.

Cavestro, G. M., et al. (2018). Early onset sporadic colorectal cancer: Worrisome trends and oncogenic features. *Digestive and Liver Disease: Official Journal of the Italian Society of Gastroenterology and the Italian Association for the Study of the Liver, 50*(6), 521−532.

Ciombor, K. K., et al. (2019). Abstract LB-235: COLOMATE: Colorectal cancer and liquid biopsy screening protocol for molecularly assigned therapy. *Cancer Research, 79*(13 Supplement), p. LB-235.

Cox, A. D., et al. (2014). Drugging the undruggable RAS: Mission possible? *Nature Reviews. Drug Discovery, 13*(11), 828−851.

Davies, H., et al. (2002). Mutations of the BRAF gene in human cancer. *Nature, 417*(6892), 949−954.

Deihimi, S., et al. (2017). BRCA2, EGFR, and NTRK mutations in mismatch repair-deficient colorectal cancers with MSH2 or MLH1 mutations. *Oncotarget, 8*(25), 39945−39962.

Doebele, R. C., et al. (2020). Entrectinib in patients with advanced or metastatic NTRK fusion-positive solid tumours: Integrated analysis of three phase 1−2 trials. *The Lancet Oncology, 21*(2), 271−282.

Douillard, J. Y., et al. (2013). Panitumumab-FOLFOX4 treatment and RAS mutations in colorectal cancer. *The New England Journal of Medicine, 369*(11), 1023−1034.

Drilon, A., et al. (2018). Efficacy of larotrectinib in TRK fusion-positive cancers in adults and children. *The New England Journal of Medicine, 378*(8), 731−739.

Entrectinib (Rozlytrek®) [package insert]. (2019). South San Francisco, CA: Genentech, Inc.

Erbitux (Cetuximab) [package insert]. (2018). New York and Princeton, NJ: ImClone Systems and Bristol-Myers Squibb.

Fearon, E. R., & Vogelstein, B. (1990). A genetic model for colorectal tumorigenesis. *Cell, 61* (5), 759−767.

Gysin, S., et al. (2011). Therapeutic strategies for targeting ras proteins. *Genes Cancer, 2*(3), 359−372.

Hampel, H., et al. (2005). Screening for the Lynch syndrome (hereditary nonpolyposis colorectal cancer). *The New England Journal of Medicine, 352*(18), 1851−1860.

Hampel, H., et al. (2008). Feasibility of screening for Lynch syndrome among patients with colorectal cancer. *Journal of Clinical Oncology: Official Journal of the American Society of Clinical Oncology, 26*(35), 5783−5788.

Hecht, J. R., et al. (2010). Lack of correlation between epidermal growth factor receptor status and response to Panitumumab monotherapy in metastatic colorectal cancer. *Clinical Cancer Research: An Official Journal of the American Association for Cancer Research, 16*(7), 2205−2213.

Heinemann, V., et al. (2014). FOLFIRI plus cetuximab vs FOLFIRI plus bevacizumab as first-line treatment for patients with metastatic colorectal cancer (FIRE-3): A randomised, open-label, phase 3 trial. *The Lancet Oncology, 15*(10), 1065−1075.

Herceptin [package insert]. (2017). South San Francisco, CA: Genentech, Inc.

Hsu, L., et al. (2015). A model to determine colorectal cancer risk using common genetic susceptibility loci. *Gastroenterology, 148*(7), 1330-9 e14.

Jasperson, K. W., et al. (2010). Hereditary and familial colon cancer. *Gastroenterology, 138*(6), 2044−2058.

Jeon, J., et al. (2018). Determining risk of colorectal cancer and starting age of screening based on lifestyle, environmental, and genetic factors. *Gastroenterology, 154*(8), 2152−2164, e19.

Jones, J. C., et al. (2017). *(Non-V600) BRAF mutations define a clinically distinct molecular subtype of metastatic colorectal cancer. Journal of Clinical Oncology: Official Journal of the American Society of Clinical Oncology, 35*(23), 2624−2630.

KADCYLA [package insert]. (2019). Genentech, Inc.

Keytruda [package insert]. (2020). Whitehouse Station, NJ: Merck Sharp & Dohme Corp.

Kopetz, S., et al. (2015). Phase II pilot study of vemurafenib in patients with metastatic BRAF-mutated colorectal cancer. *Journal of Clinical Oncology: Official Journal of the American Society of Clinical Oncology, 33*(34), 4032−4038.

Kopetz, S., et al. (2019). Encorafenib, binimetinib, and cetuximab in BRAF V600E-mutated colorectal cancer. *The New England Journal of Medicine, 381*(17), 1632−1643.

Kopetz, S., et al. (2020). Encorafenib plus cetuximab with or without binimetinib for BRAF V600E-mutant metastatic colorectal cancer: Quality-of-life results from a randomized, three-arm, phase III study vs the choice of either irinotecan or FOLFIRI plus cetuximab (BEACON CRC). *Journal of Clinical Oncology, 38*(4_suppl), p. 8-8.

Le, D. T., et al. (2015). PD-1 blockade in tumors with mismatch-repair deficiency. *The New England Journal of Medicine, 372*(26), 2509−2520.

Le, D. T., et al. (2020). Phase II open-label study of pembrolizumab in treatment-refractory, microsatellite instability-high/mismatch repair-deficient metastatic colorectal cancer: Keynote-164. *Journal of Clinical Oncology: Official Journal of the American Society of Clinical Oncology, 38*(1), 11−19.

Loree, J. M., et al. (2018). Classifying colorectal cancer by tumor location rather than sidedness highlights a continuum in mutation profiles and consensus molecular subtypes. *Clinical Cancer Research: An Official Journal of the American Association for Cancer Research, 24* (5), 1062−1072.

Lu, Y., et al. (2019). Large-scale genome-wide association study of east asians identifies loci associated with risk for colorectal cancer. *Gastroenterology, 156*(5), 1455−1466.

McKay, J. A., et al. (2002). Evaluation of the epidermal growth factor receptor (EGFR) in colorectal tumours and lymph node metastases. *European Journal of Cancer, 38*(17), 2258−2264.

Mektovi® (binimetinib). (2019). *Prescribing information.* Boulder, CO: Array BioPharma, Inc.

Mork, M. E., et al. (2015). High prevalence of hereditary cancer syndromes in adolescents and young adults with colorectal cancer. *Journal of Clinical Oncology: Official Journal of the American Society of Clinical Oncology, 33*(31), 3544−3549.

Morse, M. A., et al. (2019). Safety of nivolumab plus low-dose ipilimumab in previously treated microsatellite instability-high/mismatch repair-deficient metastatic colorectal cancer. *The Oncologist, 24*(11), 1453−1461.

Munro, A. J., Lain, S., & Lane, D. P. (2005). P53 abnormalities and outcomes in colorectal cancer: A systematic review. *British Journal of Cancer, 92*(3), 434−444.

National Comprehensive Cancer Network. (2020). *Colon cancer (Version 2.2020).* < https://www.nccn.org/professionals/physician_gls/pdf/colon.pdf>. Accessed 14.04.20.

Nunes, L., et al. (2020). Molecular characterization of a large unselected cohort of metastatic colorectal cancers in relation to primary tumor location, rare metastatic sites and prognosis. *Acta Oncologica (Stockholm, Sweden), 59*(4), 417−426.

Opdivo [package insert]. (2020). Princeton, NJ: Bristol-Myers Squibb Company.

Overman, M. J., et al. (2017). Nivolumab in patients with metastatic DNA mismatch repair-deficient or microsatellite instability-high colorectal cancer (CheckMate 142): An open-label, multicentre, phase 2 study. *The Lancet Oncology, 18*(9), 1182−1191.

Palles, C., et al. (2013). Germline mutations affecting the proofreading domains of POLE and POLD1 predispose to colorectal adenomas and carcinomas. *Nature Genetics, 45*(2), 136−144.

Pardini, B., et al. (2020). DNA repair and cancer in colon and rectum: Novel players in genetic susceptibility. *International Journal of Cancer, 146*(2), 363−372.

Perjeta [package insert]. (2017). South San Francisco, CA: Genentech, Inc.

Popat, S., Hubner, R., & Houlston, R. S. (2005). Systematic review of microsatellite instability and colorectal cancer prognosis. *Journal of Clinical Oncology: Official Journal of the American Society of Clinical Oncology, 23*(3), 609−618.

Powell, S. M., et al. (1993). Molecular diagnosis of familial adenomatous polyposis. *The New England Journal of Medicine, 329*(27), 1982−1987.

Prahallad, A., et al. (2012). Unresponsiveness of colon cancer to BRAF(V600E) inhibition through feedback activation of EGFR. *Nature, 483*(7387), 100−103.

Rex, D. K., et al. (2017). Colorectal cancer screening: Recommendations for physicians and patients from the U.S. multi-society task force on colorectal cancer. *The American Journal of Gastroenterology, 112*(7), 1016−1030.

Ryan, M. B., & Corcoran, R. B. (2018). Therapeutic strategies to target RAS-mutant cancers. *Nature Reviews Clinical Oncology, 15*(11), 709−720.

Sartore-Bianchi, A., et al. (2016). Dual-targeted therapy with trastuzumab and lapatinib in treatment-refractory, KRAS codon 12/13 wild-type, HER2-positive metastatic colorectal cancer (HERACLES): A proof-of-concept, multicentre, open-label, phase 2 trial. *The Lancet Oncology, 17*(6), 738–746.

Seo, A. N., et al. (2014). HER2 status in colorectal cancer: Its clinical significance and the relationship between HER2 gene amplification and expression. *PLoS One, 9*(5), e98528.

Siegel, R. L., et al. (2017). Colorectal cancer incidence patterns in the United States, 1974–2013. *Journal of the National Cancer Institute, 109*(8).

Siegel, R. L., et al. (2020). Colorectal cancer statistics, 2020. *CA: A Cancer Journal for Clinicians.*

Siena, S., et al. (2019). Entrectinib in NTRK-fusion positive gastrointestinal cancers: Integrated analysis of patients enrolled in three trials (STARTRK-2, STARTRK-1, and ALKA-372-001). *Annals of Oncology: Official Journal of the European Society for Medical Oncology, 30*(Suppl 4), iv134.

Solomon, J. P., et al. (2020). NTRK fusion detection across multiple assays and 33,997 cases: Diagnostic implications and pitfalls. *Modern Pathology: An Official Journal of the United States and Canadian Academy of Pathology, 33*(1), 38–46.

Sorich, M. J., et al. (2015). Extended RAS mutations and anti-EGFR monoclonal antibody survival benefit in metastatic colorectal cancer: A meta-analysis of randomized, controlled trials. *Annals of Oncology: Official Journal of the European Society for Medical Oncology, 26*(1), 13–21.

Souglakos, J., et al. (2009). Prognostic and predictive value of common mutations for treatment response and survival in patients with metastatic colorectal cancer. *British Journal of Cancer, 101*(3), 465–472.

Stoffel, E., et al. (2009). Calculation of risk of colorectal and endometrial cancer among patients with Lynch syndrome. *Gastroenterology, 137*(5), 1621–1627.

Stoffel, E. M. (2016). Heritable gastrointestinal cancer syndromes. *Gastroenterology Clinics of North America, 45*(3), 509–527.

Strickler, J. H., et al. (2017). A phase II, open label study of tucatinib (ONT-380) combined with trastuzumab in patients with HER2+ metastatic colorectal cancer (mCRC) (MOUNTAINEER). *Journal of Clinical Oncology, 35*(15_suppl), p. TPS3624.

Syngal, S., et al. (2015). ACG clinical guideline: Genetic testing and management of hereditary gastrointestinal cancer syndromes. *The American Journal of Gastroenterology, 110*(2), 223–262.

Tykerb (lapatinib) prescribing information. (2018). East Hanover, NJ: Novartis.

Vectibix (panitumumab) [package insert]. (2014). Thousand Oaks, CA: Amgen, Inc.

VITRAKVI [package insert]. (2019). Whippany, NJ: Bayer HealthCare Pharmaceuticals, Inc.

Vogelstein, B., et al. (1988). Genetic alterations during colorectal-tumor development. *The New England Journal of Medicine, 319*(9), 525–532.

Wang, F., et al. (2019). Evaluation of POLE and POLD1 mutations as biomarkers for immunotherapy outcomes across multiple cancer types. *JAMA Oncology.*

Weigl, K., et al. (2018). Genetic risk score is associated with prevalence of advanced neoplasms in a colorectal cancer screening population. *Gastroenterology, 155*(1), 88–98, e10.

Willauer, A. N., et al. (2019). Clinical and molecular characterization of early-onset colorectal cancer. *Cancer, 125*(12), 2002–2010.

Yervoy [package insert]. (2019). Princeton, NJ: Bristol Meyers Squib.

Chapter 13

Hepatic and bile duct cancers

Jingquan Jia[1] and Rachna Shroff[2]
[1]*Division of Medical Oncology, Department of Medicine, Duke University Medical Center, Durham, NC, United States,* [2]*Division of Hematology and Oncology, Department of Medicine, University of Arizona College of Medicine, Tucson, AZ, United States*

Introduction

Hepatic and bile duct cancers encompass multiple types of invasive carcinomas including hepatocellular carcinoma (HCC), cholangiocarcinoma (CCA), and gallbladder cancer (GBC). CCA and GBC are collectively called biliary tract cancers (BTCs). HCC arises from hepatocytes or hepatic stem cells while BTCs originate from bile duct epithelium. In 2018, there were 1,060,500 cases of hepatobiliary cancer diagnosed and 946,718 deaths related to hepatobiliary cancer worldwide (Bray et al., 2018). In the United States, 54,390 new cases of hepatobiliary cancer were diagnosed, while 35,740 patients died from these malignancies in 2019. Hepatobiliary cancers are rapidly becoming one of the deadliest cancers, with incidence rates more than tripling and the death rate more than doubling since 1980. Over the past decade, the incidence rate increased by about 3% per year with the mortality rate increasing by 2.4% per year (Siegel, Miller, & Jemal, 2019).

Diagnosis of HCC, CCA, or GBC is based on radiologic and/or pathologic evaluations. Treatment approaches for hepatic and bile duct cancers are quite distinct. Early-stage HCC is generally treated with surgical resection, liver transplantation, embolization, ablation, or radiotherapy, while surgical resection remains the mainstay of treatment for early-stage biliary cancers. For advanced HCC, tyrosine kinase inhibitors and immunotherapy are the current standard of care systemic therapies in both first-line and second-line settings (Abou-Alfa et al., 2018; Bruix et al., 2017; El-Khoueiry et al., 2017; Finn et al., 2020; Finn et al. 2020; Kudo et al., 2018; Llovet et al., 2008; Thomas Yau, Tae-You, & El-Khoueiry, 2019; Zhu et al., 2019). In contrast, cytotoxic chemotherapy remains the cornerstone treatment for BTCs (Lamarca et al., 2021; Shroff et al., 2019; Valle et al., 2010).

Genomic and Precision Medicine. DOI: https://doi.org/10.1016/B978-0-12-800684-9.00015-0
217

Epidemiology

Hepatocellular carcinoma

HCC is most prevalent in Eastern Asia and sub-Saharan Africa, accounting for over 85% of the incidence globally (Mittal & El-Serag 2013). Meanwhile, its incidence in the Western world is steadily rising in recent years (Center & Jemal 2011). Overall survival of patients with HCC in Taiwan and Japan is substantially better than in sub-Saharan Africa, likely as a result of nationwide intensive surveillance programs leading to diagnosis of HCC at an earlier stage (Park et al., 2015).

Risk factors for the development of HCC include viral infections, metabolic disorders, and certain environmental toxin exposures. Chronic hepatitis B virus (HBV) and/or hepatitis C virus (HCV) infection are the leading cause of HCC in Eastern Asia and Africa, whereas alcoholic cirrhosis and nonalcoholic fatty liver disease (NAFLD) are the most common risk factors in the United States and Europe. Although relatively rare, aflatoxin, aristolochic acid (AA), and other chronic liver diseases such as α-1-antitrypsin deficiency, hemochromatosis, and primary biliary cholangitis are additional risk factors for HCC (Yang et al., 2019).

Primary prevention is essential for decreasing the burden of HCC in viral hepatitis endemic areas. Most countries now conduct universal HBV vaccination for neonates and infants. Maintaining a healthy lifestyle and avoiding heavy alcohol consumption are additional strategies for HCC primary prevention.

HCC surveillance is a secondary prevention strategy aiming for early detection and diagnosis of HCC. It is indicated in patients with liver cirrhosis or chronic HBV infection with high-risk features and comprised liver ultrasonography in conjunction with serum alpha fetal protein (AFP) at regular intervals.

Biliary tract cancers

GBC is the most common BTC. The incidence of GBC increases with age and women more are likely to be affected than men (Henley, Weir, Jim, Watson, & Richardson, 2015). The incidence rate is elevated in certain areas of the world, such as Southeast Asia, Pakistan, Korea, and Japan. The highest incidence rate is seen in women in northern India at 21.5 of 100,000 (Randi, Franceschi, & La Vecchia, 2006; Sharma, Sharma, Gupta, Yadav, & Kumar, 2017). Risk factors associated with GBC include cholelithiasis with chronic gallbladder inflammation, porcelain gallbladder, gallbladder polyp, congenital biliary cysts, chronic typhoid infection, and obesity (Tazuma & Kajiyama 2001). Prophylactic cholecystectomy can be considered for patients at high risk of GBC, although there are no recommended screening guidelines for GBC.

Based on the anatomical location, CCA is further categorized into intrahepatic cholangiocarcinoma (iCCA) and extrahepatic cholangiocarcinoma (eCCA), which includes perihilar cholangiocarcinoma (pCCA) and distal cholangiocarcinoma (dCCA). eCCA is more common than iCCA. According to SEER database, the incidences of both iCCA and eCCA have increased significantly over the past several decades, albeit eCCA at a slower rate (Mukkamalla, Naseri, Kim, Katz, & Armenio, 2018). Most patients with CCA do not have identifiable risk factors, although primary sclerosing cholangitis and bile duct anomalies such as choledochal cysts are known to be predisposing factors for CCA (Tyson & El-Serag 2011). In South-East Asia, especially in Northeast Thailand, endemic liver fluke infection is a major risk factor for CCA. Interestingly, recent studies showed that many well-established risk factors for HCC, including HBV infection, cirrhosis, obesity, alcohol, and NAFLD, are also associated with the development of iCCA, suggesting certain similarity in tumorigenesis between HCC and iCCA (Welzel et al., 2007; Wongjarupong et al., 2017).

Genomic alterations

In recent years, advances in DNA sequencing technologies have enabled our understanding of the complex genomic landscape of hepatic and bile duct cancers. Genomic analysis has not only identified driver mutations responsible for tumor initiation and progression, but also provided a detailed mutational signature of HCC and BTCs.

Molecular drivers

Hepatocellular carcinoma

Each HCC has an average of 40 somatic alterations detected in the coding region of the genome. Most of these mutations occur in "passenger genes" without significant impact on carcinogenesis, but a few are considered "drivers" that are directly involved in activating key signaling pathways for HCC tumorigenesis (Llovet et al., 2016). Well-characterized driver pathways include mutations affecting telomere maintenance, WNT activation, cell-cycle control, chromatin remodeling, PIK3CA/AKT/m-TOR, RAS/RAF/MAPK, and oxidative stress pathways (Table 13.1).

Telomerase reverse transcriptase (TERT), which encodes the catalytic unit of the telomerase complex, is essential in telomere maintenance. TERT mRNA expression and telomerase activity were found in ~90% of human HCC tissues (Nagao, Tomimatsu, Endo, Hisatomi, & Hikiji, 1999). Elevated telomerase activity was significantly associated with increased risk of HCC recurrence after surgical resection (Kobayashi, Kubota, Takayama, & Makuuchi, 2001). Furthermore, TERT was not expressed in normal hepatocytes, but was present in early neoplastic lesions such as regenerative nodules and dysplastic nodules in cirrhotic liver (Kotoula et al., 2002),

TABLE 13.1 Major driver genetic alterations in HCC.

Pathway	Gene	Alteration	Frequency
Telomere maintenance	TERT	Activating mutation in promoter	54%−60%
		Amplification	1%−6%
Wnt-β catenin signaling	CTNNB1	Activating mutation	11%−37%
	AXIN1	Inactivating mutation or deletion	5%−15%
	APC	Inactivating mutation	1%−3%
Cell-cycle control	TP 53	Inactivating mutation	12%−48%
	RB1	Inactivating mutation or deletion	3%−8%
	CDKN2A	Inactivating mutation or deletion	2%−12%
	CCND1	Amplification	7%
Chromatin remodeling	ARID1A	Inactivating mutation	4%−7%
	ARID2	Inactivating mutation	3%−18%
MAPK and PIK3CA	RPS6KA3	Mutation	2%−9%
	FGF3, FGF4, and FGF19	Amplification	4%−6%
	TSC1 and TSC2	Mutation	3%−8%
	PIK3CA	Activating mutation	1%−4%
	PTEN	Inactivation mutation or deletion	1%−3%
Oxidative stress	NFE2L2	Activating mutation	3%−6%
	KEAP1	Activating mutation	2%−5%

Source: Adapted from Llovet, J. M., Zucman-Rossi, J., Pikarsky, E., Sangro, B., Schwartz, M., Sherman, M., et al. (2016). Hepatocellular carcinoma. *Nature Reviews Disease Primers, 2*, 16018. Llovet, J. M., Zucman-Rossi, J., Pikarsky, E., Sangro, B., Schwartz, M., Sherman, M., et al. (2016). Hepatocellular carcinoma. *Nature Reviews Disease Primers, 2*, 16018. Ding, X. X., Zhu, Q. G., Zhang, S. M., et al. (2017). Precision medicine for hepatocellular carcinoma: driver mutations and targeted therapy. *Oncotarget, 8*(33), 55715−55730.

suggesting that TERT activation is involved in early hepatocarcinogenesis. *TERT* promoter mutations are the most frequently found genetic alterations in HCC (Nault et al., 2013). Other mechanisms involved in telomerase reactivation in HCC include HBV insertion in the *TERT* promoter (Sung et al., 2012) and *TERT* amplification (Totoki et al., 2014).

The WNT/β-catenin pathway is a critical oncogenic pathway that is responsible for the tumorigenesis of many types of cancers. Its activation is frequently observed in HCC, especially in well-differentiated tumors (Audard et al., 2007). In 26% of human HCCs, activating mutations of *CTNNB1* lead to accumulation of β-catenin in the nucleus (de La Coste et al., 1998). Inactivating mutations or deletions of *AXIN1* (Satoh et al., 2000; Taniguchi et al., 2002), *APC* and *ZNRF3* are also observed, although the latter two are rare (Schulze et al., 2015; Totoki et al., 2014).

TP53 inactivation results in loss of cell-cycle control in HCC (Bressac, Kew, Wands, & Ozturk, 1991; Hsu et al., 1991; Schulze et al., 2015). This is detected in 24% of patients with HCC (Schulze et al., 2015). Retinoblastoma 1 (*RB1*) mutations and cyclin-dependent kinase inhibitor (*CDKN2A*) deletions have also been identified in HCC and are associated with poor prognosis, possibly by contributing to a more aggressive phenotype (Ahn et al., 2014; Schulze et al., 2015). Amplification of cyclin D1 (*CCND1*) (Wang et al., 2013) and recurrent HBV insertions in cyclin E1 (*CCNE1*) (Sung et al., 2012) are two other genetic aberrations underlying cell-cycle dysregulation in HCC.

Key subunits of chromatin remodeling SWI/SNF (switch/sucrose nonfermentable) complex encoded by *ARID1A* (AT-rich interactive domain-containing protein 1A) and *ARID2* genes are frequently inactivated in HCC, through missense, nonsense, frameshift mutations, and splice site mutations (Fujimoto et al., 2012; Li et al., 2011). ARID2 protein expression is significantly downregulated in human HCC tissue in comparison to normal liver. Furthermore, restoration of ARID2 expression inhibits tumor cell proliferation in preclinical models, highlighting the potential role of the SWI/SNF complex as a tumor suppressor (Duan et al., 2016). Somatic mutations also occur in the histone methylation writer family, mainly in *MLL* genes (Cleary et al., 2013).

The PIK3CA/AKT/m-TOR and RAS/RAF/MAPK pathways are often activated in HCC by a variety of genetic alterations, including amplification of the *FGF3*, *FGF4*, and *FGF19* locus (Sawey et al., 2011), activating mutations of *PIK3CA*, deletion of *PTEN*, and inactivating mutations of *TSC1* or *TSC2*. Activating mutations of the *RAS* gene family and inactivating mutations of genes encoding for RAS inhibitor *RSK2* are present but relatively rare in HCC (Guichard et al., 2012; Schulze et al., 2015). Recent studies suggest that the mTOR pathway plays a vital role in hepatocarcinogenesis and that activation of mTOR pathway is associated with poor prognosis (Villanueva et al., 2008; Zhou, Huang, Li, & Wang, 2010).

The oxidative stress pathway is relevant in HCC through mutations that activate nuclear factor erythroid 2−related factor 2 (*NFE2L2*) or mutations that inactivate Kelch-like ECH-associated protein 1 (*KEAP1*). Silencing of *KEAP1* prevents proteasome degradation of NFE2L2 induced by ubiquitinylation. NFE2L2 appears to play a dual role in HCC where it may protect against chronic oxidative stress on the liver and HCC development in mouse models (DeNicola et al., 2011) but constitutive activation facilitates late-stage tumor progression (Sporn & Liby 2012).

Biliary tract cancers

Similar to HCC, a number of genetic alterations in BTCs are involved in receptor tyrosine kinase signaling, cell-cycle regulation, DNA repair, and epigenetic regulation of gene expression (Table 13.2).

Large-scale whole exome sequencing (WES) from a Japanese BTC cohort discovered that the majority of the driver mutations are located in *TP53*, *KRAS*, and *ARID1A* genes. Other driver mutations occur in *SMAD4*, *BAP1*, *PIK3CA*, *ARID2*, and *GNAS*, albeit at lower frequencies. Different anatomic subtypes of BTCs seem to have both shared driver genes and subtype-specific ones. *TP53*, *BRCA1*, *BRCA2*, and *PIK3CA* mutations are common drivers across all BTCs. *KRAS*, *SMAD4*, *ARID1A*, and *GNAS* mutations are shared in iCCA and eCCA but rarely found in GBCs. *ELF3* and *ARID1B* mutations and *PRKACA* or *PRKACB* fusions are considered eCCA specific, whereas *IDH1/2* and *BAP1* mutations as well as *FGFR2* fusions are more iCCA specific. In addition to *TERT* promoter mutations, *EGFR*, *ERBB2*, *PTEN*, and *ARID2* mutations are more common in GBC (Nakamura et al., 2015).

TABLE 13.2 Major driver genetic alterations in BTCs.

Pathway	Gene	Alteration	Frequency
Cell-cycle control	TP53	Inactivating mutation	26%–32%
	CDKN2A	Inactivating mutation	47% (iCCA)
MAPK and PIK3CA	KRAS	Activating mutation	16%–18%
	FGFR2	Fusion	10%–16% (iCCA)
	PIK3CA	Activating mutation	4%–9%
	PTEN	Inactivating mutation	4%
	ERBB2	Amplification	11%–20% (eCCA)
Chromatin remodeling	ARID1A	Inactivating mutation	11%–17%
	SMAD4	Inactivating mutation	13%
	BAP1	Inactivating mutation	8.5%
Epigenetic regulation	IDH1/IDH2	Inactivating mutation	25% (iCCA)

Source: Adapted from Pellino, A., Loupakis, F., Cadamuro, M., Dadduzio, V., Fassan, M., Guido, M., Cillo, U., Indraccolo, S., & Fabris, L. (2018) Precision medicine in cholangiocarcinoma. *Translational Gastroenterology and Hepatology, 3*, 40.

Consistent with the findings from the Japanese cohort, the largest CCA dataset encompassing 489 cases including both Asian and Caucasian patients identified *TP53, ARID1A, KRAS, SMAD4,* and *BAP1* as the top five driver genes in CCA, followed by *APC, PRBM1,* and *ELF3*. Since this dataset contained the largest number of fluke-positive samples, it was able to identify *KRAS, FBXW7,* and *PTEN* mutations as well as *ERBB2* amplifications as more frequently associated with fluke-positive CCA. In contrast, *IDH1, PBRM1,* and *ACVR2A* mutations were detected in nonfluke-associated CCA (Jusakul et al., 2017). Comprehensive genomic profiling on 412 iCCA and 57 eCCA tumors by Javle et al. showed that the most frequent genomic alterations in iCCA are *TP53, CDKN2A/B, KRAS, ARID1A,* and *IDH1,* whereas *KRAS, TP53, CKDN2A/B,* and *SMAD4* are most frequently mutated in eCCA (Javle et al., 2016).

Genomic profiling data for GBC are limited. NGS profiling on advanced biliary cancers using the FoundationOne platform by Javle et al. included 85 cases of GBC. *TP53* (59%), *CDKN2A/B* (19%), *ERBB2* (16%), and *ARID1A* (13%) were the most frequently altered genes (Javle et al., 2016). In another WES analysis on 57 Chinese GBC samples, the top three driver genes were *TP53, ERBB3,* and *KRAS*. In fact, all ErbB family members, including *EGFR, ERBB2, ERBB3,* and *ERBB4,* were extensively mutated (Li et al., 2014).

Mutational signatures

Somatic mutations induced by different endogenous and exogenous mutagenic processes bear unique patterns of nucleotide changes known as "mutational signatures." Understanding such somatic mutation patterns can help us understand tumorigenesis in hepatobiliary cancers.

Hepatocellular carcinoma

Based on WES data of 503 HCC cases from Japanese and TCGA cohorts, three mutational signatures (COSMIC signatures 1, 16, and 22) were isolated. COSMIC signature 1 was common in both cohorts, whereas signature 16 was prevalent in Japanese male patients and signature 22 was preferentially found in Asian patients living in the United States (Totoki et al., 2014). Whole-genome sequencing (WGS) on 300 Japanese patients revealed seven mutational signatures, among which COSMIC signature 4 was associated with smoking and the presence of *TP53* mutations, while COSMIC signature 16 was associated with alcohol consumption, male gender, old age, and the presence of *TERT* promoter mutations and *CTNNB1* mutations (Fujimoto et al., 2012). Similar analysis was carried out in 243 cases of HCC from Europe and identified a novel COSMIC signature 24 that was associated with a *TP53* R249S mutation characteristic of aflatoxin B1 exposure

(Schulze et al., 2015). The association between COSMIC signature 16, *CTNNB1* mutations, and epidemiological factors such as alcohol intake and male gender was later confirmed in a European HCC cohort (Letouze et al., 2017). Consistent with exposure to herbal medicines, AA-associated HCC was frequently associated with COSMIC signature 22 in Taiwan, China, and Southeast Asia but was infrequently detected in Korea or Japan and even more rare in North America and Europe (Ng et al., 2017).

Biliary tract cancers

Two mutational signatures were extracted from the aforementioned Japanese BTC cohort: COSMIC signature 1 and COSMIC signature 2/13 (APOBEC signature). COSMIC signature 1 was more prevalent in iCCA, while the APOBEC signature was more frequently found in eCCA and GBC (Nakamura et al., 2015).

Integrative molecular classification

Recognizing the genetic heterogeneity of hepatic and bile duct cancers is the first step toward understanding the molecular behavior of these tumors. While tumors at the same clinical stage can be vastly different from each other biologically, the treatment of hepatic and bile duct cancers mainly relies on clinical staging. Deeper understanding of the different biological events at a molecular level is indispensable for the development of novel biomarkers and targeted therapeutics in the era of precision medicine. Integrating genomic data with transcriptomic, epigenetic, and/or proteomic profiling of resected hepatic and bile duct tumors has enabled identification of unique molecular subtypes of HCC and BTCs.

Hepatocellular carcinoma

Integrative transcriptome data isolated three HCC molecular subtypes, with the S1 subtype reflecting aberrant noncanonical WNT signaling and TGF-β activation, the S2 subtype reflecting a progenitor cell signature, and the S3 subtype reflecting normal hepatocyte differentiation (Hoshida et al., 2009). Recently, the largest collection of DNA sequencing, DNA methylation, RNA, miRNA, and proteomic expression in 196 HCC cases was assembled by The Cancer Genome Atlas HCC group. By incorporating five data platforms, three subtypes were identified (iClust1−3). The gene expression profile of iClust1 subtype resembles that of the progenitor subtype (S2), iClust2 subtype shares molecular characteristics with the S3 subtype, and iClust3 had a higher degree of chromosomal instability with a high frequency of *TP53* mutations and poor prognosis (Cancer Genome Atlas Research Network, 2017).

Based on these findings, two major molecular classifications of HCC have been proposed: a proliferation class and a nonproliferation class, each accounting for ~50% of HCC. The proliferation class is characterized by enrichment in cell proliferation signaling activation such as RAS/MAPK, AKT/mTOR, and IGF2, chromosomal instability such as *FGF19* amplification, vascular invasion gene signatures, and DNA methylation profile resembling progenitor cells. HCCs belonging to the proliferation class are often more aggressive, with higher AFP levels, are associated with HBV infection, and have poor clinical outcomes. In contrast, the nonproliferation class frequently exhibits activation of the canonical WNT signaling pathway and an immune-related gene signature. HCCs in this class are often HCV and alcohol-related, have a less aggressive phenotype, have lower AFP levels, and are prognostically favorable (Zucman-Rossi, Villanueva, Nault, & Llovet, 2015).

Since HCC develops in an inflammatory milieu, a novel immune-based molecular classification of HCC has been proposed. Of note, 30% of HCCs are classified as "immune class" with increased immune cell infiltration, IFNγ signaling activation, and high levels of PD-1/PD-L1 expression. About 25% of HCCs are classified as "immune excluded class," characterized by *CTNNB1* mutations and exclusion of T cells from the tumor microenvironment (Sia et al., 2017).

Biliary tract cancers

Clustering of global gene expression levels determined by transcriptome sequencing analysis elucidated four molecular subtypes in BTCs (Nakamura cluster 1−4). This classification was correlated with driver gene combinations and patient prognosis. Expression of metabolic genes was increased in cluster 3 where *BAP1*, *IDH1*, and *NRAS* mutations and *FGFR2* fusions were enriched. Patients in cluster 1 had the best survival and they mostly had eCCA. Patients in cluster 4 had worse survival and showed positive enrichment for immune-related genes, antiapoptotic genes, and cytokine activity. Cluster 4 tumors also had a significantly higher mutation load. Consistent with these findings, expression of nine immune checkpoint molecules, including PD-L1, was also significantly higher in this cluster (Nakamura et al., 2015).

Jusakul et al. performed WGS in combination with epigenomic analysis on 489 CCAs and revealed four distinct hierarchical clusters by different clinical features and genomic alterations (Jusakul et al., 2017). Cluster 1 consisted of mostly fluke-positive tumors, characterized by hypermethylation of promoter CpG islands, enrichment of *ARID1A* and *BRCA2* mutations, as well as high frequency of *H3K27me3* promoter mutations. Cluster 2 comprised fluke-positive and fluke-negative tumors, with upregulated CTNNB1, WNT5B, and AKT1 expression. Both Clusters 1 and 2 were enriched for

ERBB2 amplification and HER2 overexpression. In contrast to Clusters 1 and 2, Clusters 3 and 4 were mostly fluke-negative tumors. Cluster 3 exhibited the highest levels of somatic copy number alterations, particularly amplifications at chromosome 2p and 2q. In addition, high levels of immune checkpoint genes (PD-1, PD-L1, and BTLA) and genes involved in antigen cross-presentation and T-cell signal transduction were found in Cluster 3 relative to other clusters. On the other hand, Cluster 4 displayed enrichment of *BAP1* mutations, *IDH1/2* mutations, *FGFR* alterations, upregulation of PI3K pathway, and FGFR family gene expression.

Clinical applications of genomics

Prognostic value

Hepatocellular carcinoma

To date, over 40 prognostic gene signatures have been described in HCC (Hoshida, Moeini, Alsinet, Kojima, & Villanueva, 2012). While most prognostic signatures were developed from transcriptome data (Boyault et al., 2007; Lee et al., 2004; Villanueva et al., 2011), others were generated using epigenetic data (Budhu et al., 2008; Toffanin et al., 2011; Villanueva et al., 2015). HCCs with a gene expression profile resembling fetal hepatoblasts were found to have poor prognosis (Lee et al., 2006). Meanwhile, a 36-CpG DNA methylation marker signature accurately predicted poor survival in HCC (Villanueva et al., 2015).

Most of the prognostic gene signatures identify patients within the proliferation class, with the exception of a 5-gene score (Nault et al., 2013) that is also able to predict the prognosis of tumors in the nonproliferative class. The molecular 5-gene score, based on combined expression level of *HN1*, *RAN*, *RAMP3*, *KRT19*, and *TAF9*, was derived and validated from a French cohort containing 314 HCC surgical specimens. The 5-gene score was associated with disease-specific survival independent of other clinical and pathology features of HCC. It also predicted disease-specific survival when incorporated in a nomogram with data on microvascular invasion and the Barcelona Clinic Liver Cancer (BCLC) staging system.

In addition to analyzing the tumor tissue itself, efforts have also been made to characterize the gene expression profile from adjacent cirrhotic liver tissue in order to understand the impact of stroma on HCC prognosis. A 186-gene signature enriched for genes related to inflammation, including IFN signaling, NF-κB activation, and TNF-α signaling, was associated with poor prognosis (Hoshida et al., 2008). Similarly, gene expression data from human liver undergoing hepatic injury and regeneration revealed a 233-gene signature that was significantly associated with late recurrence of HCC. Of note, 4- and 20-gene predictors were further developed and validated from the full

233-gene signature and *STAT3* activation was shown to be of particular significance (Kim et al., 2014).

Interestingly, combining the 5-gene score from the tumor with the 186-gene signature from its surrounding liver tissue further enhanced its prognostic accuracy (Nault et al., 2013), suggesting that genomic data from both the tumor and the adjacent liver are complementary for optimizing prognosis.

Despite the promising findings above, none of the molecular signatures have made a meaningful impact in clinical practice. First of all, most of the molecular profiles were generated from surgically resected specimens, therefore they may not accurately reflect the molecular landscape of advanced HCCs. Secondly, these prognostic gene signatures require tissue collection, which often is not necessary for the diagnosis of HCC. In recent years, circulating tumor DNA (ctDNA) has also been investigated in various HCC cohorts but no definitive prognostic ctDNA signatures have been developed yet (Ye, Ling, Zheng, & Xu, 2019).

Biliary tract cancers

Some genetic mutations have been reported to have prognostic value in BTCs. A retrospective analysis of 183 patients with advanced BTCs found that *CDKN2A*, *TP53*, and *ARID1A* mutations were significantly associated with outcome and may predict response to gemcitabine and platinum-based chemotherapy. For example, patients with loss of *TP53* and *CDKN2A* in the absence of *ARID1A* mutations have the worst survival (Ahn et al., 2016). Similarly, genetic alterations in *KRAS*, *TP53*, *CDKN24/B*, and MAPK/ERK pathway are correlated with worse survival, independent of clinical stage or tumor location (Javle et al., 2016). Specifically in iCCA, *TP53* and *KRAS* mutations are associated with significantly worse overall survival. While some studies suggest that *FGFR2* fusions are associated with a relatively indolent disease course and better survival (Graham et al., 2014; Javle et al., 2016), the natural history of patients with *FGFR2* fusions remains largely unknown. Consistently, *KRAS*-mutated CCAs are characterized by major involvement of adjacent organs and R1 resections, resulting in shorter progression free survival (Zhu et al., 2014). In eCCA, genetic alterations in chromatin remodeling genes, *BAP1* and *PBRM1*, are associated with bone metastases and worse survival (Churi et al., 2014).

When looking at markers relevant to immune therapies in iCCA, high levels of PD-L1 expression are associated with poorly differentiated histology, higher tumor stage, and worse survival (Sabbatino et al., 2016). In a cohort of 308 European, liver fluke-negative CCAs, patients with microsatellite instability-high (MSI-H) tumors had longer survival despite more advanced stage (Goeppert et al., 2019). This finding is in contrast to an earlier report, showing MMR deficiency was associated with a trend toward worse survival (Kunk et al., 2018).

Targeted therapeutics

Even though most of the clonal mutations and prevalent driver mutations detected in HCC are not clinically actionable, novel therapeutics are being actively developed, especially for biliary cancers. Here we reviewed some of the most promising molecular-targeted therapies.

FGFR-targeted therapy

FGF19 amplification occurs in ∼10% of HCC and has been implicated in sorafenib resistance (Gao et al., 2017). FGFR4 is the predominant FGFR expressed in the liver and is therefore considered as a promising target. Multiple specific FGFR4 kinase inhibitors are currently moving through the clinical development pipeline, among which Fisogatinib is the furthest in clinical development with preliminary data showing a response rate of 17% in patients with FGF19 IHC-positive advanced HCCs (Kim et al., 2019).

In CCA, *FGFR2* fusions are seen in about 15% of patients and have become a relevant target with FDA-approved options for patients. It has been shown that patients with *FGFR* genetic alterations who are treated with FGFR-targeted therapies have superior survival compared to those who did not receive treatment with FGFR inhibitors (Javle et al., 2016). Pemigatinib, a selective tyrosine kinase inhibitor of FGFR1, 2, and 3, was approved by the US FDA for the treatment of chemotherapy-refractory unresectable or metastatic CCA with *FGFR2* fusions or other rearrangements. This approval was based on the FIGHT-202 study, which showed an overall response rate (ORR) of 36% and median duration of response (DOR) of 9.1 months (Vogel, Sahai, Hollebecque, Vaccaro, & Abou-Alfa, 2019). Infigratinib is an orally available pan-FGFR inhibitor, which showed a 19% response rate and 83% disease control rate in patients with *FGFR2* fusion positive metastatic CCA who were refractory to chemotherapy (Javle et al., 2018). The US FDA recently granted Fast Track designation to infigratinib for patients as first-line treatment for advanced or metastatic CCA with *FGFR2* gene fusions or translocations. Futibatinib, an irreversible FGFR1−4 tyrosine kinase inhibitor, was associated with a 34.3% ORR and disease control rate of 76.1%. Median time to response was 1.6 months and DOR was 6.2 months (Goyal et al., 2020).

CDK4/6-targeted therapy

CDKN2A silencing is one of the core driver pathways in HCC. The CDK4/6 inhibitors palbociclib and ribociclib are both being evaluated in phase 2 trials for their tolerability and efficacy in RB + HCC. Focal loss of *CDKN2A/2B* has been frequently identified in iCCA. It is unknown if the antitumor activity of CDK4/6 inhibitors is enhanced in CCAs with *CDKN2A/2B* alterations.

IDH1/2-targeted therapy

IDH1/2 mutations are one of the most common genetic alterations in iCCA. Ivosidenib is a selective IDH1 inhibitor that demonstrated promising efficacy with a favorable toxicity profile in patients with *IDH1*-mutated CCA (Lowery et al., 2019). In the ClarIDHy study, ivosidenib significantly improved progression-free survival compared to placebo. Although the ORR was only 2.4%, 50.8% of patients had stable disease—significantly greater than the placebo arm (Abou-Alfa et al., 2020). In addition to the selective IDH inhibitor, PARP inhibition is also being investigated as another potential therapeutic agent for IDH--mutated CCAs given that *IDH1/2* mutations can induce a "BRCAness" phenotype that renders sensitivity to PARP inhibitors in IDH-mutated cell lines (Sulkowski et al., 2017).

HER2-targeted therapy

Anti-HER2 therapy has shown promising activity in small cohorts of patients with advanced BTCs carrying *HER2* amplification, overexpression, or activating mutations. A retrospective review identified nine patients with metastatic GBC who received HER2-directed therapy, either alone or in combination with chemotherapy. One patient had complete response, four patients had partial response, and three patients had stable disease (SD) (Javle et al., 2015). In the MyPathway phase II basket study, among the 11 patients with metastatic BTC, 8 participants had *HER2* amplification/overexpression and 3 had *HER2* activating mutations. Treatment with trastuzumab plus pertuzumab was associated with a 29% ORR and three patients had SD greater than 120 days (Hainsworth et al., 2018).

As molecular profiling data on hepatobiliary cancer further accumulate, many targeted agents are actively being investigated. Numerous clinical trials are ongoing at various stages evaluating the therapeutic efficacy of targeted agents in biomarker-selected populations. Selected trials of molecularly targeted therapy are listed in Table 13.3.

Predicting response to immunotherapy

Immunotherapy is an essential component of advanced HCC systemic treatment. Bevacizumab plus atezolizumab demonstrated improved overall survival and quality of life compared to sorafenib, leading to its FDA approval as the new standard of care front-line therapy in patients with advanced and metastatic HCC (Finn et al., 2020). In addition, durvalumab plus tremelimumab combination therapy and lenvatinib plus pembrolizumab therapy have both shown promising clinical activity as first-line treatment for advanced HCC and are currently being further investigated in phase 3 trials (Finn et al., 2020a; Kelley et al., 2020). In the second-line setting, pembrolizumab and nivolumab were both approved by the FDA for patients with advanced

TABLE 13.3 Selected ongoing clinical trials of molecularly targeted therapy in hepatobiliary cancers.

Agent	Molecular target	Disease	Line/phase	Biomarker enrichment	Reference
Palbociclib	CDK4/6	HCC	2nd-L/II	RB +	NCT01356628
Capmatinib	MET	HCC	2nd-L/II	MET +	NCT01737827
Galunisertib	TGFβR1	HCC	1st-L (+sorafenib)/I-II	None	NCT02178358
SF1126	PI3K and mTOR	HCC	2nd-L/I	None	NCT03059147
Fisogatinib	FGFR4	HCC	2nd-L/I-II	FGF19 + by IHC	NCT02508467
INCB062079	FGFR4	HCC CCA	2nd-L/I-II	FGF19 amplification	NCT03144661
Pemigatinib	FGFR2	CCA	1st-L/III	FGFR2 rearrangement	NCT03656536
Infigratinib	FGFR2	CCA	1st-L/III	FGFR2 fusions/translocations	NCT03773302
Ivosidenib	IDH1	CCA	2nd-L/III	IDH1 mutation	NCT02989857
Olaparib	PARP	CCA	2nd-L/II	IDH1 or IDH2 mutation	NCT03212274
Niraparib	PARP	CCA	2nd-L/II	BAP1and other DNA damage repair pathway mutations	NCT03207347
A166	HER2	CCA	2nd-L/I-II	HER2 +	NCT03602079
Afatinib	EGFR HER2 HER4	CCA	2nd-L (+capecitabine)/I	None	NCT02451553
DKN-01	DKK1	CCA	2nd-L (+ Nivolumab)/II	None	NCT04057365
CB-103	NOTCH	CCA	2nd-L/II	None	NCT03422679
AbGn-107	AG7	CCA	2nd-L/I	AG7 high	NCT02908451

Source: Adapted from Harris, W. P., Wong, K. M., Saha, S., Dika, I. E., & Abou-Alfa, G. K. (2018). Biomarker-driven and molecular targeted therapies for hepatobiliary cancers. *Seminars in Oncology, 45*(3), 116. Llovet, J. M., Montal, R., Sia, D., & Finn, R. S. (2018). Molecular therapies and precision medicine for hepatocellular carcinoma. *Nature Reviews Clinical Oncology, 15*(10), 599.

HCC previously treated with sorafenib. The response rate for single agent anti-PD-1 therapies was modest at about 15%, but the responses were durable with a median DOR at 16.6 months. Recently, the combination of nivolumab and ipilimumab gained FDA approval as a second-line treatment of advanced HCC that has progressed on sorafenib based on CheckMate-040 data showing 31% ORR and median DOR of 17 months (He et al., 2020). Despite its activity, predictive biomarkers to immunotherapy in HCC remain elusive. There was no correlation between clinical benefit and PD-L1 expression or underlying etiology of cirrhosis (El-Khoueiry et al., 2017). In the KEYNOTE-224 study, the ORR was higher in patients with PD-L1 CPS ≥ 1 tumors compared to CPS < 1, suggesting that PDL1 CPS might be predictive of response to pembrolizumab (Zhu et al., 2018). In another study where patients with advanced HCC underwent prospective NGS profiling of their tumors, it was found that for patients treated with ICIs, activating mutations of the WNT/β-catenin signaling pathway were associated with lower disease control rate and shorter survival, suggesting that aberrant WNT/β-catenin pathway activation may confer immunotherapy resistance (Harding et al., 2019).

Aside from MSI-H or DNA mismatch repair deficient (dMMR) tumors, the efficacy of ICIs in BTCs is not well defined. However, MSI-H or dMMR BTCs are infrequent, though this varies among different studies (1%−22%) (Goeppert et al., 2019; Kunk et al., 2018; Silva et al., 2016). ICIs in nonbiomarker selected, advanced BTCs have been evaluated in a number of early phase clinical trials. ICI monotherapy has not shown clear activity in chemotherapy refractory BTCs. The same challenge remains in BTCs as in HCCs as no predictive biomarkers for immunotherapy have yet been identified.

Conclusion

Hepatic and bile duct cancers are associated with poor prognosis. Precision medicine promises treatment directed at the underlying tumor molecular biology. Despite an improved understanding of tumor heterogeneity and technologic advances in NGS, further research is needed to translate knowledge on the genomics of hepatic and bile duct cancers into predictive and prognostic biomarkers that can guide our clinical decision-making and improve patient outcomes. Progress has been made with targeted therapies and immunotherapies in hepatobiliary malignancies, but the ultimate goal will be the ability to guide treatments based wholly on molecular biomarkers, which will only occur when the drivers of these cancers are known.

References

Abou-Alfa, G. K., Macarulla, T., Javle, M. M., Kelley, R. K., Lubner, S. J., Adeva, J., ... Zhu, A. X. (2020). Ivosidenib in IDH1-mutant, chemotherapy-refractory cholangiocarcinoma

(ClarIDHy): A multicentre, randomised, double-blind, placebo-controlled, phase 3 study. *The Lancet Oncology, 21*(6), 796−807.

Abou-Alfa, G. K., Meyer, T., Cheng, A. L., El-Khoueiry, A. B., Rimassa, L., Ryoo, B. Y., . . . Kelley. (2018). Cabozantinib in patients with advanced and progressing hepatocellular carcinoma. *The New England Journal of Medicine, 379*(1), 54−63.

Ahn, D. H., Javle, M., Ahn, C. W., Jain, A., Mikhail, S., Noonan, A. M., . . . Bekaii-Saab. (2016). Next-generation sequencing survey of biliary tract cancer reveals the association between tumor somatic variants and chemotherapy resistance. *Cancer, 122*(23), 3657−3666.

Ahn, S. M., Jang, S. J., Shim, J. H., Kim, D., Hong, S. M., Sung, C. O., . . . Kong. (2014). Genomic portrait of resectable hepatocellular carcinomas: Implications of RB1 and FGF19 aberrations for patient stratification. *Hepatology, 60*(6), 1972−1982.

Audard, V., Grimber, G., Elie, C., Radenen, B., Audebourg, A., Letourneur, F., . . . Terris, B. (2007). Cholestasis is a marker for hepatocellular carcinomas displaying beta-catenin mutations. *The Journal of Pathology, 212*(3), 345−352.

Boyault, S., Rickman, D. S., de Reynies, A., Balabaud, C., Rebouissou, S., Jeannot, E., . . . Zucman-Rossi, J. (2007). Transcriptome classification of HCC is related to gene alterations and to new therapeutic targets. *Hepatology, 45*(1), 42−52.

Bray, F., Ferlay, J., Soerjomataram, I., Siegel, R. L., Torre, L. A., & Jemal, A. (2018). Global cancer statistics 2018: GLOBOCAN estimates of incidence and mortality worldwide for 36 cancers in 185 countries. *CA: A Cancer Journal for Clinicians, 68*(6), 394−424.

Bressac, B., Kew, M., Wands, J., & Ozturk, M. (1991). Selective G to T mutations of p53 gene in hepatocellular carcinoma from southern Africa. *Nature, 350*(6317), 429−431.

Bruix, J., Qin, S., Merle, P., Granito, A., Huang, Y. H., Bodoky, G., . . . Han, G.RESORCE Investigators. (2017). Regorafenib for patients with hepatocellular carcinoma who progressed on sorafenib treatment (RESORCE): A randomised, double-blind, placebo-controlled, phase 3 trial. *Lancet, 389*(10064), 56−66.

Budhu, A., Jia, H. L., Forgues, M., Liu, C. G., Goldstein, D., Lam, A., . . . Wang, X. W. (2008). Identification of metastasis-related microRNAs in hepatocellular carcinoma. *Hepatology, 47* (3), 897−907.

Cancer Genome Atlas Research Network. (2017). Comprehensive and integrative genomic characterization of hepatocellular carcinoma. *Cell, 169*(7), 1327−1341.e23.

Center, M. M., & Jemal, A. (2011). International trends in liver cancer incidence rates. *Cancer Epidemiology, Biomarkers & Prevention: A Publication of the American Association for Cancer Research, Cosponsored by the American Society of Preventive Oncology, 20*(11), 2362−2368.

Churi, C. R., Shroff, R., Wang, Y., Rashid, A., Kang, H. C., Weatherly, J., . . . Javle, M. (2014). Mutation profiling in cholangiocarcinoma: Prognostic and therapeutic implications. *PLoS One, 9*(12), e115383.

Cleary, S. P., Jeck, W. R., Zhao, X., Chen, K., Selitsky, S. R., Savich, G. L., . . . Chiang, D. Y. (2013). Identification of driver genes in hepatocellular carcinoma by exome sequencing. *Hepatology, 58*(5), 1693−1702.

de La Coste, A., Romagnolo, B., Billuart, P., Renard, C. A., Buendia, M. A., Soubrane, O., . . . Perret, C. (1998). Somatic mutations of the beta-catenin gene are frequent in mouse and human hepatocellular carcinomas. *Proceedings of the National Academy of Sciences of the United States of America, 95*(15), 8847−8851.

DeNicola, G. M., Karreth, F. A., Humpton, T. J., Gopinathan, A., Wei, C., Frese, K., . . . Tuveson, D. A. (2011). Oncogene-induced Nrf2 transcription promotes ROS detoxification and tumorigenesis. *Nature, 475*(7354), 106−109.

Duan, Y., Tian, L., Gao, Q., Liang, L., Zhang, W., Yang, Y., . . . Tang, N. (2016). Chromatin remodeling gene ARID2 targets cyclin D1 and cyclin E1 to suppress hepatoma cell progression. *Oncotarget.*, *7*(29), 45863−45875.

El-Khoueiry, A. B., Sangro, B., Yau, T., Crocenzi, T. S., Kudo, M., Hsu, C., . . . Melero, I. (2017). Nivolumab in patients with advanced hepatocellular carcinoma (CheckMate 040): An open-label, non-comparative, phase 1/2 dose escalation and expansion trial. *Lancet, 389* (10088), 2492−2502.

Finn, R. S., Ikeda, M., Zhu, A. X., Sung, M. W., Baron, A. D., Kudo, M., . . . Llovet, J. M. (2020a). Phase Ib Study of Lenvatinib Plus Pembrolizumab in Patients With Unresectable Hepatocellular Carcinoma. *Journal of Clinical Oncology*, *38*(26), 2960−2970.

Finn, R. S., Qin, S., Ikeda, M., Galle, P. R., Ducreux, M., Kim, T. Y., . . . Cheng, A.IMbrave150 Investigators. (2020). Atezolizumab plus bevacizumab in unresectable hepatocellular carcinoma. *The New England Journal of Medicine*, *382*(20), 1894−1905.

Finn, R. S., Ryoo, B. Y., Merle, P., Kudo, M., Bouattour, M., Lim, H. Y., . . . Cheng, A. KEYNOTE-240 Investigators. (2020). Pembrolizumab as second-line therapy in patients with advanced hepatocellular carcinoma in KEYNOTE-240: A randomized, double-blind, phase III trial. *Journal of Clinical Oncology: Official Journal of the American Society of Clinical Oncology*, *38*(3), 193−202.

Fujimoto, A., Totoki, Y., Abe, T., Boroevich, K. A., Hosoda, F., Nguyen, H. H., . . . Nakagawa, H. (2012). Whole-genome sequencing of liver cancers identifies etiological influences on mutation patterns and recurrent mutations in chromatin regulators. *Nature Genetics*, *44*(7), 760−764.

Gao, L., Wang, X., Tang, Y., Huang, S., Hu, C. A., & Teng, Y. (2017). FGF19/FGFR4 signaling contributes to the resistance of hepatocellular carcinoma to sorafenib. *Journal of Experimental & Clinical Cancer Research*, *36*(1), 8.

Goeppert, B., Roessler, S., Renner, M., Singer, S., Mehrabi, A., Vogel, M. N., . . . Kloor, M. (2019). Mismatch repair deficiency is a rare but putative therapeutically relevant finding in non-liver fluke associated cholangiocarcinoma. *British Journal of Cancer*, *120*(1), 109−114.

Goyal, L., Meric-Bernstam, F., Hollebecque, A., Valle, J. W., Morizane, C., Karasic, T. B., . . . Bridgewater, J. A. (2020). FOENIX-CCA2: A phase II, open-label, multicenter study of futibatinib in patients (pts) with intrahepatic cholangiocarcinoma (iCCA) harboring FGFR2 gene fusions or other rearrangements. *Journal of Clinical Oncology*, *38*(15), 108−108.

Graham, R. P., Barr Fritcher, E. G., Pestova, E., Schulz, J., Sitailo, L. A., Vasmatzis, G., . . . Kipp, B. R. (2014). Fibroblast growth factor receptor 2 translocations in intrahepatic cholangiocarcinoma. *Human Pathology*, *45*(8), 1630−1638.

Guichard, C., Amaddeo, G., Imbeaud, S., Ladeiro, Y., Pelletier, L., Maad, I. B., . . . Zucman-Rossi, J. (2012). Integrated analysis of somatic mutations and focal copy-number changes identifies key genes and pathways in hepatocellular carcinoma. *Nature Genetics*, *44*(6), 694−698.

Hainsworth, J. D., Meric-Bernstam, F., Swanton, C., Hurwitz, H., Spigel, D. R., Sweeney, C., . . . Kurzrock, R. (2018). Targeted therapy for advanced solid tumors on the basis of molecular profiles: Results from mypathway, an open-label, phase IIa multiple basket study. *Journal of Clinical Oncology: Official Journal of the American Society of Clinical Oncology*, *36*(6), 536−542.

Harding, J. J., Nandakumar, S., Armenia, J., Khalil, D. N., Albano, M., Ly, M., . . . Abou-Alfa, G. K. (2019). Prospective genotyping of hepatocellular carcinoma: Clinical implications of next-generation sequencing for matching patients to targeted and immune therapies. *Clinical*

Cancer Research: An Official Journal of the American Association for Cancer Research, 25 (7), 2116–2126.

He, A. R., Yau, T., Hsu, C., Kang, Y., Kim, T., Santoro, A., ... El-Khoueiry, A. B. (2020). Nivolumab (NIVO) + ipilimumab (IPI) combination therapy in patients (pts) with advanced hepatocellular carcinoma (aHCC): Subgroup analyses from CheckMate 040. *Journal of Clinical Oncology, 38*(4), 512–512.

Henley, S. J., Weir, H. K., Jim, M. A., Watson, M., & Richardson, L. C. (2015). Gallbladder cancer incidence and mortality, United States 1999–2011. *Cancer Epidemiology, Biomarkers & Prevention: A Publication of the American Association for Cancer Research, Cosponsored by the American Society of Preventive Oncology, 24*(9), 1319–1326.

Hoshida, Y., Moeini, A., Alsinet, C., Kojima, K., & Villanueva, A. (2012). Gene signatures in the management of hepatocellular carcinoma. *Seminars in Oncology, 39*(4), 473–485.

Hoshida, Y., Nijman, S. M., Kobayashi, M., Chan, J. A., Brunet, J. P., Chiang, D. Y., ... Golub, T. R. (2009). Integrative transcriptome analysis reveals common molecular subclasses of human hepatocellular carcinoma. *Cancer Research, 69*(18), 7385–7392.

Hoshida, Y., Villanueva, A., Kobayashi, M., Peix, J., Chiang, D. Y., Camargo, A., ... Golub, T. R. (2008). Gene expression in fixed tissues and outcome in hepatocellular carcinoma. *The New England Journal of Medicine, 359*(19), 1995–2004.

Hsu, I. C., Metcalf, R. A., Sun, T., Welsh, J. A., Wang, N. J., & Harris, C. C. (1991). Mutational hotspot in the p53 gene in human hepatocellular carcinomas. *Nature, 350*(6317), 427–428.

Javle, M., Bekaii-Saab, T., Jain, A., Wang, Y., Kelley, R. K., Wang, K., ... Ross, J. (2016). Biliary cancer: Utility of next-generation sequencing for clinical management. *Cancer, 122*(24), 3838–3847.

Javle, M., Churi, C., Kang, H. C., Shroff, R., Janku, F., Surapaneni, R., ... Siegel, A. (2015). HER2/neu-directed therapy for biliary tract cancer. *Journal of Hematology & Oncology, 8*, 58.

Javle, M., Lowery, M., Shroff, R. T., Weiss, K. H., Springfeld, C., Borad, M. J., ... Bekaii-Saab, T. (2018). Phase II study of BGJ398 in patients with FGFR-altered advanced cholangiocarcinoma. *Journal of Clinical Oncology: Official Journal of the American Society of Clinical Oncology, 36*(3), 276–282.

Jusakul, A., Cutcutache, I., Yong, C. H., Lim, J. Q., Huang, M. N., Padmanabhan, N., ... Tan, P. (2017). Whole-genome and epigenomic landscapes of etiologically distinct subtypes of cholangiocarcinoma. *Cancer Discovery, 7*(10), 1116–1135.

Kelley, R. K., Sangro, B., Harris, W. P., Ikeda, M., Okusaka, T., Kang, Y., ... Abou-Alfa, G. K. (2020). Efficacy, tolerability, and biologic activity of a novel regimen of tremelimumab (T) in combination with durvalumab (D) for patients (pts) with advanced hepatocellular carcinoma (aHCC). *Journal of Clinical Oncology, 38*(4), 4508–4508.

Kim, J. H., Sohn, B. H., Lee, H. S., Kim, S. B., Yoo, J. E., Park, Y. Y., ... Lee, J. (2014). Genomic predictors for recurrence patterns of hepatocellular carcinoma: Model derivation and validation. *PLoS Medicine, 11*(12), e1001770.

Kim, R. D., Sarker, D., Meyer, T., Yau, T., Macarulla, T., Park, J. W., ... Kang, Y. (2019). First-in-human phase I study of fisogatinib (BLU-554) validates aberrant FGF19 signaling as a driver event in hepatocellular carcinoma. *Cancer Discovery, 9*(12), 1696–1707.

Kobayashi, T., Kubota, K., Takayama, T., & Makuuchi, M. (2001). Telomerase activity as a predictive marker for recurrence of hepatocellular carcinoma after hepatectomy. *American Journal of Surgery, 181*(3), 284–288.

Kotoula, V., Hytiroglou, P., Pyrpasopoulou, A., Saxena, R., Thung, S. N., & Papadimitriou, C. S. (2002). Expression of human telomerase reverse transcriptase in regenerative and precancerous lesions of cirrhotic livers. *Liver, 22*(1), 57–69.

Kudo, M., Finn, R. S., Qin, S., Han, K. H., Ikeda, K., Piscaglia, F., . . . Cheng, A. (2018). Lenvatinib vs sorafenib in first-line treatment of patients with unresectable hepatocellular carcinoma: A randomised phase 3 non-inferiority trial. *Lancet, 391*(10126), 1163–1173.

Kunk, P. R., Obeid, J. M., Winters, K., Pramoonjago, P., Brockstedt, D. G., Giobbie-Hurder, A., . . . Rahma, O. E. (2018). Mismatch repair deficiency in cholangiocarcinoma. *Journal of Clinical Oncology: Official Journal of the American Society of Clinical Oncology, 36* (4), 269–269.

Lamarca, A., Palmer, D. H., Wasan, H. S, Ross, P. J., Ma, Y. T., Arora, A., . . . Valle, J. W. Advanced Biliary Cancer Working Group. (2021). Second-line FOLFOX chemotherapy versus active symptom control for advanced biliary tract cancer (ABC-06): a phase 3, open-label, randomised, controlled trial. *Lancet Oncology, 22*(5), 690–701.

Lee, J. S., Chu, I. S., Heo, J., Calvisi, D. F., Sun, Z., Roskams, T., . . . Thorgeirsson, S. S (2004). Classification and prediction of survival in hepatocellular carcinoma by gene expression profiling. *Hepatology, 40*(3), 667–676.

Lee, J. S., Heo, J., Libbrecht, L., Chu, I. S., Kaposi-Novak, P., Calvisi, D. F., . . . Thorgeirsson. (2006). A novel prognostic subtype of human hepatocellular carcinoma derived from hepatic progenitor cells. *Nature Medicine, 12*(4), 410–416.

Letouze, E., Shinde, J., Renault, V., Couchy, G., Blanc, J. F., Tubacher, E., . . . Zucman-Rossi. (2017). Mutational signatures reveal the dynamic interplay of risk factors and cellular processes during liver tumorigenesis. *Nature Communications, 8*(1), 1315.

Li, M., Zhang, Z., Li, X., Ye, J., Wu, X., Tan, Z., . . . Liu, Y. (2014). Whole-exome and targeted gene sequencing of gallbladder carcinoma identifies recurrent mutations in the ErbB pathway. *Nature Genetics, 46*(8), 872–876.

Li, M., Zhao, H., Zhang, X., Wood, L. D., Anders, R. A., Choti, M. A., . . . Kinzler, K. W. (2011). Inactivating mutations of the chromatin remodeling gene ARID2 in hepatocellular carcinoma. *Nature Genetics, 43*(9), 828–829.

Llovet, J. M., Ricci, S., Mazzaferro, V., Hilgard, P., Gane, E., Blanc, J. F., . . . Bruix, J.SHARP Investigator Study Group. (2008). Sorafenib in advanced hepatocellular carcinoma. *The New England Journal of Medicine, 359*(4), 378–390.

Llovet, J. M., Zucman-Rossi, J., Pikarsky, E., Sangro, B., Schwartz, M., Sherman, M., & Gores, G. (2016). Hepatocellular carcinoma. *Nature Reviews Disease Primers, 2*, 16018.

Lowery, M. A., Burris, H. A., III, Janku, F., Shroff, R. T., Cleary, J. M., Azad, N. S., . . . Abou-Alfa, G. K. (2019). Safety and activity of ivosidenib in patients with IDH1-mutant advanced cholangiocarcinoma: A phase 1 study. *The Lancet Gastroenterology and Hepatology, 4*(9), 711–720.

Mittal, S., & El-Serag, H. B. (2013). Epidemiology of hepatocellular carcinoma: Consider the population. *Journal of Clinical Gastroenterology, 47*, S2–S6, Suppl.

Mukkamalla, S. K. R., Naseri, H. M., Kim, B. M., Katz, S. C., & Armenio, V. A. (2018). Trends in incidence and factors affecting survival of patients with cholangiocarcinoma in the United States. *Journal of the National Comprehensive Cancer Network, 16*(4), 370–376.

Nagao, K., Tomimatsu, M., Endo, H., Hisatomi, H., & Hikiji, K. (1999). Telomerase reverse transcriptase mRNA expression and telomerase activity in hepatocellular carcinoma. *Journal of Gastroenterology, 34*(1), 83–87.

Nakamura, H., Arai, Y., Totoki, Y., Shirota, T., Elzawahry, A., Kato, M., . . . Shibata, T. (2015). Genomic spectra of biliary tract cancer. *Nature Genetics, 47*(9), 1003–1010.

Nault, J. C., De Reynies, A., Villanueva, A., Calderaro, J., Rebouissou, S., Couchy, G., . . . Zucman-Rossi, J. (2013). A hepatocellular carcinoma 5-gene score associated with survival of patients after liver resection. *Gastroenterology, 145*(1), 176–187.

Nault, J. C., Mallet, M., Pilati, C., Calderaro, J., Bioulac-Sage, P., Laurent, C., ... Zucman-Rossi, J. (2013). High frequency of telomerase reverse-transcriptase promoter somatic mutations in hepatocellular carcinoma and preneoplastic lesions. *Nature Communications, 4*, 2218.

Ng, A. W. T., Poon, S. L., Huang, M. N., Lim, J. Q., Boot, A., Yu, W., ... Rozen, S. G. (2017). Aristolochic acids and their derivatives are widely implicated in liver cancers in Taiwan and throughout Asia. *Science Translational Medicine, 9*, 412.

Park, J. W., Chen, M., Colombo, M., Roberts, L. R., Schwartz, M., Chen, P. J., ... Sherman, M. (2015). Global patterns of hepatocellular carcinoma management from diagnosis to death: The BRIDGE study. *Liver International: Official Journal of the International Association for the Study of the Liver, 35*(9), 2155–2166.

Randi, G., Franceschi, S., & La Vecchia, C. (2006). Gallbladder cancer worldwide: Geographical distribution and risk factors. *International Journal of Cancer. Journal International du Cancer, 118*(7), 1591–1602.

Sabbatino, F., Villani, V., Yearley, J. H., Deshpande, V., Cai, L., Konstantinidis, I. T., ... Ferrone, C. R. (2016). PD-L1 and HLA class I antigen expression and clinical course of the disease in intrahepatic cholangiocarcinoma. *Clinical Cancer Research: An Official Journal of the American Association for Cancer Research, 22*(2), 470–478.

Satoh, S., Daigo, Y., Furukawa, Y., Kato, T., Miwa, N., Nishiwaki, T., ... Nakamura, Y. (2000). AXIN1 mutations in hepatocellular carcinomas, and growth suppression in cancer cells by virus-mediated transfer of AXIN1. *Nature Genetics, 24*(3), 245–250.

Sawey, E. T., Chanrion, M., Cai, C., Wu, G., Zhang, J., Zender, L., ... Powers, S. (2011). Identification of a therapeutic strategy targeting amplified FGF19 in liver cancer by oncogenomic screening. *Cancer Cell, 19*(3), 347–358.

Schulze, K., Imbeaud, S., Letouze, E., Alexandrov, L. B., Calderaro, J., Rebouissou, S., ... Zucman-Rossi, J. (2015). Exome sequencing of hepatocellular carcinomas identifies new mutational signatures and potential therapeutic targets. *Nature Genetics, 47*(5), 505–511.

Sharma, A., Sharma, K. L., Gupta, A., Yadav, A., & Kumar, A. (2017). Gallbladder cancer epidemiology, pathogenesis and molecular genetics: Recent update. *World Journal of Gastroenterology, 23*(22), 3978–3998.

Shroff, R. T., Javle, M. M., Xiao, L., Kaseb, A. O., Varadhachary, G. R., Wolff, R. A., ... Borad, M. J. (2019). Gemcitabine, cisplatin, and nab-paclitaxel for the treatment of advanced biliary tract cancers: A phase 2 clinical trial. *JAMA Oncology, 5*(6), 824–830.

Sia, D., Jiao, Y., Martinez-Quetglas, I., Kuchuk, O., Villacorta-Martin, C., Castro de Moura, M., ... Llovet, J. M. (2017). Identification of an immune-specific class of hepatocellular carcinoma, based on molecular features. *Gastroenterology, 153*(3), 812–826.

Siegel, R. L., Miller, K. D., & Jemal, A. (2019). Cancer statistics, 2019. *CA: A Cancer Journal for Clinicians, 69*(1), 7–34.

Silva, V. W., Askan, G., Daniel, T. D., Lowery, M., Klimstra, D. S., Abou-Alfa, G. K., & Shia, J. (2016). Biliary carcinomas: pathology and the role of DNA mismatch repair deficiency. *Chinese Clinical Oncology, 5*(5), 62.

Sporn, M. B., & Liby, K. T. (2012). NRF2 and cancer: The good, the bad and the importance of context. *Nature Reviews Cancer, 12*(8), 564–571.

Sulkowski, P. L., Corso, C. D., Robinson, N. D., Scanlon, S. E., Purshouse, K. R., Bai, H., ... Bindra, R. S. (2017). 2-Hydroxyglutarate produced by neomorphic IDH mutations suppresses homologous recombination and induces PARP inhibitor sensitivity. *Science Translational Medicine, 9*(375).

Sung, W. K., Zheng, H., Li, S., Chen, R., Liu, X., Li, Y., . . . Luk, J. M. (2012). Genome-wide survey of recurrent HBV integration in hepatocellular carcinoma. *Nature Genetics*, *44*(7), 765−769.

Taniguchi, K., Roberts, L. R., Aderca, I. N., Dong, X., Qian, C., Murphy, L. M., . . . Liu, W. (2002). Mutational spectrum of beta-catenin, AXIN1, and AXIN2 in hepatocellular carcinomas and hepatoblastomas. *Oncogene*, *21*(31), 4863−4871.

Tazuma, S., & Kajiyama, G. (2001). Carcinogenesis of malignant lesions of the gall bladder. The impact of chronic inflammation and gallstones. *Langenbeck's Archives of Surgery*, *386* (3), 224−229.

Thomas Yau, Y.-K. K., Tae-You, K., & El-Khoueiry, A. B. (2019). Nivolumab (NIVO) + ipilimumab (IPI) combination therapy in patients (pts) with advanced hepatocellular carcinoma (aHCC): Results from CheckMate 040. *Journal of Clinical Oncology*, *37*(15).

Toffanin, S., Hoshida, Y., Lachenmayer, A., Villanueva, A., Cabellos, L., Minguez, B., . . . Llovet, J. M. (2011). MicroRNA-based classification of hepatocellular carcinoma and oncogenic role of miR-517a. *Gastroenterology*, *140*(5), 1618−1628, e16.

Totoki, Y., Tatsuno, K., Covington, K. R., Ueda, H., Creighton, C. J., Kato, M., . . . Shibata, T. (2014). Trans-ancestry mutational landscape of hepatocellular carcinoma genomes. *Nature Genetics*, *46*(12), 1267−1273.

Tyson, G. L., & El-Serag, H. B. (2011). Risk factors for cholangiocarcinoma. *Hepatology*, *54*(1), 173−184.

Valle, J., Wasan, H., Palmer, D. H., Cunningham, D., Anthoney, A., Maraveyas, A., . . . Bridgewater, J.ABC-02 Trial Investigators. (2010). Cisplatin plus gemcitabine vs gemcitabine for biliary tract cancer. *The New England Journal of Medicine*, *362*(14), 1273−1281.

Villanueva, A., Chiang, D. Y., Newell, P., Peix, J., Thung, S., Alsinet, C., . . . Llovet, J. M. (2008). Pivotal role of mTOR signaling in hepatocellular carcinoma. *Gastroenterology*, *135* (6), 1972−1983, 83 e1−11.

Villanueva, A., Hoshida, Y., Battiston, C., Tovar, V., Sia, D., Alsinet, C., . . . Llovet, J. M. (2011). Combining clinical, pathology, and gene expression data to predict recurrence of hepatocellular carcinoma. *Gastroenterology*, *140*(5), 1501−1512, e2.

Villanueva, A., Portela, A., Sayols, S., Battiston, C., Hoshida, Y., Mendez-Gonzalez, J., . . . Llovet, J. M.HEPTROMIC Consortium. (2015). DNA methylation-based prognosis and epi-drivers in hepatocellular carcinoma. *Hepatology*, *61*(6), 1945−1956.

Vogel, A., Sahai, V., Hollebecque, A., Vaccaro, G., & Abou-Alfa, G. K. (2019). FIGHT-202: A phase II study of pemigatinib in patients (pts) with previously treated locally advanced or metastatic cholangiocarcinoma (CCA). *Annals of Oncology*, *30*(5), Supplement.

Wang, K., Lim, H. Y., Shi, S., Lee, J., Deng, S., Xie, T., . . . Xu, J. (2013). Genomic landscape of copy number aberrations enables the identification of oncogenic drivers in hepatocellular carcinoma. *Hepatology*, *58*(2), 706−717.

Welzel, T. M., Graubard, B. I., El-Serag, H. B., Shaib, Y. H., Hsing, A. W., Davila, J. A., & McGlynn, K. A. (2007). Risk factors for intrahepatic and extrahepatic cholangiocarcinoma in the United States: A population-based case-control study. *Clinical Gastroenterology and Hepatology: The Official Clinical Practice Journal of the American Gastroenterological Association*, *5*(10), 1221−1228.

Wongjarupong, N., Assavapongpaiboon, B., Susantitaphong, P., Cheungpasitporn, W., Treeprasertsuk, S., Rerknimitr, R., & Chaiteerakij, R. (2017). Non-alcoholic fatty liver disease as a risk factor for cholangiocarcinoma: A systematic review and meta-analysis. *BMC Gastroenterology*, *17*(1), 149.

Yang, J. D., Hainaut, P., Gores, G. J., Amadou, A., Plymoth, A., & Roberts, L. R. (2019). A global view of hepatocellular carcinoma: Trends, risk, prevention and management. *Nature Reviews Gastroenterology & Hepatology, 16*(10), 589–604.

Ye, Q., Ling, S., Zheng, S., & Xu, X. (2019). Liquid biopsy in hepatocellular carcinoma: Circulating tumor cells and circulating tumor DNA. *Molecular Cancer, 18*(1), 114.

Zhou, L., Huang, Y., Li, J., & Wang, Z. (2010). The mTOR pathway is associated with the poor prognosis of human hepatocellular carcinoma. *Medical Oncology (Northwood, London, England), 27*(2), 255–261.

Zhu, A. X., Borger, D. R., Kim, Y., Cosgrove, D., Ejaz, A., Alexandrescu, S., . . . Pawlik, T. M. (2014). Genomic profiling of intrahepatic cholangiocarcinoma: Refining prognosis and identifying therapeutic targets. *Annals of Surgical Oncology, 21*(12), 3827–3834.

Zhu, A. X., Finn, R. S., Edeline, J., Cattan, S., Ogasawara, S., Palmer, D., . . . Kudo, M. KEYNOTE-224 Investigators. (2018). Pembrolizumab in patients with advanced hepatocellular carcinoma previously treated with sorafenib (KEYNOTE-224): A non-randomised, open-label phase 2 trial. *The Lancet Oncology, 19*(7), 940–952.

Zhu, A. X., Kang, Y. K., Yen, C. J., Finn, R. S., Galle, P. R., Llovet, J. M., . . . Kudo, M. REACH-2 Investigators. (2019). Ramucirumab after sorafenib in patients with advanced hepatocellular carcinoma and increased alpha-fetoprotein concentrations (REACH-2): A randomised, double-blind, placebo-controlled, phase 3 trial. *The Lancet Oncology, 20*(2), 282–296.

Zucman-Rossi, J., Villanueva, A., Nault, J. C., & Llovet, J. M. (2015). Genetic landscape and biomarkers of hepatocellular carcinoma. *Gastroenterology, 149*(5), 1226–1239, e4.

Chapter 14

Biomarker-directed therapy for pancreatic cancer

Michael J. Pishvaian[1] and Jonathan R. Brody[2]

[1]*Department of Oncology, Johns Hopkins University School of Medicine, SKCC, Washington, DC, United States,* [2]*Departments of Surgery and Cell, Developmental & Cancer Biology, Brenden-Colson Center for Pancreatic Care Knight Cancer Institute, Oregon Health and Science University, Portland, OR, United States*

Introduction

Pancreatic cancer, and specifically pancreatic ductal adenocarcinoma (PDAC), is only the 11th most common cancer in the United States, but will soon be the second leading cause of cancer-related death (Rahib et al., 2014), with 47,050 patients expected to succumb to this disease in 2020 (American Cancer Society, 2020). PDAC is typically diagnosed at a later stage, with only 10%−20% of patients having resectable disease, and at least 50% being diagnosed as metastatic and thus incurable from the outset (American Cancer Society, 2020). Chemotherapy has improved median overall survival (mOS) to a rate of ~1 year with the modern regimens of FOLFIRINOX and gemcitabine + nab-paclitaxel (Conroy et al., 2011; Von Hoff et al., 2013), as compared to the prior standard of single-agent gemcitabine (Burris et al., 1997). Furthermore, an increase in the percent of patients receiving second-line therapy is in large part responsible for the mOS improving to 14−18 months (Portal et al., 2015), including in the control arms of recently presented Phase III trials (Ramanathan et al., 2019; Tempero et al., 2020). Thus systemic chemotherapy continues to be the mainstay of the treatment of PDAC, and it is shocking that that recent publications have suggested that over 50% of PDAC patients are not offered any therapy at all (Mavros et al., 2019).

Nevertheless, these improvements in outcomes are still only incremental, and there is a great need for novel effective, personalized therapies. In this chapter, we will review the data demonstrating that a subset of PDACs harbors "actionable" mutations, and we will discuss the proven and anecdotal

Genomic and Precision Medicine. DOI: https://doi.org/10.1016/B978-0-12-800684-9.00001-0

239

data demonstrating benefit to treating PDAC patients with therapies matched to their actionable biomarker. We will also highlight several still elusive targets, and discuss some of the efforts aimed at overcoming the prior barriers to therapy.

Pancreatic cancers harbor actionable mutations

It has been over a decade since key mutations and associated pathways were identified in a large cohort of PDACs (Jones et al., 2008). Since then, multiple large-scale sequencing efforts of PDAC tumors have been performed and altogether well over 5000 individual PDAC samples have been sequenced (Aguirre et al., 2018; Bailey et al., 2016; Biankin et al., 2012; Collisson et al., 2011; Heeke et al., 2018; Lowery et al., 2017; Pishvaian et al., 2018; Singhi et al., 2019; Waddell et al., 2015; Witkiewicz et al., 2015). Consistently, ∼25% (range 17%−48%) of PDAC patients' tumors harbor actionable mutations, with "actionable" defined as pathogenic or presumed pathogenic mutations and/or gene fusions that are linked with, and have been shown to be predictors of response to specific therapies, albeit in any cancer type (see Fig. 14.1). In fact, these data influenced the National Comprehensive Cancer Network (NCCN) to update their guidelines for PDAC patients in April 2019, specifying that "tumor/somatic gene profiling is recommended for patients with locally advanced/metastatic disease who are candidates for anticancer therapy to identify uncommon but actionable mutations (Tempero et al., 2019)." These actionable mutations are linked to a few tumor agnostic, biomarker-based, FDA-approved therapies, and many more are being targeted in clinical trials in PDAC.

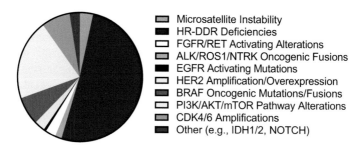

Microsatellite Instability
HR-DDR Deficiencies
FGFR/RET Activating Alterations
ALK/ROS1/NTRK Oncogenic Fusions
EGFR Activating Mutations
HER2 Amplification/Overexpression
BRAF Oncogenic Mutations/Fusions
PI3K/AKT/mTOR Pathway Alterations
CDK4/6 Amplifications
Other (e.g., IDH1/2, NOTCH)

FIGURE 14.1 **Pie chart depicting the actionable signaling pathways disrupted in PDAC.** The overall rate of actionable alterations was 26% in this publication. This pie chart demonstrates the frequency of actionable alterations within that 26%. *Extracted from Pishvaian, M. J., Blais, E. M., Brody, J. R., Lyons, E., DeArbeloa, P., Hendifar, A., et al. (2020). Overall survival in patients with pancreatic cancer receiving matched therapies following molecular profiling: A retrospective analysis of the Know Your Tumor registry trial. The Lancet Oncology, 21, 508−518.*

FDA-approved, biomarker-based therapies for pancreatic cancer

BRCA1/2 (and PALB2) mutations

DNA damage and chromosomal instability are hallmarks of PDAC and germline mutations in the Breast Cancer Associate genes 1 and 2 (*BRCA1/2*), as well as more recently the Partner and Localizer of BRCA2 (*PALB2*) gene, can predispose patients to the development of PDAC. The gene products of *BRCA1/2* and *PALB2* play a critical role in DNA damage response and repair (DDR) via homologous recombination (HR), the most error-free DNA repair mechanism. Loss of function of BRCA1/2 or PALB2 proteins can lead to ineffective repair of double-stranded DNA breaks, resulting in an accumulation of mutations that can trigger carcinogenesis (Helleday, Petermann, Lundin, Hodgson, & Sharma, 2008). Of note, 3%−5% of PDACs are associated with germline mutations in these genes (Salo-Mullen et al., 2015), and additional 2%−5% of PDACs harbor somatic *BRCA1/2* or *PALB2* mutations (Aguirre et al., 2018). The consequences of a defective HR-DDR pathway also present a therapeutic vulnerability in cancer cells that harbor *BRCA1/2* or *PALB2* mutations. Cancer cells that are HR-DDR deficient are particularly responsive to potent DNA-damaging agents such as platinum or poly(ADP-ribose) polymerase (PARP) inhibitors (Jalal, Earley, & Turchi, 2011; Pishvaian et al., 2019), as the lack of efficient DNA repair results in unrepaired double-stranded DNA breaks, leading to replication fork arrest, mitotic catastrophe, and cell death.

We and others have demonstrated that PDAC patients whose tumors harbor HR-DDR mutations have an improved mOS with platinum-based chemotherapy (Golan et al., 2017; O'Reilly et al., 2020; Pishvaian et al., 2019). Interestingly, the presence of an HR-DDR mutation was not *prognostically* favorable in platinum-naïve patients. Patients whose tumors harbored HR-DDR mutations had a mOS of 0.76 years compared to 1.13 years for DDR wild-type (DDR^{WT}) patients. By contrast, the presence of an HR-DDR mutation was strongly *predictive* of an improved outcome with platinum-based therapy. Platinum-treated patients whose tumors harbored HR-DDR mutations had a mOS of 2.37 years, compared to only 0.76 years for similar platinum-naïve patients, and compared to 1.45 years for platinum-treated, DDR^{WT} patients.

The benefits of PARP inhibitors in *BRCA1/2*-mutated were first noted by Lowery et al., who described one partial response (PR), and three stable diseases upon treatment with olaparib (Lowery et al., 2011). Subsequently, Kaufman et al. revealed a 22% objective response rate (ORR), including 1 complete response (CR) and 4 PRs among 23 *BRCA1/2*-mutated PDAC patients enrolled as part of a larger trial of olaparib in *BRCA1/2*-mutated cancers (Kaufman et al., 2015). In addition, Shroff et al.

demonstrated a 16% ORR in 16 *BRCA1/2*-mutated PDAC patients treated with rucaparib (Shroff et al., 2018). Importantly, Shroff's trial included patients with somatic-only *BRCA1/2* mutations, and at least one patient with a somatic-only mutation had a CR on rucaparib—raising the possibility that PARP inhibitors might not need to be restricted to germline *BRCA1/2* mutation carriers only. Finally, based on the positive results of the randomized Phase III POLO trial, olaparib was approved as maintenance therapy in germline *BRCA1/2*-mutated PDAC patients, whose disease was stable on or responding to platinum-based chemotherapy (Golan, Locker, & Kindler, 2019). In this study, olaparib improved progression-free survival (PFS) over placebo (7.4 vs. 3.8 months, $P = 0.004$). These results were supported by similar outcomes in a trial of maintenance rucaparib (Reiss Binder et al., 2019).

Despite the excitement of the first-targeted therapy ever approved for PDAC patients, not all patients with *BRCA1/2*-mutated PDAC benefit from PARP inhibitors. We and others have demonstrated that patients whose tumors were refractory to platinums (i.e., not just platinum "exposed" but their disease had grown while actively on platinum-based therapy) did not receive any benefit from PARP inhibitor-based therapy (Pishvaian et al., 2020; Shroff et al., 2018). We demonstrated for the first time in a patient with PDAC that a secondary *BRCA2* mutation occurred under PARP inhibitor treatment, which rescued the function of the BRCA enzyme and lead to a loss of responsiveness to therapy (Pishvaian et al., 2017; Shroff et al., 2018). Finally, there has been a growing appreciation that up to 1/3 of *BRCA1/2*-mutated PDAC patients exhibit innate resistance to platinums and PARP inhibitors, for reasons that are yet to be understood.

MSI-high tumors

In 2016, pembrolizumab was the first agent to receive a disease-type agnostic FDA label indication, being approved for the treatment of tumors with a high degree of microsatellite instability (MSI-high) of any type. Nivolumab was similarly approved for the treatment of MSI-high colorectal cancer in 2017. These two anti-PD-1 agents serve to remove the immune checkpoints that cancer cells have harnessed to dampen the anticancer immune response, and have proven to be highly effective for MSI-high tumors which harbor a high tumor mutational burden, which is highly immunogenic. MSI-high PDAC is quite uncommon, with published rates of less than 1% (Lowery et al., 2017). Interestingly, in the Memorial Sloan Kettering IMPACT experience, in which tumor molecular profiling was accompanied by germline tumor testing, all MSI-high PDAC patients identified were found to have a germline mutation in one of the MMR genes (Lynch syndrome). Thus somatic MSI-high PDAC appears to be very rare.

With regards to the clinical experience of pembrolizumab for MSI-high PDAC, Le et al. included six PDAC patients in their MSI-high pembrolizumab trial (Le et al., 2017), and all patients had some degree of tumor reduction (4/6 had a RECIST criteria response). However, an update on these data by Marabelle et al. demonstrated, disappointingly, that in the 22 PDAC patients, the ORR for pembrolizumab was only 18%, and the responses were relatively short lived (relative to MSI-high tumors from other organs), with a median PFS of only 2.1 months, and a median duration of response (DOR) of 13.4 months (the DOR was unreached for all other highlighted disease cohorts) (Marabelle et al., 2020). This highlights the need for effective combination therapies for MSI-high PDAC to increase the ORR and extend the DOR.

Neurotrophic tyrosine receptor kinase fusions

The only other agents to receive FDA approval based on a molecular abnormality (and irrespective of disease type) are the tropomyosin receptor kinase (TRK) inhibitors, larotrectinib and entrectinib. Larotrectinib was approved by FDA in 2018, and entrectinib in 2019 based on a very high response rate (75%) (Drilon et al., 2018). In the initial trials, only one PDAC patient was treated with larotrectinib, and three were treated with entrectinib, but all patients experienced some degree of benefit (with a true RECIST criteria response in 2/3 patients) (Pishvaian, Liu, Multani, Chow-Maneval, & Rolfo, 2018). Thus while identification of neurotrophic tyrosine receptor kinase (NTRK)-fusion positive PDAC is a rare event, the disproportionate benefit of the TRK inhibitors justifies screening all PDAC patients.

Additional promising targets

Other DDR mutations

The DDR pathway is a highly complex orchestrated system that exists to maintain the fidelity of the cell's original DNA sequence. While HR is critical for effective DNA repair, multiple other aspects of the DDR pathway are required for DNA repair. For example, the ataxia-telangiesctasia mutated (ATM) and ataxia telangiectasia and Rad3 related (ATR) proteins are essential for recognition of a DS-DNA break, and for coating the DNA strand for recognition for repair by the HR machinery. Cell-cycle arrest is required to allow time for effective DNA repair. Alternative pathways of DNA repair exist, and are controlled by DNA-PK. We and others have shown that multiple germline and/or somatic mutations in the broader DDR pathway exist in PDAC, and, with *BRCA1/2/PALB2* mutations, 17%−25% of all PDACs exhibit some DDR defect. What is unknown is the degree to which these non-*BRCA/PALB* DDR mutations predict for a response to therapy.

Early phase trials have suggested that the non-*BRCA/PALB* DDR mutations may be predictive of other DDR inhibitors, outside of platinums, and PARP inhibitors. For example, multiple studies have demonstrated that *ATM*-mutated tumors are more responsive to the combination of an ATR inhibitor and chemotherapy (Min et al., 2017; Perkhofer et al., 2017; Reaper et al., 2011; Schmitt et al., 2017; Shi et al., 2018; Vendetti et al., 2015), and *ATM* mutations are among the most common DDR mutation identified in PDAC, occurring in ~6% of all PDAC patients.

KRAS wild-type PDAC: mutations in *BRAF*, and other receptor tyrosine kinases

The hallmark of PDACs is the presence of an activating *KRAS* mutation, which occurs in 90%−95% of all PDACs. Recent data suggest that different *KRAS* mutations may have different effects on cell metabolism (Hobbs et al., 2020). However, in the 5%−10% of *KRAS* wild-type tumors ($KRAS^{WT}$), other "drivers" exist, several of which are targetable with currently available therapies, either off-label, or in the context of clinical trials. For example, 4% of PDACs harbor a mutation in *BRAF*, half of which are the classically activating $BRAF^{V600E}$ mutation. A clinical trial targeting this specific sub-group of patients for treatment with the combination of the raf inhibitor, encorafenib, and the MEK inhibitor, binimetinib, has recently been initiated. Anecdotal data suggest that the combination may be effective for $BRAF^{V600E}$-mutated PDAC patients, with reports of PRs with treatment (Aguirre et al., 2018; Guan et al., 2018). In addition, there have been case reports demonstrating significant benefit with therapies targeted to mutations or fusions in receptor tyrosine kinases. For example, Jones et al. reported on three patients with $KRAS^{WT}$ tumors that harbored a fusion involving neuregulin 1 (NRG1), which is a HER3 ligand (Jones et al., 2019). NRG1 activates HER3, resulting in receptor heterodimerization with other HER-family receptors leading to downstream signaling pathway activation (Jones et al., 2017). Two of the three patients were treated with the pan-HER inhibitor, afatinib, both of whom benefitted with a significant and rapid response to therapy. A similar experience was published by Heining et al. in their 3/17 PDAC patients with NRG1 fusions (Heining et al., 2018).

Similarly, fusions in the anaplastic lymphoma kinase (ALK) and receptor tyrosine kinase (ROS) genes can occur in $KRAS^{WT}$ PDAC, and appropriately targeted therapy can lead to prolonged benefit. We published our experience with one patient whose tumor harbored a SLC4A4-ROS1 gene fusion, and who experienced 7 months of stable disease with entrectinib (Pishvaian et al., 2018). Interestingly, as proof of principle that his ROS1 fusion was a driving mutation, he then had prolonged disease control for more than 1 year on the second-generation ROS1 inhibitor, brigatinib (personal communication). Singhi et al. published the Foundation Medicine experience with sequencing over 3000 PDAC patients, and identified 5 patients with ALK gene rearrangements (Singhi et al., 2017). Four of the five patients were treated with an ALK inhibitor, three of whom had prolonged stable disease (with one highlighted patient treated with ceritinib who had prolonged disease control for >17 months).

Targeting the WNT pathway: RNF43 mutations and RSPO2/3 fusions

WNT pathway activation is triggered when neighboring cells secrete WNT proteins, in a process that requires the palmitoylation of WNT by the acyltransferase, porcupine (Lum & Clevers, 2012). Upon binding to coreceptors Frizzled and LRP5/6 on the target cell, WNT pathway activation involves a shift in the equilibrium between free β-catenin, and β-catenin destined for proteosomal degradation, leading to a greater cytosolic pool of free β-catenin, translocation to the nucleus, and interaction with T cell factor (TCF) to mediate transcription of multiple oncogenic pathways (Nusse & Clevers, 2017; Pishvaian & Byers, 2007). Recently, it has been discovered that alterations in proteins functioning in the WNT pathway upstream of APC and β-catenin can increase WNT pathway activation, including activating fusions in rombospondin (*RSPO*)-2 or -3, or inactivating mutations in the RSPO coreceptor, *RNF43* (Hao et al., 2012; Jiang et al., 2013; Koo et al., 2012; Seshagiri et al., 2012; Tsukiyama et al., 2015; Wang et al., 2016). Inactivating mutations in *RNF43* occur in 5%−7% of pancreatic cancers (Cancer Genome Atlas Research Network, 2017; Waddell et al., 2015), and frequently occur in premalignant lesions of pancreas, such as intraductal papillary mucinous neoplasms and mucinous cystic neoplasms (Furukawa et al., 2011; Macgregor-Das & Iacobuzio-Donahue, 2013; Wu, Chen, Aloysius, & Hu, 2011). Occurrence of *RSPO3* fusions in pancreatic tumors has also been reported (Kleeman et al., 2019).

While previous attempts to block WNT pathway signaling for cancer therapy have been unsuccessful, currently, the most promising approach is to prevent the paracrine or autocrine secretion of WNT proteins by blocking porcupine (Lum & Clevers, 2012). Preclinical studies have confirmed that the porcupine inhibitors LGK974, etc-159, and CGX1321 reduce WNT secretion, and inhibited the growth of tumors harboring mutations in *RNF43*, including *RNF43*-mutated pancreatic cancer xenograft models (Jiang et al., 2013; Li et al., 2018; Liu et al., 2013; Madan et al., 2016).

Survival benefit—a national registry for precision medicine (the Know Your Tumor program)

We demonstrated that there is a survival benefit for patients with advanced PDAC who undergo tumor molecular profiling, and receive appropriately matched therapy (Pishvaian et al., 2020). Through a prospectively enrolled registry trial, PDAC patients were offered molecular profiling of their tumors. Based on the testing results, and their prior treatment history, patients were offered treatment options tailored to their molecular profile, which could include off-label and clinical trial considerations. Patients were then followed longitudinally, tracking the therapies they received, and for survival. In a retrospective analysis of over 1000 patients, those whose tumors harbored actionable molecular alterations and who received therapy appropriately matched to their molecular alteration ($n = 46$) lived 1 year longer than similar patients who did not receive appropriately matched therapy [$n = 143$; 2.58 years (95% CI: 2.39 to not reached) vs. 1.51 years (1.33−1.87); hazard ratio 0.42 (95% CI: 0.26−0.68), $P = 0.0004$]. Survival was also more than 1 year longer than patients whose tumors did not harbor any actionable molecular alteration [$n = 488$; 2.58 years (95% CI: 2.39 to not reached) vs. 1.32 years (1.25−1.47); HR 0.34 (95% CI: 0.22−0.53), $P < 0.0001$]. These data represent clearly, albeit through a retrospective analysis of a registry trial, the value of molecular profiling, and working toward access to appropriately matched therapy for PDAC patients.

Elusive drivers: KRAS, TP53, CDKN2A/2B, and SMAD4

The classic molecular alterations in PDAC [and their associated percent prevalence (Bailey et al., 2016; Waddell et al., 2015)], which are typically felt to be the drivers of pancreatic carcinogenesis include *KRAS* (92%), *TP53* (70%), *CDKN2A/2B* (35%), and *SMAD4* (31%). Unfortunately, to date, none of these molecular alterations have been successfully therapeutically targeted. There have been no trials that have definitively targeted TP53 nor SMAD4 loss, and even trials of CDK4/6 inhibitors, while highly successful in, for example, breast cancer, and also with published promising preclinical data in PDAC (Dhir et al., 2019), have not demonstrated that the presence of a CDKN2A/2B mutation is predictive of a response to therapy. KRAS in particular has been extensively studied, but early attempts to target farnesylation, and thus, in theory anchoring of KRAS into the membrane, were unsuccessful. Recently, several groups have looked at targeting signaling pathways downstream of KRAS, but trials of single-agent MEK inhibitors have not improved outcomes, not even when combined with chemotherapy (Infante et al., 2014). The most promising recent KRAS-targeted strategy has been targeting the $KRAS^{G12C}$ mutation specifically. AMG 510 (Amgen) can induce responses in patients whose tumors harbor a $KRAS^{G12C}$ mutation (Canon et al., 2019), which occur more frequently in nonsmall cell lung cancer and colorectal cancer. Unfortunately, this specific mutation is only found in <1% of PDACs.

Future considerations

Others have applied high-throughput sequencing and RNA analysis to subtyping and classifying PDACs into categories that presumably will aid in tailoring therapies (for a review of this work, please refer to Pishvaian and Brody (2017)); these subcategories include classical-, immunogenic-, and squamous-type PDACs. More recent, sophisticated studies are consistently refining subtypes and attempting to classify PDAC tumors (Chan-Seng-Yue et al., 2020; Martens et al., 2019; O'Kane et al., 2020). As better therapies emerge and we further identify facile methods to classify these tumors, this subtyping work should add value to precision therapy—based trials for PDAC. Other important elements of the PDAC tumor that are being exploited and explored are the critical tumor microenvironment that surrounds the epithelial PDAC cells. Not only the amount of stroma content but also the elements in the microenvironment such as low glucose and hypoxia (Blanco et al., 2016; Zarei et al., 2017) could dramatically disrupt the efficacy of targeted and nontargeted therapeutic strategies. Being able to rigorously quantitate these elements in the tumor microenvironment may factor in our decision-making for choosing certain therapies (e.g., immunotherapies and therapies targeting the metabolic milieu of the tumor).

Finally, the future of precision medicine may incorporate serial biopsies, liquid biopsies, and development of patient-derived models of cancer (PDMCs), which could all be used as companion diagnostics to guide and adjust therapies to patients in real time (Fig. 14.2). For example, our best targeted strategies typically only extend OS, and unfortunately, either many PDACs ultimately become resistant or an emerging resistant clone emerges.

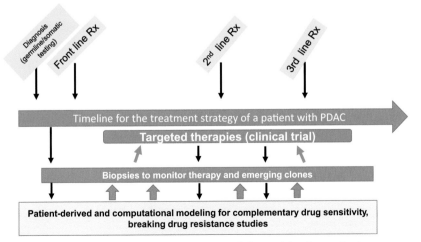

FIGURE 14.2   The future: a multidisciplinary approach for precision therapy for the treatment of PDAC.

By developing and validating PDMCs and a liquid biopsy strategy, we may be able to detect emerging clones sooner, and at the same time, use PDMCs to predict what therapies to pivot to in an effort to break acquired resistance mechanisms (Fig. 14.2).

Concluding remarks

We have demonstrated that PDAC patients who are profiled (as recommended by the NCCN) and treated with appropriately matched therapies have a significantly improved survival as compared to similar patients who were not treated with matched therapy. Yet, one of the understandable criticisms is that PDAC patients do not have adequate access to matched therapy if molecular profiling reveals an actionable mutation. For this reason, we and multiple other groups have been working to change the clinical trial landscape for PDAC patients, and to focus on biomarker-driven trials focused on small subgroups of patients. These trials are intended to demonstrate a high degree of activity, as demonstrated a clinically meaningful enhanced ORR; and/or a clinically meaningful prolongation of response. The best examples to date, albeit in a small number of patients in pancreatic cancer, continue to be PARP inhibitor-based therapy for *BRCA1/2*-mutated PDAC (Golan et al., 2019; Kaufman et al., 2015; Lowery et al., 2011; Shroff et al., 2018), and MSI-high patients treated with pembrolizumab (Marabelle et al., 2020).

Developing such biomarker trials one trial, and one target at a time is inefficient and slows down any momentum the field might have. Thus we are working toward developing an umbrella protocol through which PDAC patients anywhere in the United States can undergo molecular screening, and then be directed to one of a series of biomarker-specific trials. This effort, called TARGET-Panc, is anticipated to launch in 2020, and parallels a similar effort in the United Kingdom known as Precision Panc. TARGET-Panc may also be a feeder into the US-based Precision Promise (Pancreatic Cancer Action Network), which will sponsor a number of Phase III trials that, if successful, would lead to FDA approval/label indication for a biomarker-specific subgroup of PDAC patients. We are hopeful to see the future approval of several agents for biomarker-specific subgroups of PDAC patients.

References

Aguirre, A. J., Nowak, J. A., Camarda, N. D., Moffitt, R. A., Ghazani, A. A., Hazar-Rethinam, M., ... Wolpin, B. M. (2018). Real-time genomic characterization of advanced pancreatic cancer to enable precision medicine. *Cancer Discovery*, 8(9), 1096−1111.

American Cancer Society. (2020). *Cancer facts & figures*.

Bailey, P., Chang, D. K., Nones, K., Johns, A. L., Patch, A. M., Gingras, M. C., ... Grimmond, S. M. (2016). Genomic analyses identify molecular subtypes of pancreatic cancer. *Nature*, 531(7592), 47−52.

Biankin, A. V., Waddell, N., Kassahn, K. S., Gingras, M. C., Muthuswamy, L. B., Johns, A. L., ... Grimmond, S. M. (2012). Pancreatic cancer genomes reveal aberrations in axon guidance pathway genes. *Nature, 491*(7424), 399−405.

Blanco, F. F., Preet, R., Aguado, A., Vishwakarma, V., Stevens, L. E., Vyas, A., ... Dixon, D. A. (2016). Impact of HuR inhibition by the small molecule MS-444 on colorectal cancer cell tumorigenesis. *Oncotarget., 7*(45), 74043−74058.

Burris, H. A., 3rd, Moore, M. J., Andersen, J., Green, M. R., Rothenberg, M. L., Modiano, M. R., ... Von Hoff, D. D. (1997). Improvements in survival and clinical benefit with gemcitabine as first-line therapy for patients with advanced pancreas cancer: a randomized trial. *Journal of Clinical Oncology: Official Journal of the American Society of Clinical Oncology, 15*(6), 2403−2413.

Cancer Genome Atlas Research Network. (2017). Integrated genomic characterization of pancreatic ductal adenocarcinoma. *Cancer Cell, 32*(2), 185−203., e13.

Canon, J., Rex, K., Saiki, A. Y., Mohr, C., Cooke, K., Bagal, D., ... Lipford, J. R. (2019). The clinical KRAS (G12C) inhibitor AMG 510 drives anti-tumour immunity. *Nature, 575*(7781), 217−223.

Chan-Seng-Yue, M., Kim, J. C., Wilson, G. W., Ng, K., Figueroa, E. F., O'Kane, G. M., ... Notta, F. (2020). Transcription phenotypes of pancreatic cancer are driven by genomic events during tumor evolution. *Nature Genetics, 52*(2), 231−240.

Collisson, E. A., Sadanandam, A., Olson, P., Gibb, W. J., Truitt, M., Gu, S., ... Gray, J. W. (2011). Subtypes of pancreatic ductal adenocarcinoma and their differing responses to therapy. *Nature Medicine, 17*(4), 500−503.

Conroy, T., Desseigne, F., Ychou, M., Bouche, O., Guimbaud, R., Becouarn, Y., ... PRODIGE Intergroup. (2011). FOLFIRINOX vs gemcitabine for metastatic pancreatic cancer. *The New England Journal of Medicine, 364*(19), 1817−1825.

Dhir, T., Schultz, C. W., Jain, A., Brown, S. Z., Haber, A., Goetz, A., ... Brody, J. R. (2019). Abemaciclib is effective against pancreatic cancer cells and synergizes with HuR and YAP1 inhibition. *Molecular Cancer Research, 17*(10), 2029−2041.

Drilon, A., Laetsch, T. W., Kummar, S., DuBois, S. G., Lassen, U. N., Demetri, G. D., ... Hyman, D. M. (2018). Efficacy of larotrectinib in TRK fusion-positive cancers in adults and children. *The New England Journal of Medicine, 378*(8), 731−739.

Furukawa, T., Kuboki, Y., Tanji, E., Yoshida, S., Hatori, T., Yamamoto, M., ... Shiratori, K. (2011). Whole-exome sequencing uncovers frequent GNAS mutations in intraductal papillary mucinous neoplasms of the pancreas. *Scientific Reports, 1*, 161.

Golan, T., Locker, G. Y., & Kindler, H. L. (2019). Maintenance olaparib for metastatic pancreatic cancer. Reply. *The New England Journal of Medicine, 381*(15), 1492−1493.

Golan, T., Sella, T., O'Reilly, E. M., Katz, M. H., Epelbaum, R., Kelsen, D. P., ... Gallinger, S. (2017). Overall survival and clinical characteristics of BRCA mutation carriers with stage I/II pancreatic cancer. *British Journal of Cancer, 116*(6), 697−702.

Guan, M. B. R., Pishvaian, M. J., Halverson, D. C., Tuli, R., Klempner, S. J., Wainberg, Z. A., ... Hendifar, A. E. (2018). Molecular and clinical characterization of BRAF mutations in pancreatic ductal adenocarcinomas (PDACs). *Journal of Clinical Oncology: Official Journal of the American Society of Clinical Oncology, 36*(suppl 4S; abstr 214), 2018.

Hao, H. X., Xie, Y., Zhang, Y., Charlat, O., Oster, E., Avello, M., ... Cong, F. (2012). ZNRF3 promotes Wnt receptor turnover in an R-spondin-sensitive manner. *Nature, 485*(7397), 195−200.

Heeke, A. L., Pishvaian, M. J., Lynce, F., Xiu, J., Brody, J. R., Chen, W. J., ... Isaacs, C. (2018). Prevalence of homologous recombination-related gene mutations across multiple cancer types. *JCO Precision Oncology, 2018.*

Heining, C., Horak, P., Uhrig, S., Codo, P. L., Klink, B., Hutter, B., ... Glimm, H. (2018). NRG1 fusions in KRAS wild-type pancreatic cancer. *Cancer Discovery, 8*(9), 1087−1095.

Helleday, T., Petermann, E., Lundin, C., Hodgson, B., & Sharma, R. A. (2008). DNA repair pathways as targets for cancer therapy. *Nature Reviews Cancer, 8*(3), 193−204.

Hobbs, G. A., Baker, N. M., Miermont, A. M., Thurman, R. D., Pierobon, M., Tran, T. H., ... Der, C. J. (2020). Atypical KRAS (G12R) mutant is impaired in PI3K signaling and macropinocytosis in pancreatic cancer. *Cancer Discovery, 10*(1), 104−123.

Min, A., Im, S. A., Jang, H., Kim, S., Lee, M., Kim, D. K., ... Bang, Y.-J. (2017). AZD6738, a novel oral inhibitor of ATR, induces synthetic lethality with ATM deficiency in gastric cancer cells. *Molecular Cancer Therapeutics, 16*(4), 566−577.

Infante, J. R., Somer, B. G., Park, J. O., Li, C. P., Scheulen, M. E., Kasubhai, S. M., ... Le, N. (2014). A randomised, double-blind, placebo-controlled trial of trametinib, an oral MEK inhibitor, in combination with gemcitabine for patients with untreated metastatic adenocarcinoma of the pancreas. *European Journal of Cancer, 50*(12), 2072−2081.

Jalal, S., Earley, J. N., & Turchi, J. J. (2011). DNA repair: From genome maintenance to biomarker and therapeutic target. *Clinical Cancer Research: An Official Journal of the American Association for Cancer Research, 17*(22), 6973−6984.

Jiang, X., Hao, H. X., Growney, J. D., Woolfenden, S., Bottiglio, C., Ng, N., ... Cong, F. (2013). Inactivating mutations of RNF43 confer Wnt dependency in pancreatic ductal adenocarcinoma. *Proceedings of the National Academy of Sciences of the United States of America, 110*(31), 12649−12654.

Jones, M. R., Lim, H., Shen, Y., Pleasance, E., Ch'ng, C., Reisle, C., ... Laskin, J. (2017). Successful targeting of the NRG1 pathway indicates novel treatment strategy for metastatic cancer. *Annals of Oncology: Official Journal of the European Society for Medical Oncology, 28*(12), 3092−3097.

Jones, M. R., Williamson, L. M., Topham, J. T., Lee, M. K. C., Goytain, A., Ho, J., ... Renouf, D. J. (2019). NRG1 gene fusions are recurrent, clinically actionable gene rearrangements in KRAS wild-type pancreatic ductal adenocarcinoma. *Clinical Cancer Research: An Official Journal of the American Association for Cancer Research, 25*(15), 4674−4681.

Jones, S., Zhang, X., Parsons, D. W., Lin, J. C., Leary, R. J., Angenendt, P., ... Kinzler, K. W. (2008). Core signaling pathways in human pancreatic cancers revealed by global genomic analyses. *Science (New York, N.Y.), 321*(5897), 1801−1806.

Kaufman, B., Shapira-Frommer, R., Schmutzler, R. K., Audeh, M. W., Friedlander, M., Balmana, J., ... Domchek, S. M (2015). Olaparib monotherapy in patients with advanced cancer and a germline BRCA1/2 mutation. *Journal of Clinical Oncology: Official Journal of the American Society of Clinical Oncology, 33*(3), 244−250.

Kleeman, S. O., Koelzer, V. H., Jones, H. J., Vazquez, E. G., Davis, H., East, J. E., ... Leedham, S. J. (2019). Exploiting differential Wnt target gene expression to generate a molecular biomarker for colorectal cancer stratification. *Gut, 69*, 1092−1103.

Koo, B. K., Spit, M., Jordens, I., Low, T. Y., Stange, D. E., van de Wetering, M., ... Clevers, H. (2012). Tumour suppressor RNF43 is a stem-cell E3 ligase that induces endocytosis of Wnt receptors. *Nature, 488*(7413), 665−669.

Le, D. T., Durham, J. N., Smith, K. N., Wang, H., Bartlett, B. R., Aulakh, L. K., ... Diaz, L. A. (2017). Mismatch repair deficiency predicts response of solid tumors to PD-1 blockade. *Science (New York, N.Y.), 357*(6349), 409−413.

Li, C., Cao, J., Zhang, N., Tu, M., Xu, F., Wei, S., ... Xu, Y. (2018). Identification of RSPO2 fusion mutations and target therapy using a porcupine inhibitor. *Scientific Reports, 8*(1), 14244.

Liu, J., Pan, S., Hsieh, M. H., Ng, N., Sun, F., Wang, T., ... Harris, J. L. (2013). Targeting Wnt-driven cancer through the inhibition of Porcupine by LGK974. *Proceedings of the National Academy of Sciences of the United States of America, 110*(50), 20224−20229.

Lowery, M. A., Jordan, E. J., Basturk, O., Ptashkin, R. N., Zehir, A., Berger, M. F., ... O'Reilly, E. M. (2017). Real-time genomic profiling of pancreatic ductal adenocarcinoma: Potential actionability and correlation with clinical phenotype. *Clinical Cancer Research: An Official Journal of the American Association for Cancer Research, 23*(20), 6094−6100.

Lowery, M. A., Kelsen, D. P., Stadler, Z. K., Yu, K. H., Janjigian, Y. Y., Ludwig, E., ... O'Reilly, E. M. (2011). An emerging entity: Pancreatic adenocarcinoma associated with a known BRCA mutation: Clinical descriptors, treatment implications, and future directions. *The Oncologist, 16*(10), 1397−1402.

Lum, L., & Clevers, H. (2012). Cell biology. The unusual case of Porcupine. *Science (New York, N.Y.), 337*(6097), 922−923.

Macgregor-Das, A. M., & Iacobuzio-Donahue, C. A. (2013). Molecular pathways in pancreatic carcinogenesis. *Journal of Surgical Oncology, 107*(1), 8−14.

Madan, B., Ke, Z., Harmston, N., Ho, S. Y., Frois, A. O., Alam, J., ... Virshup, D. M. (2016). Wnt addiction of genetically defined cancers reversed by PORCN inhibition. *Oncogene, 35*(17), 2197−2207.

Marabelle, A., Le, D. T., Ascierto, P. A., Di Giacomo, A. M., De Jesus-Acosta, A., Delord, J. P., & Diaz, L. A. (2020). Efficacy of pembrolizumab in patients with noncolorectal high microsatellite instability/mismatch repair-deficient cancer: Results from the phase II KEYNOTE-158 study. *Journal of Clinical Oncology: Official Journal of the American Society of Clinical Oncology, 38*(1), 1−10.

Martens, S., Lefesvre, P., Nicolle, R., Biankin, A. V., Puleo, F., Van Laethem, J. L., ... Rooman, I. (2019). Different shades of pancreatic ductal adenocarcinoma, different paths towards precision therapeutic applications. *Annals of Oncology: Official Journal of the European Society for Medical Oncology, 30*(9), 1428−1436.

Mavros, M. N., Coburn, N. G., Davis, L. E., Mahar, A. L., Liu, Y., Beyfuss, K., ... Hallet, J. (2019). Low rates of specialized cancer consultation and cancer-directed therapy for noncurable pancreatic adenocarcinoma: A population-based analysis. *Canadian Medical Association Journal = Journal de l'Association Medicale Canadienne, 191*(21), E574−E580.

Nusse, R., & Clevers, H. (2017). Wnt/beta-catenin signaling, disease, and emerging therapeutic modalities. *Cell, 169*(6), 985−999.

O'Kane, G. M., Grunwald, B. T., Jang, G. H., Masoomian, M., Picardo, S., Grant, R. C., ... Fischer, S. E. (2020). GATA6 expression distinguishes classical and basal-like subtypes in advanced pancreatic cancer. *Clinical Cancer Research: An Official Journal of the American Association for Cancer Research, 26*, 4901−4910.

O'Reilly, E. M., Lee, J. W., Zalupski, M., Capanu, M., Park, J., Golan, T., ... Kelsen, D. P. (2020). Randomized, multicenter, phase II trial of gemcitabine and cisplatin with or without veliparib in patients with pancreas adenocarcinoma and a germline BRCA/PALB2 mutation. *Journal of Clinical Oncology: Official Journal of the American Society of Clinical Oncology*, JCO1902931.

Perkhofer, L., Schmitt, A., Romero Carrasco, M. C., Ihle, M., Hampp, S., Ruess, D. A., ... Kleger, A. (2017). ATM deficiency generating genomic instability sensitizes pancreatic ductal adenocarcinoma cells to therapy-induced DNA damage. *Cancer Research, 77*(20), 5576−5590.

Pishvaian, M. J., Bender, R. J., Halverson, D., Rahib, L., Hendifar, A. E., Mikhail, S., ... Petricoin, E. F. (2018). Molecular profiling of patients with pancreatic cancer: Initial results from the know your tumor initiative. *Clinical Cancer Research: An Official Journal of the American Association for Cancer Research, 24*(20), 5018–5027.

Pishvaian, M. J., Biankin, A. V., Bailey, P., Chang, D. K., Laheru, D., Wolfgang, C. L., ... Brody, J. R. (2017). BRCA2 secondary mutation-mediated resistance to platinum and PARP inhibitor-based therapy in pancreatic cancer. *British Journal of Cancer, 116*(8), 1021–1026.

Pishvaian, M. J., Blais, E., Brody, J. R., Rahib, L., Lyons, E., De Arbeloa, P., ... Petricoin, E., III (2019). Outcomes in patients with pancreatic adenocarcinoma with genetic mutations in DNA damage response pathways: Results from the know your tumor program. *JCO Precision Oncology*.

Pishvaian, M. J., Blais, E. M., Brody, J. R., Lyons, E., DeArbeloa, P., Hendifar, A., ... Petricoin, E. F. (2020). Overall survival in patients with pancreatic cancer receiving matched therapies following molecular profiling: A retrospective analysis of the Know Your Tumor registry trial. *The Lancet Oncology, 21*, 508–518.

Pishvaian, M. J., & Brody, J. R. (2017). Therapeutic implications of molecular subtyping for pancreatic cancer. *Oncology (Williston Park, N.Y.), 31*(3), 159–166, 68.

Pishvaian, M. J., & Byers, S. W. (2007). Biomarkers of WNT signaling. *Cancer Biomarkers: Section A of Disease Markers, 3*(4–5), 263–274.

Pishvaian, M. J., Wang, H., He, A. R., Hwang, J. J., Smaglo, B. G., Kim, S. S., ... Brody, J. R. (2020). A phase I/II study of veliparib (ABT-888) in combination with 5-fluorouracil and oxaliplatin in patients with metastatic pancreatic cancer. *Clinical Cancer Research: An Official Journal of the American Association for Cancer Research, 26*, 5092–5101.

Pishvaian, M. J., Garrido-Laguna, I., Liu, S. V., Multani, P. S., Chow-Maneval, E., & Rolfo, C. (2018). Entrectinib in TRK and ROS1 fusion-positive metastatic pancreatic cancer. *JCO Precision Oncology*, Epub ahead of print.

Portal, A., Pernot, S., Tougeron, D., Arbaud, C., Bidault, A. T., de la Fouchardiere, C., ... Taieb, J. (2015). Nab-paclitaxel plus gemcitabine for metastatic pancreatic adenocarcinoma after Folfirinox failure: An AGEO prospective multicentre cohort. *British Journal of Cancer, 113*(7), 989–995.

Rahib, L., Smith, B. D., Aizenberg, R., Rosenzweig, A. B., Fleshman, J. M., & Matrisian, L. M. (2014). Projecting cancer incidence and deaths to 2030: The unexpected burden of thyroid, liver, and pancreas cancers in the United States. *Cancer Research, 74*(11), 2913–2921.

Ramanathan, R. K. , McDonough, S. L., Philip, P. A., Hingorani, S. R., Lacy, J., Kortmansky, J. S., & Hochster, H. S. (2019). Phase IB/II randomized study of FOLFIRINOX plus pegylated recombinant human hyaluronidase vs FOLFIRINOX alone in patients with metastatic pancreatic adenocarcinoma: SWOG S1313. *Journal of Clinical Oncology: Official Journal of the American Society of Clinical Oncology, 37*(13), 1062–1069.

Reaper, P. M., Griffiths, M. R., Long, J. M., Charrier, J. D., Maccormick, S., Charlton, P. A., & John, R. P. (2011). Selective killing of ATM- or p53-deficient cancer cells through inhibition of ATR. *Nature Chemical Biology, 7*(7), 428–430.

Reiss Binder, K. M. R., O'Hara, M., Teitelbaum, U., Karasic, T., Schneider, C., O'Dwyer, P. J., ... Domchek, S. (2019). Phase II, single arm study of maintenance rucaparib in patients with platinum-sensitive advanced pancreatic cancer and a pathogenic germline or somatic mutation in BRCA1, BRCA2 or PALB2. *AACR annual meeting, Abstract CT234.*

Salo-Mullen, E. E., O'Reilly, E. M., Kelsen, D. P., Ashraf, A. M., Lowery, M. A., Yu, K. H., ... Stadler, Z. K. (2015). Identification of germline genetic mutations in patients with pancreatic cancer. *Cancer, 121*(24), 4382–4388.

Schmitt, A., Knittel, G., Welcker, D., Yang, T. P., George, J., Nowak, M., ... Reinhardt, H. C. (2017). ATM deficiency is associated with sensitivity to PARP1- and ATR inhibitors in lung adenocarcinoma. *Cancer Research, 77*(11), 3040−3056.

Seshagiri, S., Stawiski, E. W., Durinck, S., Modrusan, Z., Storm, E. E., Conboy, C. B., ... de Sauvage, F. J. (2012). Recurrent R-spondin fusions in colon cancer. *Nature, 488*(7413), 660−664.

Shi, Q., Shen, L. Y., Dong, B., Fu, H., Kang, X. Z., Yang, Y. B., ... Chen, K. N. (2018). The identification of the ATR inhibitor VE-822 as a therapeutic strategy for enhancing cisplatin chemosensitivity in esophageal squamous cell carcinoma. *Cancer Letters, 432*, 56−68.

Shroff, R. T., Hendifar, A., McWilliams, R. R., Geva, R., Epelbaum, R., Rolfe, L., ... Domchek, S. M. (2018). Rucaparib monotherapy in patients with pancreatic cancer and a known deleterious BRCA mutation. *JCO Precision Oncology, 2018*.

Singhi, A. D., Ali, S. M., Lacy, J., Hendifar, A., Nguyen, K., Koo, J., ... Bahary, N. (2017). Identification of targetable ALK rearrangements in pancreatic ductal adenocarcinoma. *Journal of the National Comprehensive Cancer Network, 15*(5), 555−562.

Singhi, A. D., George, B., Greenbowe, J. R., Chung, J., Suh, J., Maitra, A., ... Bahary, N. (2019). Real-time targeted genome profile analysis of pancreatic ductal adenocarcinomas identifies genetic alterations that might be targeted with existing drugs or used as biomarkers. *Gastroenterology, 156*(8), 2242−2253, e4.

Tempero, M. A., Malafa, M. P., Al-Hawary, M., Behrman, S. W., Al Benson, B., Cardin, D. B., ... George, G. V. (2019). *NCCN clinical practice guidelines in oncology pancreatic adenocarcinoma version 2*.

Tempero, M. A., Cutsem, E. V., Sigal, D., Oh, D. Y., Fazio, N., Macarulla, T., ... Bullock, A. J. (2020). HALO 109−301: A randomized, double-blind, placebo-controlled, phase 3 study of pegvorhyaluronidase alfa (PEGPH20) + nab-paclitaxel/gemcitabine (AG) in patients (pts) with previously untreated hyaluronan (HA)-high metastatic pancreatic ductal adenocarcinoma (mPDA). *Journal of Clinical Oncology: Official Journal of the American Society of Clinical Oncology, 38*(suppl 4; abstr 638).

Tsukiyama, T., Fukui, A., Terai, S., Fujioka, Y., Shinada, K., Takahashi, H., ... Hatakeyama, S. (2015). Molecular role of RNF43 in canonical and noncanonical Wnt signaling. *Molecular and Cellular Biology, 35*(11), 2007−2023.

Vendetti, F. P., Lau, A., Schamus, S., Conrads, T. P., O'Connor, M. J., & Bakkenist, C. J. (2015). The orally active and bioavailable ATR kinase inhibitor AZD6738 potentiates the anti-tumor effects of cisplatin to resolve ATM-deficient non-small cell lung cancer in vivo. *Oncotarget, 6*(42), 44289−44305.

Von Hoff, D. D., Ervin, T., Arena, F. P., Chiorean, E. G., Infante, J., Moore, M., ... Renschler, M. F. (2013). Increased survival in pancreatic cancer with nab-paclitaxel plus gemcitabine. *The New England Journal of Medicine, 369*(18), 1691−1703.

Waddell, N., Pajic, M., Patch, A. M., Chang, D. K., Kassahn, K. S., Bailey, P., ... Grimmond, S. M. (2015). Whole genomes redefine the mutational landscape of pancreatic cancer. *Nature, 518*(7540), 495−501.

Wang, D., Tan, J., Xu, Y., Han, M., Tu, Y., Zhu, Z., ... Liu, Z. (2016). The ubiquitin ligase RNF43 downregulation increases membrane expression of frizzled receptor in pancreatic ductal adenocarcinoma. *Tumour Biology: The Journal of the International Society for Oncodevelopmental Biology and Medicine, 37*(1), 627−631.

Witkiewicz, A. K., McMillan, E. A., Balaji, U., Baek, G., Lin, W. C., Mansour, J., ... Knudsen, E. S. (2015). Whole-exome sequencing of pancreatic cancer defines genetic diversity and therapeutic targets. *Nature Communications, 6*, 6744.

Wu, X., Chen, Y., Aloysius, H., & Hu, L. (2011). A novel high-yield synthesis of aminoacyl p-nitroanilines and aminoacyl 7-amino-4-methylcoumarins: Important synthons for the synthesis of chromogenic/fluorogenic protease substrates. *Beilstein Journal of Organic Chemistry*, 7, 1030–1035.

Zarei, M., Lal, S., Parker, S. J., Nevler, A., Vaziri-Gohar, A., Dukleska, K., ... Winter, J. M. (2017). Posttranscriptional upregulation of IDH1 by HuR establishes a powerful survival phenotype in pancreatic cancer cells. *Cancer Research*, *77*(16), 4460–4471.

Chapter 15

Esophageal and gastric cancer

Matthew Emmet[1], Arnav Mehta[1] and Samuel Klempner[2]

[1]*Department of Medicine, Massachusetts General Hospital, Boston, MA, United States,*
[2]*Harvard Medical School, Massachusetts General Hospital, Boston, MA, United States*

Introduction

In 2018 alone there were roughly 572,000 cases of esophageal cancer and 1 million cases of gastric cancer, together accounting for 9% of all cancer diagnoses (Bray et al., 2018). Only lung, breast, and colorectal cancers accounted for more new cancer diagnoses. In a reflection of their lethality, esophageal and gastric cancers recorded nearly 1.3 million deaths, or 13.5% of all cancer deaths, in 2018 (Bray et al., 2018).

Epidemiology and risk factors

Esophageal squamous cell

The incidence rates for esophageal squamous cell (ESCC) vary up to 16-fold with age standardized rates (per 100,000 people) approaching 18 in Eastern Asia and as low as 1.6 in Western Africa (Bray et al., 2018). In North America, the incidence is roughly 5.6 cases per 100,000 in men and 1.2 per 100,000 in women, highlighting the significant gender differences. Similar to head and neck cancers, smoking and alcohol consumption are known risk factors in the United States and several Western Countries (Abnet, Arnold, & Wei, 2018; Pandeya, Williams, Green, Webb, & Whiteman, 2009). In the highest incidence areas, the relative contribution of smoking and alcohol is lower, but chronic high-temperature beverages, nutritional status, and dietary patterns are major contributors (Islami et al., 2020; Yu et al., 2018). Other ESCC risk factors include underlying esophageal disease (achalasia), prior gastrectomy, and atrophic gastritis (Islami, Sheikhattari, Ren, & Kamangar, 2011; Sandler et al., 1995). Despite the association between HPV and oropharyngeal cancers, HPV is not considered a significant risk factor for ESCC (Islami et al., 2011; Petrick et al., 2014; Sandler et al., 1995).

Genomic and Precision Medicine. DOI: https://doi.org/10.1016/B978-0-12-800684-9.00009-5

Esophageal and gastroesophageal junction adenocarcinomas

Although ESCC remains the predominant overall histologic subtype, the rates of esophageal and gastroesophageal junction (GEJ) adenocarcinomas have been increasing over the last 30 years, primarily in Western countries. The consequences of chronic acid reflux, namely Barrett's metaplasia (BE), are the dominant risk factor for esophageal and GEJ adenocarcinomas (Hvid-Jensen, Pedersen, Drewes, Sørensen, & Funch-Jensen, 2011; Lagergren, Bergström, Lindgren, & Nyrén, 1999). BE is itself a heterogeneous condition, and patients with low-grade dysplasia have an annual cancer risk of ~0.5% per year, whereas those with high-grade dysplasia have cancer risk of 4%−8% per year (de Jonge et al., 2010; Pohl et al., 2016). Consequently, risk factors for BE, including obesity and metabolic syndrome, are also associated with increased adenocarcinoma risk (Thrift et al., 2014).

Gastric cancer

Nearly 10-fold differences exist between the highest incidence areas in Eastern Asia and lower incidence areas such as North America (Jemal et al., 2011). The ubiquitous bacteria *Helicobacter pylori* (*H. pylori*) is defined as a group 1 carcinogen and is estimated to account for up to 60% of all gastric cancers (Polk, Brent Polk, & Peek, 2010). A detailed discussion of *H. pylori* is beyond the scope of this chapter and readers are referred to several excellent reviews (Polk et al., 2010; Tan & Yeoh, 2015). Other gastric cancer risk factors include family history, diets high in salt and/or salt-preserved foods, diets high in nitroso compound generation, Epstein-Barr virus (EBV), and obesity (Boysen et al., 2009; Chang et al., 2006; González et al., 2003, 2006; Peleteiro, Lopes, Figueiredo, & Lunet, 2011; Turati, Tramacere, La Vecchia, & Negri, 2013). Although accounting for only a small number of global gastric cancer cases, inherited alterations in the E-cadherin encoding gene *CDH1* are responsible for hereditary diffuse gastric cancer (DGC), which is over-represented in younger patients (Stjepanovic et al., 2019; van der Post et al., 2015).

Current classification

Histological classification

True esophageal cancers are categorized as ESCC or adenocarcinomas. The vast majority of GEJ and gastric cancers are adenocarcinomas. Among gastric cancers, there are two major types defined by histomorphologic appearance: diffuse type and intestinal type (Lauren, 1965). The intestinal type of gastric cancer predominates and morphologically resembles other adenocarcinomas of the tubular gastrointestinal tract. DGC accounts for approximately 30%−35% of all cases and is characterized by an infiltrative pattern with loss of intercellular adhesions leading to a lack of the glandular structures seen in the intestinal type (Graziano, Humar, & Guilford, 2003; Lauren, 1965). DGC more commonly spreads to the peritoneal cavity and has historically been associated with decreased chemotherapy and radiotherapy responsiveness and a worse overall prognosis (Chon et al., 2017; Graziano et al., 2003; Lauren, 1965; Taghavi, Jayarajan, Davey, & Willis, 2012).

Molecular classifications: TCGA and ACRG

The improvement in higher throughput genomic technologies coupled to large collaborative efforts has led to significant advances in biologic understanding across cancer types. Our discussion is restricted to adenocarcinomas and the genomics of ESCC are reviewed elsewhere (Cancer Genome Atlas Research Network et al., 2017; Gao et al., 2014; Hu et al., 2016; Lin et al., 2014, 2018; Song et al., 2014; Wang et al., 2015).

Our molecular understanding of gastric and esophageal/GEJ adenocarcinomas has advanced substantially over the last 10 years, following publications from The Cancer Genome Atlas (TCGA) and the Asian Cancer Research Group (ACRG). The TCGA collected a cohort ($n = 295$) of previously untreated resected gastric adenocarcinomas and performed multiomic profiling including whole exome sequencing, genome-wide copy number analysis, methylation profiling, messenger RNA, micro-RNA, and reverse-phase protein array to develop distinct molecular subgroups (The Cancer Genome Atlas Research Network, 2014). Analyses ultimately defined four distinct subgroups: EBV positive (9% cases), microsatellite instable (MSI, 22% cases), chromosomally instable (CIN, 50% cases), and genomically stable (GS, 20% cases). The molecular subtypes show some anatomic clustering with CIN tumors predominating more proximally (GEJ, cardia) and MSI and GS tumors more distally (body and antrum). Contemporaneous with the TCGA gastric molecular analysis was publication from the ACRG. Using a cohort of 300 Asian gastric cancers subjected to gene expression profiling, genome-wide copy number analysis, and large panel–targeted sequencing, the ACRG identified four distinct molecular subtypes: MSI, microsatellite stable (MSS) and TP53 functional loss (MSS/TP53-), MSS with intact TP53 activity (MSS/TP53 +), and a MSS with epithelial-to-mesenchymal (EMT) signature (MSS/EMT) (Cristescu et al., 2015). In addition to distinct molecular features, these subgroups are associated with differences in anatomic location and prognosis with the MSI subgroup having the best overall survival and the MSS/EMT and GS groups having the shortest survival (Ali et al., 2015; Chao, Lee, & Klempner, 2016; Hur et al., 2020; Klempner et al., 2019; Nagaraja, Kikuchi, & Bass, 2019; Sohn et al., 2017). A summary of the key clinicopathologic and molecular features of the TCGA and ACRG analysis are summarized in Tables 15.1 and 15.2.

TABLE 15.1 Recurrent molecular features in Asian gastric cancer from the Asian Cancer Research Group analyses.

Clinicopathologic feature	MSS/TP53 − ($n = 107$)	MSS/TP53 + ($n = 79$)	MSI ($n = 68$)	MSS/EMT ($n = 46$)
Median age	65	64	66	53
Male (%)	65	72	66	58
Female (%)	35	28	34	42
Intestinal (%)	54	48	62	17
Diffuse (%)	39	46	30	80
Tumor location				
Antrum (%)	57	33	75	37
Body (%)	35	46	19	47
Cardia (%)	8	18	6	11
Molecular feature				
Gene amplifications				
HER2 (%)	18	3	0	0
EGFR (%)	7	3	2	0
FGFR2 (%)	1	3	0	5
MET (%)	4	3	2	0
KRAS (%)	5	8	2	2
PIK3CA (%)	1	0	0	0
CCND1 (%)	5	3	3	7
CCNE1 (%)	18	15	5	12
Somatic mutations				
ARID1A (%)	6	19	44	14
CDH1 (%)	4	2	7	3
RHOA (%)	4	7	0	3
PIK3CA (%)	5	17	33	8
PTEN (%)	4	3	14	6
TP53 (%)	60	24	26	33
APC (%)	8	15	16	3

APC, APC regulator of WNT signaling pathway; *EGFR*, epidermal growth factor receptor; *KRAS*, KRAS proto-oncogene, GTPase; *MSI*, microsatellite instable; *MSS*, microsatellite stable; *PTEN*, phosphatase and tensin homolog; *RHOA*, ras homolog family member A.

TABLE 15.2 Key molecular features gastric cancer from the The Cancer Genome Atlas analyses.

Clinicopathologic feature	EBV ($n = 26$) (%)	MSI ($n = 64$) (%)	GS ($n = 58$) (%)	CIN ($n = 147$) (%)
Male	81	44	62	66
Female	19	56	38	34
Intestinal	58	75	26	80
Diffuse	19	9	69	12
Tumor location				
Antrum	23	48	48	33
Body	62	39	31	39
Cardia	15	8	19	25
Molecular feature				
Gene amplifications				
HER2	12	0	4	23
EGFR	4	2	4	9
FGFR2	0	0	7	7
MET	0	2	0	7
KRAS	0	0	4	14
PIK3CA	8	3	2	8
CCND1	0	0	7	10
CCNE1	0	0	2	20
Somatic mutations				
ARID1A	54	84	16	9
CDH1	0	8	35	3
RHOA	8	5	14	2
PIK3CA	77	42	10	3
PTEN	15	25	2	1
TP53	4	39	14	71
APC	0	36	3	12

EGFR, Epidermal growth factor receptor; MSS, microsatellite stable; MSI, microsatellite instable.

Molecularly targeted therapies in advanced disease

Cytotoxic chemotherapies remain the backbone of treatment for the vast majority of patients and initial treatment with fluoropyrimidine and platinum is the global standard for frontline therapy. The taxane paclitaxel is the most commonly utilized second-line chemotherapy agent. However, the median survival for advanced gastroesophageal adenocarcinoma (GEA) patients remains poor, ranging from 10−14 months from large phase III trials (Bang et al., 2010; Cunningham et al., 2008; Fuchs et al., 2019). Thus there is a clear need to exploit biologic understanding to enhance patient outcomes.

HER2

As noted above, the transmembrane receptor tyrosine kinase (RTK) HER2 is amplified in 15%−20% of all GEA, most commonly in the CIN and MSS/TP53- subtypes. Following the observation of *HER2* amplification in GEA, there was a rapid push to exploit these therapeutically following multiple positive trials in *HER2*-amplified breast cancer. The phase III ToGA trial demonstrated a 2.7-month absolute improvement in survival with the addition of the anti-HER2 antibody trastuzumab to standard fluoropyrimidine and platinum chemotherapy. Similar to the breast cancer literature, there was an "oncogene dose" relationship with patients harboring a higher level of amplification deriving the greatest benefit. Analogous to breast cancer there was an immediate assumption that continuing HER2 pathway blockade into second line (2L) and beyond would translate to further benefit. Strategies involved dual HER2/HER3 blockade with trastuzumab + pertuzumab (JACOB trial) (Tabernero et al., 2018), intracellular blockade with the small molecule inhibitor lapatinib (TyTAN nad LOGiC trials) (Hecht et al., 2016; Satoh et al., 2014), and the antibody-drug conjugate T-DM1 combining trastuzumab and a microtubule inhibitor (GATSBY trial) (Thuss-Patience et al., 2017). All of these trials failed to demonstrate benefit. How can we explain these divergent findings? The failure to translate benefit in HER2 encapsulates several key issues in GEA, including spatial and temporal heterogeneity. GEA, in particularly CIN tumors, are known to harbor significant intratumoral and intertumoral heterogeneity at baseline. In intratumoral heterogeneity, individual oncogenes like *HER2* may be uniquely amplified in subclones of the same anatomic tumor location, such that a *HER2* nonamplified population exists at baseline (Janjigian et al., 2018; Pectasides et al., 2018; Wang et al., 2019). Furthermore, existence of coamplified oncogenes within the same cell (*EGFR* and *HER2* for example) has also been described in GEA and may blunt the effect of agents only targeting one of the coamplified targets. In cases of intertumoral heterogeneity, patients with advanced disease have differences between anatomic sites (stomach and liver for example). In a series of paired pretreatment samples, there was discordance between HER2 overexpression/amplification in up to 30% of cases at baseline (Pectasides et al., 2018). The therapeutic implications of intra- and intertumoral heterogeneity were initially brought to light from analyzing posttrastuzumab tissue samples where the initial *HER2* amplification was no longer observed. This temporal HER2 "loss" likely reflects the outgrowth of the initial *HER2* nonamplified subclonal population under therapeutic pressure of HER2-directed therapies. This biology likely partly underlies the lack of clinical benefit with continued HER2-blockade in HER2 positive patients (Shah et al., 2019). In fact, the response rates and benefits seem to be greatest in patients with concordant HER2 status between primary and metastatic or between tissue and circulating tumor DNA (ctDNA) analyses (Kim et al., 2018; Klempner & Chao 2018; Wang et al., 2019). In addition to heterogeneity of HER2 itself, genomic analyses have revealed that up to 50% of *HER2*-amplified GEA cases harbor concurrent genomic alterations that may modulate innate trastuzumab resistance include *CCNE1* amplification, *CDK6* amplification, PI3-kinase pathway mutations, as well as *MET* (*proto-oncogene, receptor tyrosine kinase*) and *EGFR* amplification (Kim et al., 2014; Klempner & Chao 2018; Lee et al., 2015; Zhao, Klempner, & Chao, 2019).

Despite a list of negative trials directed at HER2, newer strategies are showing promise. The antibody drug conjugate DS-8201 combines trastuzumab with a potent topoisomerase I inhibitor payload, with a significantly higher payload per antibody than T-DM1. In a randomized phase II trial conducted in Asia, this agent demonstrated a robust 50% response rate versus investigator choice chemotherapy in HER2 + GEA patients who had progressed after trastuzumab containing first line therapy. The overall survival of 12.5 months in the DS-8201-treated patients versus 8.4 months among investigator choice chemotherapy was highly encouraging (Shitara et al., 2020). The high drug to antibody ratio of DS-8201 reduces the dependency on higher HER2 expression and may lead to additional "bystander" effects against HER2 negative subclones and explain why DS-8201 appears to have succeeded where T-DM1 did not (Ogitani et al., 2016; Ogitani, Hagihara, Oitate, Naito, & Agatsuma, 2016; Shitara et al., 2019).

It is known that trastuzumab partly works via antibody-dependent cellular cytotoxicity and preclinical evidence has supported increased activity in combination with immune checkpoint blockade (Park et al., 2010; Taylor et al., 2007; Varadan et al., 2016). This strategy has produced promising early clinical results and is currently being explored in two ongoing phase III trials (Catenacci et al., 2020; Janjigian et al., 2020). The story of HER2 development in GEA highlights lineage-dependent differences in biology between breast cancer and GEA and provides a cautionary tale for precision medicine without clear biologic understanding.

MET

Amplification of *MET* is seen in 5%−7% of all GEA and protein overexpression is reported in up to 60% of patients. Following successes with trastuzumab in HER2 + GEA, a series of early-phase clinical trials demonstrated promising response data in MET positive patients (Lee et al., 2019; Shah et al., 2016; Van Cutsem et al., 2019). Unfortunately, many MET-directed clinical trials were negatively impacted by issues surrounding patient selection. For example, in preclinical models, activity was greatest in *MET*-amplified models, which did not clearly correlate with MET protein expression by IHC. *MET* amplification also demonstrates intra- and intertumoral heterogeneity, which leads to innate MET-directed resistance in several cases. Using MET protein expression by IHC, the METGastric trial failed to demonstrate a survival improvement with the addition of onartuzumab to chemotherapy (Shah et al., 2017). Similarly, the RILOMET-1 trial did not improve outcomes with the addition of the monoclonal antibody rilotumumab to first-line chemotherapy (Catenacci et al., 2017). Both of these trials suffered from inclusion criteria that may not identify GEA patients whose tumors are more biologically dependent on MET signaling (high-level focal *MET* amplification without other coamplification and limited heterogeneity) (Kwak et al., 2015; Sanchez-Vega et al., 2019). Although ongoing anecdotal cases and small series clearly highlight a subset of patients with exceptional response to MET inhibitors, conducting clinical trials in these subsets is practically difficult with a need to screen many patients to identify the optimal group for MET-directed therapies. Like other RTK targets in GEA, key considerations for future MET-directed approaches include biomarker thresholds (amplifications only vs. IHC +), exclusion of more heterogeneous tumors (coexisting *EGFR* or *HER2* amplification), and use of new rather than archival tissue samples to limit temporal heterogeneity and minimize the impact of prior therapies on MET biomarker status (Catenacci, 2019; Klempner & Catenacci, 2019).

Epidermal growth factor receptor

The HER-family transmembrane RTK epidermal growth factor receptor (EGFR) is a validated target across multiple tumor types. While activating mutations are common to non-small cell lung cancer, amplification predominates in GEA. Paralleling observations with *MET*, *EGFR* is amplified in roughly 5% of GEA patients but protein overexpression is reported in nearly 50% of patients (Kim et al., 2008; Langer et al., 2006; Maron et al., 2018). Given the high prevalence of protein expression and existence of approved EGFR-directed antibodies and small molecule inhibitors, several large clinical efforts to target EGFR were undertaken. In two similar phase III frontline trials, the antibodies cetuximab (EXPAND trial) and panitumumab (REAL3 trial) were added to standard chemotherapy. In a pretreated population, the COG group failed to show benefit for erlotinib over placebo in advanced GEA. Notably, none of these trials utilized a biomarker enrichment strategy for EGFR. However, of equal importance is that in planned subset analyses patients with *EGFR* amplification and/or highest protein expression by IHC, a trend toward improved outcomes was seen. These subset analyses again point to the need to incorporate biologically relevant information into targeted trial design in GEA. It is tempting to think that some surrogate measure of EGFR-dependency coupled to the heterogeneity assessments would have led to different results. In orthogonal support for the importance of optimal biomarker, selection in GEA is the continued reporting of exceptional responses to EGFR-directed therapies in these thoughtfully selected patients (Adashek, Arroyo-Martinez, Menta, Kurzrock, & Kato, 2020; Kato et al., 2019; Maron et al., 2018).

FGFR2

In a large series of GEA samples, amplifications in *FGFR2* were seen in 3%−6% of patients (Ali et al., 2015; Klempner et al., 2019). Mutations and fusions are reported but quite rare. High-level *FGFR2* amplification is associated with aggressive disease and a poor prognosis (Hur et al., 2020). Preclinical studies with homogeneous *FGFR2*-amplified cell lines reliably show robust activity with FGFR2 inhibitors, but as discussed above do not accurately recapitulate the heterogeneity seen in patients (Catenacci et al., 2020; Pearson et al., 2016; Xie et al., 2013). Again, the importance of heterogeneity was highlighted in a phase II trial of the FGFR inhibitor AZD4547 (Van Cutsem et al., 2017; Cunningham et al., 2008; Tabernero et al., 2018) in advanced GEA. The trial failed to show improved response or progression-free survival versus paclitaxel in the overall population (Van Cutsem et al., 2017). Other attempts at targeting FGFR2 with monoclonal antibodies such as bemarituzumab have shown single-agent activity in a highly selected FGFR2 + population, though the phase III FIGHT trial was recently terminated (Catenacci et al., 2019; Tejani et al., 2019).

Additional targets

The tight junction protein Claudin 18 isoform 2 (CLDN18.2) is a lineage marker for differentiated gastric mucosal epithelial cells (Nagaraja et al., 2019; Sahin et al., 2008). Up to 30%−35% of gastric cancers retain strong CLDN18.2 protein expression providing and antibodies against CLDN18.2 have shown activity in several early trials (Sahin et al., 2018; Türeci et al., 2019). Although the monotherapy activity is moderate, these trials spawned two ongoing phase III trials (NCT03504397, NCT03653507), which will explore the addition of zolbetuximab to standard first-line platinum and

5-fluorouracil in patients with previously untreated advanced gastric and GEJ adenocarcinomas with CLDN18.2 expression in at least 70% of cells. Similar to other surface proteins, there are also strategies incorporating CLDN18.2 chimeric antigen receptor T cells, and CLDN18.2-CD3 bispecific antibodies as well as antibody-drug conjugates using CLDN18.2 to deliver a cytotoxic payload. If positive, the CLDN18.2 phase III trials would be important as the selection biomarker is relatively common and would offer therapy to a large group of patients, including DGC.

Other MAPK-pathway amplifications are frequently observed, including *KRAS* amplification in up to 14% of samples (Wong et al., 2018). Preclinical work examining MEK inhibition in *KRAS*-amplified tumors demonstrated limited activity, but MEK inhibition was substantially potentiated by combined blockade of SHP2, a protein tyrosine phosphatase. Trials examining the combination of SHP2 and MEK inhibition with the SHP2 inhibitors TNO155 and RMC-4630 (NCT03114319, NCT03634982) are ongoing. In addition to MAPK-pathway alterations, a significant number of GEA tumors harbor recurrent alterations in cell-cycle regulators, particularly in CIN tumors (Alexandrov, Nik-Zainal, Siu, Leung, & Stratton, 2015; Frankell et al., 2019). Deletion of *CDK2NA* and amplification of *CCNE1*, *CCND1*, and *CDK6* are observed frequently and are potentially actionable. Monotherapy data for CDK4/6 inhibitors are limited, and significant preclinical work is needed to understand the cellular dependencies in this subgroup. In a small phase II trial, the CDK4/6 inhibitor palbociclib demonstrated discouraging results, and it is expected that combination strategies will be needed to capitalize on the therapeutic potential (Karasic et al., 2020).

Immunotherapy in esophageal and gastric cancers

Modulation of the antitumor immune response through blockade of key immune checkpoints including programmed cell death protein 1 (PD-1), programmed death ligand 1 (PD-L1), and cytotoxic T-lymphocyte-associated protein 4 (CTLA4) has revolutionized cancer care across tumor types. A comprehensive review of immune checkpoint inhibitors (ICI) is beyond the scope of this chapter and reviewed elsewhere (Chuang, Chao, Hendifar, Klempner, & Gong, 2019; Cohen, Strong, & Janjigian, 2018; Terrero & Lockhart 2020). Enriching patient selection for optimal responders based on biologic tumor information parallels several of the concepts described above with regards to RTK-directed agents (Zhou et al., 2020). Tumors with higher level PD-L1 expression by IHC—particularly those with a combined positive score (CPS) of 10 or greater—have higher response rates and derive greater benefit from immune checkpoint blockade (Shen et al., 2018; Shitara et al., 2018; Yamamoto & Kato, 2020). Similarly, molecularly defined subsets including MSI-high (MSI-H) and EBV-associated GEA have achieved some of the greatest benefit from ICI and warrant specific mention. Much of the ongoing work is centered around expanding the portion of patients who benefit, largely through PD-1 combinations with other checkpoint inhibitors (TIM3, LAG3, VISTA, etc.), as well as cytotoxic agents, though early results are mixed (Boku et al., 2019; Shitara et al., 2020; Xu et al., 2020). Similarly, neo-angiogenesis plays an important role in gastric and esophageal cancers and may drive an immunosuppressive microenvironment influencing intrinsic PD-1 resistance. Combinations of PD-1 agents with antiangiogenic strategies including the VEGFR2 antibody ramucirumab as well as small molecule inhibitors regorafenib and lenvatininb have shown early promise and are moving forward in development (Fukuoka et al., 2020; Herbst et al., 2019; Kawazoe et al., 2020; Klempner & Wainberg, 2019).

Epstein-Barr virus

EBV is a ubiquitous herpesvirus and among the most common viruses found in humans. TCGA and ACRG efforts identified EBV + gastric cancers and defined several hallmarks of potential ICI sensitivity. First, these tumors are associated with frequent amplification of the chromosome 9 locus containing the PD-1 and PD-L1 genes and are associated with high levels of PD-L1 expression in tumor and adjacent immune cells. Second, EBV + GEA tumors are defined histologically by a peritumoral immune cell infiltration suggesting a level of immune recognition. The increased sensitivity was validated in a phase II trial of Korean gastric cancer patients (Kim et al., 2018). In this trial, the response rate was 100% among the six patients with known EBV + tumors. In another retrospective analysis of Japanese patients with EBV + GEA ($n = 6$), the later line response rate to anti-PD-1 immunotherapy was 33%, somewhat tempering the prior data (Kim et al., 2018; Kubota et al., 2020). The incidence of EBV + GEA varies geographically with estimates of up to 10% in some Asian series, but closer to 3%−5% in Western populations.

Microsatellite instable

MSI-H and/or mismatch repair deficiency (dMMR) is a well-described genomic correlate of immunotherapy response and previously reviewed elsewhere (Le et al., 2015, 2017). Briefly, these tumors are characteristically defined by the accumulation of somatic mutations leading to an increased burden of potentially immunogenic neoantigens. Interestingly, anecdotal cases have demonstrated intratumoral heterogeneity of MMR protein expression in gastric cancer and may underlie the lack of response in some MSI-H patients (Kim et al., 2018). However, the overall trend toward increased immune checkpoint response in dMMR and MSI-H tumors extends to gastric cancer as well. In pooled analyses from the Keynote-059, Keynote-061, and Keynote-062 trials, clinical outcomes were significantly improved in the small numbers of dMMR/MSI-H patients (Fuchs et al., 2018; Shitara et al., 2018, 2020). Similar to EBV, dMMR/MSH-H GEA represent only a minor fraction of GEA, roughly 3%−6% of all advanced patients.

Key barriers to genomic medicine in esophageal and gastric cancers

In the above sections, we have touched on several potential barriers to precision medicine in gastric and esophageal cancers, but it is worth summarizing again here. The CIN subgroup is the most well studied and most affected by spatial and temporal biomarker heterogeneity (Kim et al., 2014; Kwak et al., 2015; Lee et al., 2019; Maron et al., 2018; Pectasides et al., 2018; Zhou et al., 2020). Understanding baseline heterogeneity is important for both optimizing drug development and patient selection for targeted therapies, particularly RTK-directed approaches. The importance of heterogeneity extends beyond tumoral features to the composition of the tumor immune microenvironment as well. Here, much work is needed to better understand the interplay between the tumor genomics and the immune composition of the microenvironment. One could envision the potential ability to individualize immunotherapy approaches based on the ability to understand the primary immunosuppressive mechanisms in the tumor microenvironment (Derks et al., 2020; Gullo et al., 2019; Kim, da Silva, Coit, & Tang, 2019). The GS subgroup is enriched for DGC and characterized by a paucity of genomic alterations. This group repeatedly has worse outcomes with immunotherapy and chemotherapy. Recently, model systems have suggested a role for focal adhesion kinase activation in DGC, and this may represent a potential therapeutic target (Zhang et al., 2020). Even among the highly immune responsive MSI-H patients, the response rates are only 50%−60% suggesting a subset with intrinsic resistance. Interestingly, there is preclinical work suggesting that both immune editing and extensive subclonal heterogeneity in MSI-H tumors may underlie the observation that not all MSI-H tumors respond to ICIs (Trinh & Polyak, 2019; Wolf et al., 2019).

Conclusion and future directions

Genomic blueprints have enabled reliable classification into molecular subsets with distinct biological features and clinical behavior. To move beyond the genomic barriers to targeted therapies, we need to incorporate expanding tools, including serial ctDNA analyses, and repeated tissue profiling. Early work suggests that we may be able to refine patient selection based on capturing the degree of heterogeneity present within a given patient's tumor, with more homogenous tumors relying more heavily on a given pathway. Further laboratory investigations are needed to identify genetic dependencies that can inform novel clinical combination approaches. We remain optimistic for the future of gastric and esophageal cancers.

References

Abnet, C. C., Arnold, M., & Wei, W.-Q. (2018). Epidemiology of esophageal squamous cell carcinoma. *Gastroenterology, 154*(2), 360–373, Jan.

Adashek, J. J., Arroyo-Martinez, Y., Menta, A. K., Kurzrock, R., & Kato, S. (2020). Therapeutic implications of epidermal growth factor receptor (EGFR) in the treatment of metastatic gastric/GEJ cancer. *Frontiers in Oncology, 10*, 1312, Aug 4.

Alexandrov, L. B., Nik-Zainal, S., Siu, H. C., Leung, S. Y., & Stratton, M. R. (2015). A mutational signature in gastric cancer suggests therapeutic strategies. *Nature Communications, 6*, 8683, Oct 29.

Ali, S. M., Sanford, E. M., Klempner, S. J., Rubinson, D. A., Wang, K., Palma, N. A., et al. (2015). Prospective comprehensive genomic profiling of advanced gastric carcinoma cases reveals frequent clinically relevant genomic alterations and new routes for targeted therapies. *The Oncologist, 20*(5), 499–507, May.

Bang, Y.-J., Van Cutsem, E., Feyereislova, A., Chung, H. C., Shen, L., Sawaki, A., et al. (2010). Trastuzumab in combination with chemotherapy vs chemotherapy alone for treatment of HER2-positive advanced gastric or gastro-oesophageal junction cancer (ToGA): A phase 3, open-label, randomised controlled trial. *Lancet, 376*(9742), 687–697, Aug 28.

Boku, N., Ryu, M.-H., Kato, K., Chung, H. C., Minashi, K., Lee, K.-W., et al. (2019). Safety and efficacy of nivolumab in combination with S-1/capecitabine plus oxaliplatin in patients with previously untreated, unresectable, advanced, or recurrent gastric/gastroesophageal junction cancer: Interim results of a randomized, phase II trial (ATTRACTION-4). *Annals of Oncology: Official Journal of the European Society for Medical Oncology, 30*(2), 250–258, Feb 1.

Boysen, T., Mohammadi, M., Melbye, M., Hamilton-Dutoit, S., Vainer, B., Hansen, A. V., et al. (2009). EBV-associated gastric carcinoma in high- and low-incidence areas for nasopharyngeal carcinoma. *British Journal of Cancer, 101*(3), 530–533, Aug 4.

Bray, F., Ferlay, J., Soerjomataram, I., Siegel, R. L., Torre, L. A., & Jemal, A. (2018). Global cancer statistics 2018: GLOBOCAN estimates of incidence and mortality worldwide for 36 cancers in 185 countries. *CA: A Cancer Journal for Clinicians, 68*(6), 394–424, Nov.

Cancer Genome Atlas Research Network, Analysis Working Group: Asan University, BC Cancer Agency, Brigham and Women's Hospital, Broad Institute, Brown University, et al. (2017). Integrated genomic characterization of oesophageal carcinoma. *Nature, 541*(7636), 169–175, Jan 12.

Catenacci, D. V., Tesfaye, A., Tejani, M., Cheung, E., Eisenberg, P., Scott, A. J., et al. (2019). Bemarituzumab with modified FOLFOX6 for advanced FGFR2-positive gastroesophageal cancer: FIGHT Phase III study design. *Future Oncology (London, England), 15*(18), 2073–2082, Jun.

Catenacci, D. V. T. (2019). When inhibitor MET biomarker: Postmortem or initium novum. *JCO Precision Oncology [Internet], 3.* Available from https://doi.org/10.1200/PO.18.00359, Mar 8.

Catenacci, D. V. T., Kang, Y.-K., Park, H., Uronis, H. E., Lee, K.-W., Ng, M. C. H., et al. (2020). Margetuximab plus pembrolizumab in patients with previously treated, HER2-positive gastro-oesophageal adenocarcinoma (CP-MGAH22-05): A single-arm, phase 1b-2 trial. *The Lancet Oncology, 21*(8), 1066–1076, Aug.

Catenacci, D. V. T., Rasco, D., Lee, J., Rha, S. Y., Lee, K.-W., Bang, Y. J., et al. (2020). Phase I escalation and expansion study of bemarituzumab (FPA144) in patients with advanced solid tumors and FGFR2b-selected gastroesophageal adenocarcinoma. *Journal of Clinical Oncology: Official Journal of the American Society of Clinical Oncology, 38*(21), 2418–2426, Jul 20.

Catenacci, D. V. T., Tebbutt, N. C., Davidenko, I., Murad, A. M., Al-Batran, S.-E., Ilson, D. H., et al. (2017). Rilotumumab plus epirubicin, cisplatin, and capecitabine as first-line therapy in advanced MET-positive gastric or gastro-oesophageal junction cancer (RILOMET-1): A randomised, double-blind, placebo-controlled, phase 3 trial. *The Lancet Oncology, 18*(11), 1467–1482, Nov.

Chang, M.-S., Uozaki, H., Chong, J.-M., Ushiku, T., Sakuma, K., Ishikawa, S., et al. (2006). CpG island methylation status in gastric carcinoma with and without infection of Epstein-Barr virus. *Clinical Cancer Research: An Official Journal of the American Association for Cancer Research, 12*(10), 2995–3002, May 15.

Chao, J., Lee, J., & Klempner, S. J. (2016). Moving molecular subtypes to the clinic in gastric cancer. *Translational Cancer Research, 5*(1), S25–S30, Jun.

Chon, H. J., Hyung, W. J., Kim, C., Park, S., Kim, J.-H., Park, C. H., et al. (2017). Differential prognostic implications of gastric signet ring cell carcinoma: Stage adjusted analysis from a single high-volume center in Asia. *Annals of Surgery, 265*(5), 946–953, May.

Chuang, J., Chao, J., Hendifar, A., Klempner, S. J., & Gong, J. (2019). Checkpoint inhibition in advanced gastroesophageal cancer: Clinical trial data, molecular subtyping, predictive biomarkers, and the potential of combination therapies. *Translational Gastroenterology and Hepatology, 4*, 63, Aug 27.

Cohen, N. A., Strong, V. E., & Janjigian, Y. Y. (2018). Checkpoint blockade in esophagogastric cancer. *Journal of Surgical Oncology, 118*(1), 77–85, Jul.

Cristescu, R., Lee, J., Nebozhyn, M., Kim, K.-M., Ting, J. C., Wong, S. S., et al. (2015). Molecular analysis of gastric cancer identifies subtypes associated with distinct clinical outcomes. *Nature Medicine, 21*(5), 449–456, May.

Cunningham, D., Starling, N., Rao, S., Iveson, T., Nicolson, M., Coxon, F., et al. (2008). Capecitabine and oxaliplatin for advanced esophagogastric cancer. *The New England Journal of Medicine, 358*(1), 36–46, Jan 3.

de Jonge, P. J. F., van Blankenstein, M., Looman, C. W. N., Casparie, M. K., Meijer, G. A., & Kuipers, E. J. (2010). Risk of malignant progression in patients with Barrett's oesophagus: A Dutch nationwide cohort study. *Gut, 59*(8), 1030–1036, Aug.

Derks, S., de Klerk, L. K., Xu, X., Fleitas, T., Liu, K. X., Liu, Y., et al. (2020). Characterizing diversity in the tumor-immune microenvironment of distinct subclasses of gastroesophageal adenocarcinomas. *Annals of Oncology: Official Journal of the European Society for Medical Oncology, 31*(8), 1011–1020, Aug.

Frankell, A. M., Jammula, S., Li, X., Contino, G., Killcoyne, S., Abbas, S., et al. (2019). The landscape of selection in 551 esophageal adenocarcinomas defines genomic biomarkers for the clinic. *Nature Genetics, 51*(3), 506–516, Mar.

Fuchs, C. S., Doi, T., Jang, R. W., Muro, K., Satoh, T., Machado, M., et al. (2018). Safety and efficacy of pembrolizumab monotherapy in patients with previously treated advanced gastric and gastroesophageal junction cancer: Phase 2 clinical KEYNOTE-059 trial. *JAMA Oncology, 4*(5), e180013, May 10.

Fuchs, C. S., Shitara, K., Di Bartolomeo, M., Lonardi, S., Al-Batran, S.-E., Van Cutsem, E., et al. (2019). Ramucirumab with cisplatin and fluoropyrimidine as first-line therapy in patients with metastatic gastric or junctional adenocarcinoma (RAINFALL): A double-blind, randomised, placebo-controlled, phase 3 trial. *The Lancet Oncology, 20*(3), 420–435, Mar.

Fukuoka, S., Hara, H., Takahashi, N., Kojima, T., Kawazoe, A., Asayama, M., et al. (2020). Regorafenib plus nivolumab in patients with advanced gastric or colorectal cancer:

An open-label, dose-escalation, and dose-expansion phase Ib trial (REGONIVO, EPOC1603). *Journal of Clinical Oncology: Official Journal of the American Society of Clinical Oncology, 38*(18), 2053−2061, Jun 20.

Gao, Y.-B., Chen, Z.-L., Li, J.-G., Hu, X.-D., Shi, X.-J., Sun, Z.-M., et al. (2014). Genetic landscape of esophageal squamous cell carcinoma. *Nature Genetics, 46*(10), 1097−1102, Oct.

González, C. A., Jakszyn, P., Pera, G., Agudo, A., Bingham, S., Palli, D., et al. (2006). Meat intake and risk of stomach and esophageal adenocarcinoma within the European Prospective Investigation Into Cancer and Nutrition (EPIC). *Journal of the National Cancer Institute, 98* (5), 345−354, Mar 1.

González, C. A., Pera, G., Agudo, A., Palli, D., Krogh, V., Vineis, P., et al. (2003). Smoking and the risk of gastric cancer in the European Prospective Investigation Into Cancer and Nutrition (EPIC). *International Journal of Cancer. Journal International du Cancer, 107*(4), 629−634, Nov 20.

Graziano, F., Humar, B., & Guilford, P. (2003). The role of the E-cadherin gene (CDH1) in diffuse gastric cancer susceptibility: From the laboratory to clinical practice. *Annals of Oncology: Official Journal of the European Society for Medical Oncology, 14*(12), 1705−1713, Dec.

Gullo, I., Oliveira, P., Athelogou, M., Gonçalves, G., Pinto, M. L., Carvalho, J., et al. (2019). New insights into the inflamed tumor immune microenvironment of gastric cancer with lymphoid stroma: From morphology and digital analysis to gene expression. *Gastric Cancer: Official Journal of the International Gastric Cancer Association and the Japanese Gastric Cancer Association, 22*(1), 77−90, Jan.

Hecht, J. R., Randolph Hecht, J., Bang, Y.-J., Qin, S. K., Chung, H. C., Xu, J. M., et al. (2016). Lapatinib in combination with capecitabine plus oxaliplatin in human epidermal growth factor receptor 2−positive advanced or metastatic gastric, esophageal, or gastroesophageal adenocarcinoma: TRIO-013/LOGiC—A randomized phase III trial [Internet]. *Journal of Clinical Oncology*, 443−451. Available from https://doi.org/10.1200/jco.2015.62.6598.

Herbst, R. S., Arkenau, H.-T., Santana-Davila, R., Calvo, E., Paz-Ares, L., Cassier, P. A., et al. (2019). Ramucirumab plus pembrolizumab in patients with previously treated advanced non-small-cell lung cancer, gastro-oesophageal cancer, or urothelial carcinomas (JVDF): A multicohort, non-randomised, open-label, phase 1a/b trial. *The Lancet Oncology, 20*(8), 1109−1123, Aug.

Hu, N., Kadota, M., Liu, H., Abnet, C. C., Su, H., Wu, H., et al. (2016). Genomic landscape of somatic alterations in esophageal squamous cell carcinoma and gastric cancer. *Cancer Research, 76*(7), 1714−1723, Apr 1.

Hur, J. Y., Chao, J., Kim, K., Kim, S. T., Kim, K.-M., Klempner, S. J., et al. (2020). High-level FGFR2 amplification is associated with poor prognosis and lower response to chemotherapy in gastric cancers. *Pathology, Research and Practice, 216*(4), 152878, Apr.

Hvid-Jensen, F., Pedersen, L., Drewes, A. M., Sørensen, H. T., & Funch-Jensen, P. (2011). Incidence of adenocarcinoma among patients with Barrett's esophagus. *The New England Journal of Medicine, 365*(15), 1375−1383, Oct 13.

Islami, F., Poustchi, H., Pourshams, A., Khoshnia, M., Gharavi, A., Kamangar, F., et al. (2020). A prospective study of tea drinking temperature and risk of esophageal squamous cell carcinoma. *International Journal of Cancer, 146*(1), 18−25, Jan 1.

Islami, F., Sheikhattari, P., Ren, J. S., & Kamangar, F. (2011). Gastric atrophy and risk of oesophageal cancer and gastric cardia adenocarcinoma—a systematic review and *meta*-analysis. *Annals of Oncology: Official Journal of the European Society for Medical Oncology, 22*(4), 754−760, Apr.

Janjigian, Y. Y., Maron, S. B., Chatila, W. K., Millang, B., Chavan, S. S., Alterman, C., et al. (2020). First-line pembrolizumab and trastuzumab in HER2-positive oesophageal, gastric, or gastro-oesophageal junction cancer: An open-label, single-arm, phase 2 trial. *The Lancet Oncology*, *21*(6), 821−831, Jun.

Janjigian, Y. Y., Sanchez-Vega, F., Jonsson, P., Chatila, W. K., Hechtman, J. F., Ku, G. Y., et al. (2018). Genetic predictors of response to systemic therapy in esophagogastric cancer. *Cancer Discovery*, *8*(1), 49−58, Jan.

Jemal, A., Bray, F., Center, M. M., Ferlay, J., Ward, E., & Forman, D. (2011). Global cancer statistics. *CA: A Cancer Journal for Clinicians*, *61*(2), 69−90, Mar.

Karasic, T. B., O'Hara, M. H., Teitelbaum, U. R., Damjanov, N., Giantonio, B. J., d'Entremont, T. S., et al. (2020). Phase II trial of palbociclib in patients with advanced esophageal or gastric cancer. *Oncologist [Internet]*. Available from https://doi.org/10.1634/theoncologist.2020-0681.

Kato, S., Okamura, R., Mareboina, M., Lee, S., Goodman, A., Patel, S. P., et al. (2019). Revisiting epidermal growth factor receptor amplification as a target for anti-EGFR therapy: Analysis of cell-free circulating tumor DNA in patients with advanced malignancies. *JCO Precision Oncology [Internet]*, *3*. Available from https://doi.org/10.1200/PO.18.00180, Jan 22.

Kawazoe, A., Fukuoka, S., Nakamura, Y., Kuboki, Y., Wakabayashi, M., Nomura, S., et al. (2020). Lenvatinib plus pembrolizumab in patients with advanced gastric cancer in the first-line or second-line setting (EPOC1706): An open-label, single-arm, phase 2 trial. *The Lancet Oncology*, *21*(8), 1057−1065, Aug.

Kim, J., Fox, C., Peng, S., Pusung, M., Pectasides, E., Matthee, E., et al. (2014). Preexisting oncogenic events impact trastuzumab sensitivity in ERBB2-amplified gastroesophageal adenocarcinoma. *The Journal of Clinical Investigation*, *124*(12), 5145−5158, Dec.

Kim, M. A., Lee, H. S., Lee, H. E., Jeon, Y. K., Yang, H. K., & Kim, W. H. (2008). EGFR in gastric carcinomas: Prognostic significance of protein overexpression and high gene copy number. *Histopathology*, *52*(6), 738−746, May.

Kim, S. T., Banks, K. C., Pectasides, E., Kim, S. Y., Kim, K., Lanman, R. B., et al. (2018). Impact of genomic alterations on lapatinib treatment outcome and cell-free genomic landscape during HER2 therapy in HER2 gastric cancer patients [Internet], Vol*Annals of Oncology*, 1037−1048. Available from https://doi.org/10.1093/annonc/mdy034.

Kim, S. T., Cristescu, R., Bass, A. J., Kim, K.-M., Odegaard, J. I., Kim, K., et al. (2018). Comprehensive molecular characterization of clinical responses to PD-1 inhibition in metastatic gastric cancer. *Nature Medicine*, *24*(9), 1449−1458, Sep.

Kim, T. S., da Silva, E., Coit, D. G., & Tang, L. H. (2019). Intratumoral immune response to gastric cancer varies by molecular and histologic subtype. *The American Journal of Surgical Pathology*, *43*(6), 851−860, Jun.

Klempner, S. J., & Catenacci, D. V. T. (2019). Variety is the spice of life, but maybe not in gastroesophageal adenocarcinomas. *Cancer Discovery*, *9*(2), 166−168, Feb.

Klempner, S. J., & Chao, J. (2018). Toward optimizing outcomes in Her2-positive gastric cancer: Timing and genomic context matter. *Annals of Oncology: Official Journal of the European Society for Medical Oncology*, *29*(4), 801−802, Apr 1.

Klempner, S. J., Madison, R., Pujara, V., Ross, J. S., Miller, V. A., Ali, S. M., et al. (2019). Altered gastroesophageal adenocarcinomas are an uncommon clinicopathologic entity with a distinct genomic landscape. *The Oncologist*, *24*(11), 1462−1468, Nov.

Klempner, S. J., & Wainberg, Z. A. (2019). Ramucirumab plus pembrolizumab: Can we make the maths work. *The Lancet Oncology, 20*(8), 1041−1043, Aug.

Kubota, Y., Kawazoe, A., Sasaki, A., Mishima, S., Sawada, K., Nakamura, Y., et al. (2020). The impact of molecular subtype on efficacy of chemotherapy and checkpoint inhibition in advanced gastric cancer. *Clinical Cancer Research: An Official Journal of the American Association for Cancer Research, 26*(14), 3784−3790, Jul 15.

Kwak, E. L., Ahronian, L. G., Siravegna, G., Mussolin, B., Borger, D. R., Godfrey, J. T., et al. (2015). Molecular heterogeneity and receptor coamplification drive resistance to targeted therapy in MET-amplified esophagogastric cancer. *Cancer Discovery, 5*(12), 1271−1281, Dec.

Lagergren, J., Bergström, R., Lindgren, A., & Nyrén, O. (1999). Symptomatic gastroesophageal reflux as a risk factor for esophageal adenocarcinoma. *The New England Journal of Medicine, 340*(11), 825−831, Mar 18.

Langer, R., Von Rahden, B. H. A., Nahrig, J., Von Weyhern, C., Reiter, R., Feith, M., et al. (2006). Prognostic significance of expression patterns of c-erbB-2, p53, p16INK4A, p27KIP1, cyclin D1 and epidermal growth factor receptor in oesophageal adenocarcinoma: A tissue microarray study. *Journal of Clinical Pathology, 59*(6), 631−634, Jun.

Lauren, P. (1965). The two histological main types of gastric carcinoma: Diffuse and so-called intestinal-type carcinoma. An attempt at a histo-clinical classification. *Acta Pathologica et Microbiologica Scandinavica, 64*, 31−49.

Le, D. T., Durham, J. N., Smith, K. N., Wang, H., Bartlett, B. R., Aulakh, L. K., et al. (2017). Mismatch repair deficiency predicts response of solid tumors to PD-1 blockade. *Science (New York, N.Y.), 357*(6349), 409−413, Jul 28.

Le, D. T., Uram, J. N., Wang, H., Bartlett, B. R., Kemberling, H., Eyring, A. D., et al. (2015). PD-1 Blockade in tumors with mismatch-repair deficiency. *The New England Journal of Medicine, 372*(26), 2509−2520, Jun 25.

Lee, J., Kim, S. T., Kim, K., Lee, H., Kozarewa, I., Mortimer, P. G. S., et al. (2019). Tumor genomic profiling guides patients with metastatic gastric cancer to targeted treatment: The VIKTORY Umbrella Trial. *Cancer Discovery, 9*(10), 1388−1405, Oct.

Lee, J. Y., Hong, M., Kim, S. T., Park, S. H., Kang, W. K., Kim, K.-M., et al. (2015). The impact of concomitant genomic alterations on treatment outcome for trastuzumab therapy in HER2-positive gastric cancer. *Scientific Report, 5*, 9289, Mar 19.

Lin, D.-C., Dinh, H. Q., Xie, J.-J., Mayakonda, A., Silva, T. C., Jiang, Y.-Y., et al. (2018). Identification of distinct mutational patterns and new driver genes in oesophageal squamous cell carcinomas and adenocarcinomas. *Gut, 67*(10), 1769−1779, Oct.

Lin, D.-C., Hao, J.-J., Nagata, Y., Xu, L., Shang, L., Meng, X., et al. (2014). Genomic and molecular characterization of esophageal squamous cell carcinoma. *Nature Genetics, 46*(5), 467−473, May.

Maron, S. B., Alpert, L., Kwak, H. A., Lomnicki, S., Chase, L., Xu, D., et al. (2018). Targeted therapies for targeted populations: Anti-EGFR treatment for -amplified gastroesophageal adenocarcinoma. *Cancer Discovert, 8*(6), 696−713, Jun.

Nagaraja, A. K., Kikuchi, O., & Bass, A. J. (2019). Genomics and targeted therapies in gastro-esophageal adenocarcinoma. *Cancer Discovery, 9*(12), 1656−1672, Dec.

The Cancer Genome Atlas Research Network. (2014). Comprehensive molecular characterization of gastric adenocarcinoma [Internet]. *Nature*, 202−209. Available from https://doi.org/10.1038/nature13480.

Ogitani, Y., Aida, T., Hagihara, K., Yamaguchi, J., Ishii, C., Harada, N., et al. (2016). DS-8201a, a novel HER2-targeting ADC with a novel DNA topoisomerase I inhibitor, demonstrates a promising antitumor efficacy with differentiation from T-DM1. *Clinical Cancer*

Research: An Official Journal of the American Association for Cancer Research, 22(20), 5097−5108, Oct 15.

Ogitani, Y., Hagihara, K., Oitate, M., Naito, H., & Agatsuma, T. (2016). Bystander killing effect of DS-8201a, a novel anti-human epidermal growth factor receptor 2 antibody-drug conjugate, in tumors with human epidermal growth factor receptor 2 heterogeneity [Internet]. *Cancer Science*, 1039−1046. Available from https://doi.org/10.1111/cas.12966.

Pandeya, N., Williams, G., Green, A. C., Webb, P. M., & Whiteman, D. C. (2009). Australian cancer study. Alcohol consumption and the risks of adenocarcinoma and squamous cell carcinoma of the esophagus. *Gastroenterology, 136*(4), 1215−1224, Apr.

Park, S., Jiang, Z., Mortenson, E. D., Deng, L., Radkevich-Brown, O., Yang, X., et al. (2010). The therapeutic effect of anti-HER2/neu antibody depends on both innate and adaptive immunity. *Cancer Cell, 18*(2), 160−170, Aug 9.

Pearson, A., Smyth, E., Babina, I. S., Herrera-Abreu, M. T., Tarazona, N., Peckitt, C., et al. (2016). High-level clonal FGFR amplification and response to FGFR inhibition in a translational clinical trial [Internet]. *Cancer Discovery*, 838−851. Available from https://doi.org/10.1158/2159-8290.cd-15-1246.

Pectasides, E., Stachler, M. D., Derks, S., Liu, Y., Maron, S., Islam, M., et al. (2018). Genomic heterogeneity as a barrier to precision medicine in gastroesophageal adenocarcinoma. *Cancer Discovery, 8*(1), 37−48, Jan.

Peleteiro, B., Lopes, C., Figueiredo, C., & Lunet, N. (2011). Salt intake and gastric cancer risk according to *Helicobacter pylori* infection, smoking, tumour site and histological type. *British Journal of Cancer, 104*(1), 198−207, Jan 4.

Petrick, J. L., Wyss, A. B., Butler, A. M., Cummings, C., Sun, X., Poole, C., et al. (2014). Prevalence of human papillomavirus among oesophageal squamous cell carcinoma cases: Systematic review and *meta*-analysis. *British Journal of Cancer, 110*(9), 2369−2377, Apr 29.

Pohl, H., Pech, O., Arash, H., Stolte, M., Manner, H., May, A., et al. (2016). Length of Barrett's oesophagus and cancer risk: Implications from a large sample of patients with early oesophageal adenocarcinoma. *Gut, 65*(2), 196−201, Feb.

Polk, D. B., Brent Polk, D., & Peek, R. M. (2010). *Helicobacter pylori*: Gastric cancer and beyond [Internet]. *Nature Reviews Cancer*, 403−414. Available from https://doi.org/10.1038/nrc2857. Available from.

Sahin, U., Koslowski, M., Dhaene, K., Usener, D., Brandenburg, G., Seitz, G., et al. (2008). Claudin-18 splice variant 2 is a pan-cancer target suitable for therapeutic antibody development. *Clinical Cancer Research: An Official Journal of the American Association for Cancer Research, 14*(23), 7624−7634, Dec 1.

Sahin, U., Schuler, M., Richly, H., Bauer, S., Krilova, A., Dechow, T., et al. (2018). A phase I dose-escalation study of IMAB362 (Zolbetuximab) in patients with advanced gastric and gastro-oesophageal junction cancer. *European Journal of Cancer, 100*, 17−26, Sep.

Sanchez-Vega, F., Hechtman, J. F., Castel, P., Ku, G. Y., Tuvy, Y., Won, H., et al. (2019). EGFR and MET amplifications determine response to HER2 inhibition in ERBB2-amplified esophagogastric cancer. *Cancer Discovery, 9*(2), 199−209, Feb.

Sandler, R. S., Nyrén, O., Ekbom, A., Eisen, G. M., Yuen, J., & Josefsson, S. (1995). The risk of esophageal cancer in patients with achalasia. A population-based study. *JAMA: The Journal of the American Medical Association, 274*(17), 1359−1362, Nov 1.

Satoh, T., Xu, R.-H., Chung, H. C., Sun, G.-P., Doi, T., Xu, J.-M., et al. (2014). Lapatinib plus paclitaxel vs paclitaxel alone in the second-line treatment of HER2-amplified advanced gastric cancer in Asian populations: TyTAN−a randomized, phase III study. *Journal of*

Clinical Oncology: Official Journal of the American Society of Clinical Oncology, *32*(19), 2039–2049, Jul 1.

Shah, M. A., Bang, Y.-J., Lordick, F., Alsina, M., Chen, M., Hack, S. P., et al. (2017). Effect of fluorouracil, leucovorin, and oxaliplatin with or without onartuzumab in HER2-negative, MET-positive gastroesophageal adenocarcinoma: The METGastric randomized clinical trial. *JAMA Oncology*, *3*(5), 620–627, May 1.

Shah, M. A., Cho, J., Tan, I. B., Tebbutt, N. C., Yen, C., Kang, A., et al. (2016). A randomized phase II study of FOLFOX with or without the MET inhibitor onartuzumab in advanced adenocarcinoma of the stomach and gastroesophageal junction [Internet]. *The Oncologist*, 1085–1090. Available from https://doi.org/10.1634/theoncologist.2016-0038.

Shah, M. A., Kang, Y.-K., Thuss-Patience, P. C., Ohtsu, A., Ajani, J. A., Van Cutsem, E., et al. (2019). Biomarker analysis of the GATSBY study of trastuzumab emtansine vs a taxane in previously treated HER2-positive advanced gastric/gastroesophageal junction cancer [Internet]. *Gastric Cancer: Official Journal of the International Gastric Cancer Association and the Japanese Gastric Cancer Association*, 803–816. Available from https://doi.org/10.1007/s10120-018-00923-7.

Shen, L., Ajani, J. A., Kim, S.-B., Van Cutsem, E., Guo, B., Song, J., et al. (2018). A phase 3, randomized, open-label study to compare the efficacy of tislelizumab (BGB-A317) vs chemotherapy as second-line therapy for advanced unresectable/metastatic esophageal squamous cell carcinoma (ESCC) [Internet]. *Journal of Clinical Oncology*, TPS3111. Available from https://doi.org/10.1200/jco.2018.36.15_suppl.tps3111.

Shitara, K., Bang, Y.-J., Iwasa, S., Sugimoto, N., Ryu, M.-H., Sakai, D., et al. (2020). Trastuzumab deruxtecan in previously treated HER2-positive gastric cancer. *The New England Journal of Medicine*, *382*(25), 2419–2430, Jun 18.

Shitara, K., Iwata, H., Takahashi, S., Tamura, K., Park, H., Modi, S., et al. (2019). Trastuzumab deruxtecan (DS-8201a) in patients with advanced HER2-positive gastric cancer: A dose-expansion, phase 1 study. *The Lancet Oncology*, *20*(6), 827–836, Jun.

Shitara, K., Özgüroğlu, M., Bang, Y.-J., Di Bartolomeo, M., Mandalà, M., Ryu, M.-H., et al. (2018). Pembrolizumab vs paclitaxel for previously treated, advanced gastric or gastro-oesophageal junction cancer (KEYNOTE-061): A randomised, open-label, controlled, phase 3 trial. *Lancet*, *392*(10142), 123–133, Jul 14.

Shitara, K., Van Cutsem, E., Bang, Y.-J., Fuchs, C., Wyrwicz, L., Lee, K.-W., et al. (2020). Efficacy and safety of pembrolizumab or pembrolizumab plus chemotherapy vs chemotherapy alone for patients with first-line, advanced gastric cancer: The KEYNOTE-062 phase 3 randomized clinical trial. *JAMA Oncology [Internet]*. Available from https://doi.org/10.1001/jamaoncol.2020.3370, Sep 3; Available from.

Sohn, B. H., Hwang, J.-E., Jang, H.-J., Lee, H.-S., Oh, S. C., Shim, J.-J., et al. (2017). Clinical significance of four molecular subtypes of gastric cancer identified by the cancer genome atlas project. *Clinical Cancer Research [Internet]*. Available from https://doi.org/10.1158/1078-0432.CCR-16-2211, Jul 26; Available from.

Song, Y., Li, L., Ou, Y., Gao, Z., Li, E., Li, X., et al. (2014). Identification of genomic alterations in oesophageal squamous cell cancer. *Nature*, *509*(7498), 91–95, May 1.

Stjepanovic, N., Moreira, L., Carneiro, F., Balaguer, F., Cervantes, A., Balmaña, J., et al. (2019). Hereditary gastrointestinal cancers: ESMO Clinical Practice Guidelines for diagnosis, treatment and follow-up. *Annals of Oncology: Official Journal of the European Society for Medical Oncology*, *30*(10), 1558–1571, Oct 1.

Tabernero, J., Hoff, P. M., Shen, L., Ohtsu, A., Shah, M. A., Cheng, K., et al. (2018). Pertuzumab plus trastuzumab and chemotherapy for HER2-positive metastatic gastric or gastro-oesophageal junction cancer (JACOB): Final analysis of a double-blind, randomised, placebo-controlled phase 3 study. *The Lancet Oncology, 19*(10), 1372−1384, Oct.

Taghavi, S., Jayarajan, S. N., Davey, A., & Willis, A. I. (2012). Prognostic significance of signet ring gastric cancer. *Journal of Clinical Oncology: Official Journal of the American Society of Clinical Oncology, 30*(28), 3493−3498, Oct 1.

Tan, P., & Yeoh, K.-G. (2015). Genetics and molecular pathogenesis of gastric adenocarcinoma [Internet]. *Gastroenterology*, 1153−1162.e3. Available from https://doi.org/10.1053/j.gastro.2015.05.059. Available from.

Taylor, C., Hershman, D., Shah, N., Suciu-Foca, N., Petrylak, D. P., Taub, R., et al. (2007). Augmented HER-2 specific immunity during treatment with trastuzumab and chemotherapy. *Clinical Cancer Research: An Official Journal of the American Association for Cancer Research, 13*(17), 5133−5143, Sep 1.

Tejani, M. A., Cheung, E., Eisenberg, P. D., Scott, A. J., Tesfaye, A. A., Dreiling, L., et al. (2019). Phase I results from the phase 1/3 FIGHT study evaluating bemarituzumab and mFOLFOX6 in advanced gastric/GEJ cancer (GC) [Internet]. *Journal of Clinical Oncology*, 91. Available from https://doi.org/10.1200/jco.2019.37.4_suppl.91.

Terrero, G., & Lockhart, A. C. (2020). Role of immunotherapy in advanced gastroesophageal cancer. *Current Oncology Reports, 22*(11), 112, Aug 17.

Thrift, A. P., Shaheen, N. J., Gammon, M. D., Bernstein, L., Reid, B. J., Onstad, L., et al. (2014). Obesity and risk of esophageal adenocarcinoma and Barrett's esophagus: A Mendelian randomization study. *Journal of National Cancer Institute [Internet], 106*(11). Available from https://doi.org/10.1093/jnci/dju252, Nov.

Thuss-Patience, P. C., Shah, M. A., Ohtsu, A., Van Cutsem, E., Ajani, J. A., Castro, H., et al. (2017). Trastuzumab emtansine vs taxane use for previously treated HER2-positive locally advanced or metastatic gastric or gastro-oesophageal junction adenocarcinoma (GATSBY): An international randomised, open-label, adaptive, phase 2/3 study [Internet]. *The Lancet Oncology*, 640−653. Available from https://doi.org/10.1016/s1470-2045(17)30111-0.

Trinh, A., & Polyak, K. (2019). Tumor neoantigens: When too much of a good thing is bad. *Cancer Cell, 36*(5), 466−467, Nov 11.

Turati, F., Tramacere, I., La Vecchia, C., & Negri, E. (2013). A meta-analysis of body mass index and esophageal and gastric cardia adenocarcinoma. *Annals of Oncology: Official Journal of the European Society for Medical Oncology, 24*(3), 609−617, Mar.

Türeci, O., Sahin, U., Schulze-Bergkamen, H., Zvirbule, Z., Lordick, F., Koeberle, D., et al. (2019). A multicentre, phase IIa study of zolbetuximab as a single agent in patients with recurrent or refractory advanced adenocarcinoma of the stomach or lower oesophagus: The MONO study. *Annals of Oncology: Official Journal of the European Society for Medical Oncology, 30*(9), 1487−1495, Sep 1.

Van Cutsem, E., Bang, Y.-J., Mansoor, W., Petty, R. D., Chao, Y., Cunningham, D., Ferry, D. R., Smith, N. R., Frewer, P., Ratnayake, J., Stockman, P. K., Kilgour, E., & Landers, D. (2017). A randomized, open-label study of the efficacy and safety of AZD4547 monotherapy versus paclitaxel for the treatment of advanced gastric adenocarcinoma with FGFR2 polysomy or gene amplification. *Ann Oncol., 28*(6), 1316−1324, Jun 1. Available from https://doi.org/10.1093/annonc/mdx107, PMID: 29177434.

Van Cutsem, E., Karaszewska, B., Kang, Y.-K., Chung, H. C., Shankaran, V., Siena, S., et al. (2019). A multicenter phase II study of AMG 337 in patients with *MET*-amplified gastric/

gastroesophageal junction/esophageal adenocarcinoma and other *MET*-amplified solid tumors. *Clinical Cancer Research: An Official Journal of the American Association for Cancer Research, 25*(8), 2414−2423, Apr 15.

van der Post, R. S., Vogelaar, I. P., Carneiro, F., Guilford, P., Huntsman, D., Hoogerbrugge, N., et al. (2015). Hereditary diffuse gastric cancer: Updated clinical guidelines with an emphasis on germline CDH1 mutation carriers. *Journal of Medical Genetics, 52*(6), 361−374, Jun.

Varadan, V., Gilmore, H., Miskimen, K. L. S., Tuck, D., Parsai, S., Awadallah, A., et al. (2016). Immune signatures following single dose trastuzumab predict pathologic response to preoperative trastuzumab and chemotherapy in HER2-positive early breast cancer. *Clinical Cancer Research: An Official Journal of the American Association for Cancer Research, 22* (13), 3249−3259, Jul 1.

Wang, D.-S., Liu, Z.-X., Lu, Y.-X., Bao, H., Wu, X., Zeng, Z.-L., et al. (2019). Liquid biopsies to track trastuzumab resistance in metastatic HER2-positive gastric cancer. *Gut, 68*(7), 1152−1161, Jul.

Wang, K., Johnson, A., Ali, S. M., Klempner, S. J., Bekaii-Saab, T., Vacirca, J. L., et al. (2015). Comprehensive genomic profiling of advanced esophageal squamous cell carcinomas and esophageal adenocarcinomas reveals similarities and differences. *The Oncologist, 20*(10), 1132−1139, Oct.

Wolf, Y., Bartok, O., Patkar, S., Eli, G. B., Cohen, S., Litchfield, K., et al. (2019). UVB-induced tumor heterogeneity diminishes immune response in melanoma. *Cell, 179*(1), 219−235, Sep 19.

Wong, G. S., Zhou, J., Liu, J. B., Wu, Z., Xu, X., Li, T., et al. (2018). Targeting wild-type KRAS-amplified gastroesophageal cancer through combined MEK and SHP2 inhibition. *Nature Medicine, 24*(7), 968−977, Jul.

Xie, L., Su, X., Zhang, L., Yin, X., Tang, L., Zhang, X., et al. (2013). FGFR2 gene amplification in gastric cancer predicts sensitivity to the selective FGFR inhibitor AZD4547. *Clinical Cancer Research: An Official Journal of the American Association for Cancer Research, 19* (9), 2572−2583, May 1.

Xu, J., Bai, Y., Xu, N., Li, E., Wang, B., Wang, J., et al. (2020). Tislelizumab plus chemotherapy as first-line treatment for advanced esophageal squamous cell carcinoma and gastric/ gastroesophageal junction adenocarcinoma. *Clinical Cancer Research: An Official Journal of the American Association for Cancer Research, 26*(17), 4542−4550, Sep 1.

Yamamoto, S., & Kato, K. (2020). Immuno-oncology for esophageal cancer. *Future Oncology [Internet]*. Available from https://doi.org/10.2217/fon-2020-0545, Aug 11; Available from.

Yu, C., Tang, H., Guo, Y., Bian, Z., Yang, L., Chen, Y., et al. (2018). Hot tea consumption and its interactions with alcohol and tobacco use on the risk for esophageal cancer: A population-based cohort study. *Annals of Internal Medicine, 168*(7), 489−497, Apr 3.

Zhang, H., Schaefer, A., Wang, Y., Hodge, R. G., Blake, D. R., Diehl, J. N., et al. (2020). Gain-of-function mutations promote focal adhesion kinase activation and dependency in diffuse gastric cancer. *Cancer Discovery, 10*(2), 288−305, Feb.

Zhao, D., Klempner, S. J., & Chao, J. (2019). Progress and challenges in HER2-positive gastro-esophageal adenocarcinoma. *Journal of Hematology & Oncology, 12*(1), 50, May 17.

Zhou, K. I., Peterson, B. F., Serritella, A., Thomas, J., Reizine, N., Moya, S., et al. (2020). Spatial and temporal heterogeneity of PD-L1 expression and tumor mutational burden in gastroesophageal adenocarcinoma at baseline diagnosis and after chemotherapy. *Clinical Cancer Research [Internet]*. Available from https://doi.org/10.1158/1078-0432.CCR-20-2085, Aug 20; Available from.

Chapter 16

Prostate Cancer

Jason Zhu
Levine Cancer Institute, Charlotte, NC, United States

Introduction

Prostate cancer is the most common cancer diagnosis in men in the United States with estimated 191,930 new cases and 33,330 deaths in 2020 (Siegel, Miller, & Jemal, 2020). Localized prostate cancer is highly treatable and curable, with nearly 100% survival at 5 years (https://www.cancer.org/cancer/prostate-cancer/detection-diagnosis-staging/survival-rates.html). However, for men with metastatic disease, long-term survival remains poor, with only 31% survival at 5 years. Over the past decade, advances in prostate cancer genomics have introduced the field to biomarker-based therapies, providing patients with targeted treatment options that may improve both quality of life and overall survival. In this chapter, we will discuss the genomic medicine approaches regarding screening and treatment of prostate cancer.

Precision prostate cancer screening

Prostate cancer is the most heritable common cancer—in the landmark paper by Lichtenstein et al., 42% of the variation underlying prostate cancer could be explained by genetic factors (Lichtenstein et al., 2000). This heritable risk stays constant across age through late life (Hjelmborg et al., 2014). Prostate cancer screening recommendations by the US Preventive Services Task Force have changed twice in the past decade, from recommending against screening in 2012, to recommending shared decision making in 2018. Genomic germline information may be useful in this discussion (Jemal et al., 2015).

Many genes have been identified, which may increase one's risk of developing prostate cancer, and a precision screening approach may be appropriate for patients with known mutations (they may have been tested if a family member had tested positive for a pathogenic mutation). The first gene which was implicated in hereditary prostate cancer was *HOXB13* G84E, discovered after targeted sequencing exons on chromosome

Genomic and Precision Medicine. DOI: https://doi.org/10.1016/B978-0-12-800684-9.00012-5

17q21−22 resulted in the identification of a recurrent mutation, G84E in *HOXB13*, in four probands of families with a history of prostate cancer (Ewing et al., 2012). For patients with an inherited mutation in *HOXB13*, experts agree that screening for prostate cancer should begin at age 40, or 10 years prior to the youngest family member diagnosed with prostate cancer (Giri et al., 2018). Similar recommendations have been made for men with known *BRCA2* mutations. *BRCA2* carriers are known to have a higher incidence of prostate cancer, younger age at diagnosis, as well as more aggressive primary tumors (Page, Braun, Partin, Caporaso, & Walsh, 1997). A recent prospective study of 447 *BRCA2* carriers found that *BRCA2* carriers had a stronger association with Gleason ≥ 7 and a higher risk of death from prostate cancer with a standardized mortality ratio of 3.85 (95% CI: 1.44−10.3) (Nyberg et al., 2020). Additionally, the location of the mutation within the *BRCA2* gene was important—men with BRCA2 mutations in the ovarian cancer cluster region (OCCR, c. 2831−6401) had a lower risk of prostate cancer than men with mutations outside of the OCCR. However, regardless of *BRCA2* mutation location, men with any *BRCA2* mutation had a higher risk for prostate cancer than the general population. A precision medicine approach to screening patients with high-risk germline variants may enable detection of patients who have high risk disease at a young age.

Management of localized prostate cancer

After the diagnosis of localized prostate cancer, decisions regarding definitive local treatment (surgery, radiation, hormonal therapy) or active surveillance are nuanced by several factors including patient preferences and pathologic characteristics (Gleason grade, perineural invasion, extraprostatic extension). For patients with low risk disease, defined as clinical stage $\leq T2$, prostate specific antigen (PSA) ≤ 10 ng/mL, Gleason score $\leq 3 + 3$, ≤ 2 to 3 positive cores, $\leq 50\%$ single core involvement, an active surveillance strategy may be appropriate, but these guidelines may not capture some patients who have aggressive cancer. Several molecular and genomic assays have been developed to aid in a precision medicine approach (Reichard, Stephenson, & Klein, 2015). Some assays are gene expression panels that provide a score based on a set RNA quantification and gene-expression profile (Oncotype DX, Prolaris), while others evaluate immuno-histochemical protein quantification (ProMark). These tests provide additional data for clinicians and patients deciding between definitive management or active surveillance and prospective data have shown increases in active surveillance with the use of these tests (Badani et al., 2015). In the postprostatectomy setting, genomic tests have also been developed to help identify which men may benefit from treatment intensification or deescalation. Decipher genomic classifier (GC) utilizes a 22-gene panel to risk stratify patients into low (5.5%), intermediate (15%), or high risk (26%)

of metastases at 10 years (Spratt et al., 2017). Most recently, the Decipher GC was found to be predictive of metastatic disease for men with NCCN high risk disease (Tosoian et al., 2020).

There are limitations to these tests, especially those utilizing prostate biopsy specimens. For patients with multifocal prostate cancer, sampling bias may limit the utility of these tests (Salami et al., 2018). To overcome sampling bias, one novel method involves interrogation of small noncoding RNAs (sncRNA) isolated from urinary exosomes (Klotz et al., 2020). In the prospective validation cohort ($n = 329$), this test demonstrated a sensitivity of 98% and specificity of 96% of the detection of prostate cancer, and was also able to distinguish between grade group 1 cancer (Gleason 6) and grade group 2−5 cancer (Gleason ≥ 7) with a sensitivity of 100% and specific of 99%. This test is now undergoing additional validation. Tumor heterogeneity and multifocality represent a significant challenge in the management of men with localized prostate cancer, and noninvasive blood or urine-based biomarkers are necessary to overcome these obstacles.

Precision oncology advances in metastatic prostate cancer

Prior to 2017, all treatments for metastatic, castration-resistant prostate cancer were agnostic of tumor or germline mutations. Treatment options included second-generation androgen receptor antagonists such as enzalutamide or abiraterone, chemotherapy (docetaxel and cabazitaxel), a bone targeting agent (radium-223), and an immunotherapy (sipuleucel-T). However, in 2017, the U.S. Food and Drug Administration (FDA)-approved pembrolizumab for the treatment of all patients with unresectable or metastatic, microsatellite instability-high (MSI-H) or mismatch repair deficient solid tumors that have progressed following prior treatment (Marcus, Lemery, Keegan, & Pazdur, 2019). Thus pembrolizumab became the first biomarker-guided therapy for patients withmetastatic castration resistant prostate cancer (mCRPC). In the largest case series of patients with MSI-high prostate cancer on immune checkpoint inhibitors ($n = 11$), 45% (5/11) of patients were still on therapy, including one patient who continued to be free from progression at 7 years (Abida et al., 2019).

While the use of pembrolizumab under this specific indication is exciting, MSI high status is rare in prostate cancer (3%), compared with uterine (31%) and colon cancer (20%) (Abida et al., 2019; Bonneville et al., 2017). Additionally, in unselected patients, immune checkpoint inhibition is largely ineffective, with objective response rates <5% (De Bono et al., 2018). Thus additional approaches for precision medicine are necessary to identify patients who may respond to immunotherapy. *CDK12* is one such biomarker currently under evaluation. Biallelic loss of *CDK12* is associated with genomic instability and focal tandem duplications, and may lead to increased neoantigen burden as well as T cell infiltration (Wu et al., 2018). In the pilot clinical study of immune checkpoint inhibition for patients with *CDK12* mutated prostate cancer, two of four heavily pretreated patients with mCRPC demonstrated marked PSA decline with anti-PD1 therapy (Wu et al., 2018). Several clinical trials are underway, evaluating the role of immunotherapy for tumors with *CDK12* mutations (NCT03570619, NCT04104893).

A second precision medicine approach to the treatment of mCRPC involves targeting the genes regulating the DNA repair pathway. A landmark study by Robinson et al. found that 23% of patients with mCRPC harbor mutations in the DNA repair pathway, and 8% of all patients had germline mutations. A precision oncology strategy has been developed for men harboring pathogenic variants in *BRCA1/2* or *ATM*-mutated mCRPC (Hussain et al., 2019). *BRCA1* and *BRCA2* are critical for DNA double-strand break repair by homologous recombination, and when cells lack these critical functions, they become sensitive to PARP inhibition, resulting in cell cycle arrest and apoptosis (Farmer et al., 2005). This hypothesis was tested in a Phase II study by Mateo et al., which reported that 88% (14 of 16) of biomarker positive (a homozygous deletion or deleterious mutation was identified in a gene associated with DNA damage repair or sensitivity to PARP inhibition) patients had a response to olaparib compared to only 6% (2 of 33) of biomarker negative patients (Mateo et al., 2015). Patients in the biomarker positive group had improved radiographic progression free survival and overall survival (13.8 months vs 7.5 months). Subsequent data from PROfound, a Phase III study of olaparib versus enzalutamide or abiraterone for patients with mCRPC and homologous recombination repair gene alterations validated the utility of this precision medicine approach (Hussain et al., 2019). Patients were divided into two cohorts, cohort A (BRCA1, BRCA2, ATM alterations) and cohort B (other alterations). In cohort A, receipt of olaparib was associated with improved progression-free survival [7.4 months vs 3.9 months, hazard ratio (HR): 0.34, 95% confidence interval (CI): 0.25−0.47] as well as objective overall response (33.3% vs 2.3%, odds ratio (OR): 20.86, 95% CI: 4.18−379.18). Based on PROfound, the FDA approved olaparib for the treatment of mCRPC with recombination repair gene mutations. There are a number of ongoing PARP inhibitors trials, both single agent as well as in combination with other therapies. Given that there is a now a biomarker-specific treatment for these men, national guidelines including NCCN and American Urological Association (AUA) recommend screening all men with advanced prostate cancer for germline alterations (Mohler et al., 2019).

In addition to biomarkers that predict for efficacy, a genomic biomarker has been validated as a marker of resistance for androgen receptor−targeted therapies (abiraterone and enzalutamide): the presence of androgen-receptor splice variant 7 messenger RNA (AR-V7) (Antonarakis et al., 2014). Enzalutamide and abiraterone are second-generation androgen antagonists that inhibit the androgen receptor signaling pathway and deplete adrenal and intratumoral androgens, respectively. Both therapies have been shown to improve overall survival for men with mCRPC, as well as metastatic castration-sensitive prostate cancer (mCSPC) (Davis et al., 2019; Fizazi et al., 2017; Ryan et al., 2013; Scher et al., 2012). However, sequencing these therapies is challenging as most patients who have an initial response to either abiraterone or enzalutamide do not respond to the other therapy at the time of progression (Noonan et al., 2013; Schrader et al., 2014). One explanation for the development of primary resistance involves the presence of truncated androgen receptor proteins that lack a C-terminal ligand binding domain but retain the activating N-terminal domain. There are many androgen-receptor variants but AR-V7 has two characteristics that make it favorable as a biomarker. First, it is detectable in clinical specimens, and second, it encodes a functional protein product. In a prospective study of men with mCRPC who were initiating treatment with enzalutamide or abiraterone, AR-V7 positive patients had a 0% response rate to either agent, compared with 50%−60% of patients who were AR-V7 negative (Antonarakis et al., 2014). This has subsequently been validated in a prospective multicenter study with 118 men, which found that detection of AR-V7 by circulating tumor cells (CTCs) is independently associated with shorter progression-free survival and overall survival with abiraterone or enzalutamide (Armstrong et al., 2019). One limitation of this study is that the AR-V7 assays currently require detectable CTCs, which are typically only present in patients with high-volume, poor risk disease. For newly diagnosed patients with mCSPC, especially if they are treatment-naïve, AR-V7 is unlikely to be detected or affect management decisions. In the second line setting, for patients who are chemotherapy-naïve, you could potentially use detection of AR-V7 to help decide between a second androgen receptor antagonist versus chemotherapy (docetaxel or cabazitaxel).

Challenges

While significant progress has been made in the understanding of prostate cancer genomics, a number of limitations still exist, which prevent us from treating all patients with a precision oncology approach. The most glaring limitation is the lack of biomarker-guided therapies. While immunotherapy for patients with MSI high tumors and PARP inhibitors for patients with pathogenic *BRCA1/BRCA2/ATM* variants have become part of standard practice, this represents less than 10% of all patients with metastatic prostate cancer. In one retrospective study of 77 patients with mCRPC who underwent tumor next generation sequencing (NGS) sequencing, 9% of patients had a treatment change based on the targeted gene panel and only 5% of patients derived clinical benefit (Zhu et al., 2019). The current lack of targeted therapies is the most important limitation and significant efforts should be placed toward novel drug discovery.

From a technical standpoint, another limitation for men with prostate cancer is having adequate tissue to sample. For most men with mCRPC, the burden of disease resides in the bone, with few men having visceral metastases. Bone biopsies are technically challenging and often result in insufficient tumor tissue for biopsy. Certain institutions have instituted prostate cancer–specific bone biopsy protocols that have shown to improve the yield to upward of 80% (McKay et al., 2014; Sailer et al., 2018). One method to overcome this challenge is collection of circulating tumor DNA (ctDNA). This approach has been successful in the setting metastatic lung cancer for patients without accessible tissue, and is currently under evaluation for men with metastatic prostate cancer (Normanno, Denis, Thress, Ratcliffe, Reck., 2017; Vandekerkhove et al., 2019).

Conclusions

Precision oncology approaches to prostate cancer may impact all stages of the disease, from screening in asymptomatic patients with known high-risk germline mutations to treatment selection for men with mCRPC. Given the available biomarker-selected treatments for men with prostate cancer, it is important to obtain tumor genomic profiling, if feasible, on all men with mCRPC. Additionally, all men with advanced prostate cancer should undergo germline testing, which may impact both cancer screening and prevention.

References

Abida, W., Cheng, M. L., Armenia, J., et al. (2019). Analysis of the prevalence of microsatellite instability in prostate cancer and response to immune checkpoint blockade. *JAMA Oncology, 5*, 471–478.

Antonarakis, E. S., Lu, C., Wang, H., et al. (2014). AR-V7 and resistance to enzalutamide and abiraterone in prostate cancer. *New England Journal of Medicine, 371*, 1028–1038.

Armstrong, A. J., Halabi, S., Luo, J., et al. (2019). Prospective multicenter validation of androgen receptor splice variant 7 and hormone therapy resistance in high-risk castration-resistant prostate cancer: The PROPHECY Study. *Journal of Clinical Oncology, 37*, 1120–1129.

Badani, K. K., Kemeter, M. J., Febbo, P. G., et al. (2015). The impact of a biopsy based 17-gene genomic prostate score on treatment recommendations in men with newly diagnosed clinically prostate cancer who are candidates for active surveillance. *Urology Practice, 2*, 181–189.

Bonneville, R., Krook, M. A., Kautto, E. A., et al. (2017). Landscape of microsatellite instability across 39 cancer types. *JCO Precision Oncology, 1*, 1–15.

Davis, I. D., Martin, A. J., Stockler, M. R., et al. (2019). Enzalutamide with standard first-line therapy in metastatic prostate cancer. *New England Journal of Medicine, 381*, 121–131.

De Bono, J. S., Goh, J. C., Ojamaa, K., et al. (2018). KEYNOTE-199: Pembrolizumab (pembro) for docetaxel-refractory metastatic castration-resistant prostate cancer (mCRPC). *Journal of Clinical Oncology.*

Ewing, C. M., Ray, A. M., Lange, E. M., et al. (2012). Germline mutations in HOXB13 and prostate-cancer risk. *New England Journal of Medicine, 366,* 141–149.

Farmer, H., McCabe, N., Lord, C. J., et al. (2005). Targeting the DNA repair defect in BRCA mutant cells as a therapeutic strategy. *Nature, 434,* 917–921.

Fizazi, K., Tran, N., Fein, L., et al. (2017). Abiraterone plus prednisone in metastatic, castration-sensitive prostate cancer. *New England Journal of Medicine, 377,* 352–360.

Giri, V. N., Knudsen, K. E., Kelly, W. K., et al. (2018). Role of genetic testing for inherited prostate cancer risk: Philadelphia Prostate Cancer Consensus Conference 2017. *Journal of Clinical Oncology, 36,* 414.

Hjelmborg, J. B., Scheike, T., Holst, K., et al. (2014). The Heritability of Prostate Cancer in the Nordic Twin Study of Cancer. *Cancer Epidemiology, Biomarkers & Prevention: A publication of the American Association for Cancer Research, cosponsored by the American Society of Preventive Oncology, 23,* 2303–2310.

Hussain, M., Mateo, J., Fizazi, K., et al. (2019). LBA12_PRPROfound: Phase III study of olaparib vs enzalutamide or abiraterone for metastatic castration-resistant prostate cancer (mCRPC) with homologous recombination repair (HRR) gene alterations. *Annals of Oncology, 30.*

Jemal, A., Fedewa, S. A., Ma, J., et al. (2015). Prostate cancer incidence and PSA testing patterns in relation to USPSTF screening recommendations. *JAMA, 314,* 2054–2061.

Klotz, L., Tilki, D., Fleshner, N. E., DiRienzo, A. G., Wang, W.-L. W., & Tenniswood, M. (2020). Analysis of small non-coding RNAs in urinary exosomes to classify prostate cancer into low-grade (GG1) and higher-grade (GG2-5). *Journal of Clinical Oncology.*

Lichtenstein, P., Holm, N. V., Verkasalo, P. K., et al. (2000). Environmental and heritable factors in the causation of cancer—Analyses of cohorts of twins from Sweden, Denmark, and Finland. *New England Journal of Medicine, 343,* 78–85.

Marcus, L., Lemery, S. J., Keegan, P., & Pazdur, R. (2019). FDA approval summary: Pembrolizumab for the treatment of microsatellite instability-high solid tumors. *Clinical Cancer Research,* clincanres.4070.2018.

Mateo, J., Carreira, S., Sandhu, S., et al. (2015). DNA-repair defects and olaparib in metastatic prostate cancer. *New England Journal of Medicine, 373,* 1697–1708.

McKay, R. R., Zukotynski, K. A., Werner, L., et al. (2014). Imaging, procedural and clinical variables associated with tumor yield on bone biopsy in metastatic castration-resistant prostate cancer. *Prostate Cancer and Prostatic Diseases, 17,* 325–331.

Mohler, J. L., Antonarakis, E. S., Armstrong, A. J., et al. (2019). Prostate cancer, version 2.2019, NCCN clinical practice guidelines in oncology. *Journal of the National Comprehensive Cancer Network, 17,* 479–505.

Noonan, K., North, S., Bitting, R., Armstrong, A., Ellard, S., & Chi, K. (2013). Clinical activity of abiraterone acetate in patients with metastatic castration-resistant prostate cancer progressing after enzalutamide. *Annals of Oncology, 24,* 1802–1807.

Normanno, N., Denis, M. G., Thress, K. S., Ratcliffe, M., & Reck, M. (2017). Guide to detecting epidermal growth factor receptor (EGFR) mutations in ctDNA of patients with advanced non-small-cell lung cancer. *Oncotarget, 8,* 12501.

Nyberg, T., Frost, D., Barrowdale, D., et al. (2020). Prostate cancer risks for male BRCA1 and BRCA2 mutation carriers: A prospective cohort study. *European Urology, 77,* 24–35.

Page, W. F., Braun, M. M., Partin, A. W., Caporaso, N., & Walsh, P. (1997). Heredity and prostate cancer: A study of World War II veteran twins. *The Prostate, 33,* 240–245.

Reichard, C. A., Stephenson, A. J., & Klein, E. A. (2015). Applying precision medicine to the active surveillance of prostate cancer. *Cancer, 121,* 3403–3411.

Ryan, C. J., Smith, M. R., De Bono, J. S., et al. (2013). Abiraterone in metastatic prostate cancer without previous chemotherapy. *New England Journal of Medicine, 368*, 138−148.

Sailer, V., Schiffman, M. H., Kossai, M., et al. (2018). Bone biopsy protocol for advanced prostate cancer in the era of precision medicine. *Cancer, 124*, 1008−1015.

Salami, S. S., Hovelson, D. H., Kaplan, J. B., et al. (2018). Transcriptomic heterogeneity in multifocal prostate cancer. *JCI Insight, 3*.

Scher, H. I., Fizazi, K., Saad, F., et al. (2012). Increased survival with enzalutamide in prostate cancer after chemotherapy. *New England Journal of Medicine, 367*, 1187−1197.

Schrader, A. J., Boegemann, M., Ohlmann, C.-H., et al. (2014). Enzalutamide in castration-resistant prostate cancer patients progressing after docetaxel and abiraterone. *European Urology, 65*, 30−36.

Siegel, R. L., Miller, K. D., & Jemal, A. (2020). Cancer statistics, 2020. *CA: A Cancer Journal for Clinicians, 70*, 7−30.

Spratt, D. E., Yousefi, K., Deheshi, S., et al. (2017). Individual patient-level *meta*-analysis of the performance of the decipher genomic classifier in high-risk men after prostatectomy to predict development of metastatic disease. *Journal of Clinical Oncology: Official Journal of the American Society of Clinical Oncology, 35*, 1991.

Tosoian, J. J., Birer, S. R., Karnes, R. J., et al. (2020). Performance of clinicopathologic models in men with high risk localized prostate cancer: Impact of a 22-gene genomic classifier. *Prostate Cancer and Prostatic Diseases*, 1−8.

Vandekerkhove, G., Struss, W. J., Annala, M., et al. (2019). Circulating tumor DNA abundance and potential utility in de novo metastatic prostate cancer. *European Urology, 75*, 667−675.

Wu, Y.-M., Cieślik, M., Lonigro, R. J., et al. (2018). Inactivation of CDK12 delineates a distinct immunogenic class of advanced prostate cancer. *Cell, 173*, 1770−1782.e14.

Zhu, J., Tucker, M., Marin, D., et al. (2019). *Clinical utility of FoundationOne tissue molecular profiling in men with metastatic prostate cancer. Urologic oncology: Seminars and original investigations* (pp. 813.e1−813.e9). Elsevier.

Chapter 17

Genomics and precision medicine: kidney cancer

Ankit Madan[1] and David M. Nanus[2]

[1]SOVAH Cancer Center, Danville, VA, United States, [2]Division of Hematology and Medical Oncology, Weill Cornell Medical College, New York, NY, United States

Introduction

In the United States approximately 73,750 new cases of kidney cancer will be diagnosed in 2020 and over 14,830 people are anticipated to die from this disease. Globally, in 2018 there were over 400,000 new cases of renal cell carcinoma (RCC) and over 175,000 deaths attributable to this malignancy (Bray et al., 2018). The classification of kidney tumors has evolved over time. It has been classified based on anatomical origin into four types: RCC (renal cortex involvement), collecting duct carcinoma (renal medulla), renal medullary carcinoma (renal medulla), and papillary urothelial carcinoma (transitional epithelium lining renal pelvis and ureter) (Eble, Sauter, & Epstein, 2004). Papillary urothelial carcinoma is more similar to transitional carcinoma of bladder and ureter than other types of renal tumors (Solomon & Hansel, 2015). RCC has further been classified on the basis of histopathology into three major types: clear cell RCC (ccRCC), papillary RCC (pRCC), and chromophobe RCC (chRCC). ccRCC is the most common accounting for 65%−70% of cases. It is characterized histologically by the presence of clear lipid and glycogen-rich cytoplasm. pRCC accounts for 15%−20% of adult RCCs and is further divided into two types. Type 1 pRCC (p1RCC) is characterized by papillae covered by small cells in a single layer of papillary basement membrane along with scant cytoplasm. Type 2 pRCC (p2RCC) is characterized by abundant eosinophilic cytoplasm and pseudostratified nuclei of higher nuclear grade. chRCC comprises larger polygonal cells containing atypical nuclei and smaller granular cells in a solid growth pattern (Hsieh, Le, Cao, Cheng, & Creighton, 2018). Other rare subtypes of RCC include sarcomatoid RCC, oncocytoma, and unclassified RCC. In 2016 World Health Organization updated the RCC classification and included other subtypes based on genomic and molecular profiling done by The Cancer Genomic Atlas (TCGA) investigators and individual groups (Moch, Humphrey, Ulbright, & Reuter, 2016).

Genomic and Precision Medicine. DOI: https://doi.org/10.1016/B978-0-12-800684-9.00004-6

Genomic and molecular profiling of renal cell carcinoma subtypes

Clear cell renal cell carcinoma

The role of the Von Hippel−Lindau (*VHL*) gene and hypoxia inducible factors 1α and 2α (HIF-1α and HIF-2α) is well described in the pathogenesis of sporadic and hereditary ccRCC. The *VHL* gene is located on chromosome 3p and encodes pVHL. pVHL along with elongin C and other components form a E3 ubiquitin ligase complex called VEC. pVHL has an α domain that binds with elongin C and a β-domain to bind with HIF-1α and HIF-2α. The VEC complex is responsible for ubiquitination of HIF-1α and HIF-2α through proteasomal degradation. Under normoxia, HIF-1α and HIF-2α are hydroxylated, then bind with VEC and undergo degradation. In hypoxic conditions, hydroxylation does not occur. This results in accumulation of HIF transcript proteins, which leads to transcriptional activation of HIF-target genes [such as vascular endothelial growth factor (VEGF), GLUT-1] which are involved in angiogenesis, metabolism, and cell death. HIF-1α is a key transcription factor for glycolysis and HIF-2α is involved in erythropoiesis. In VHL disease where one allele of VHL is mutated, conditions similar to hypoxia prevail following loss of the one functioning VHL allele (Hsieh et al., 2018; Hu, Wang, Chodosh, Keith, & Simon, 2003; Sufan, Jewett, & Ohh, 2004).

The TCGA evaluated somatic alterations from 417 primary nephrectomy samples. Somatic copy number alterations revealed loss of chromosome 3p in 91% of samples. Chromosome 3p houses *VHL*, polybromo 1 (*PBRM1*), *SETD2*, and BRCA1 associated protein-1 (*BAP-1*) genes. Gain of chromosome 5q (67% samples) and arm-level losses of chromosome 14q (45% samples) were also noted. Whole-exome sequencing (WES) revealed 19 significantly mutated genes, with the most common being *VHL, PBRM1, SETD2, KDM5C, PTEN, BAP1, mTOR*, and *TP53*. There was an average of 1.1 ± 0.5 nonsilent mutations per megabase (Cancer Genome Atlas Research Network, 2013). In 2010 Bannon et al. who had reported transcriptomic data on 47 ccRCC cases from unsupervised clustering analysis showed two ccRCC clusters, clear cell types A and B (ccA and ccB), each with a different prognosis [overall survival (OS): 59 months vs 36 months, respectively] (Brannon et al., 2010). TCGA analysis also revealed four distinct clusters: m1, m2, m3, and m4. The m1 cluster was similar to ccA. Clusters m2 and m3 corresponded to ccB and the m4 cluster was unclassifiable as compared to the ccA and ccB clusters. Cluster m1 had an improved survival compared with the other clusters. Cluster m1 also had a greater frequency of the *PBRM-1* mutation. Cluster m3 contained a higher number of *PTEN* mutations and *CDKN2A* deletions. Cluster m4 demonstrated a higher number of *BAP-1* and *mTOR* mutations (Cancer Genome Atlas Research Network, 2013).

Papillary renal cell carcinoma

One hundred and sixty-one pRCCs were analyzed by the TCGA Consortium by WES. Copy number alterations revealed one subgroup (predominantly type 1 pRCCs) with almost universal gain of chromosomes 7 and 17 and less frequent gain of chromosomes 2, 3, 12, 16, and 20. The other subgroup (predominantly type 2 tumors) was characterized by a high degree of aneuploidy with multiple chromosomal losses, including frequent loss of chromosome 9p. The second subgroup was associated with an inferior survival. WES analysis reported five genes that were frequently mutated: *MET*, *SETD2*, *NF2*, *KDM6A*, and *SMARCB1*. Further analysis revealed six additional significantly mutated genes consisting of *FAT1*, *BAP1*, *PBRM1*, *STAG2*, *NFE2L2*, and *TP53*. *TFE3* or *TFEB* gene fusions were noticed in 10.6% of tumors. An average of 1.45 nonsilent mutations per megabase was identified. Seventeen percent of type 1 pRCC tumors had characteristic *MET* gene mutations. Alterations in *CDKN2A* which encodes p16 (INK4A) were more common amongst type 2 pRCC samples. A CpG Island Methylator Phenotype (CIMP) subgroup was identified in type 2 pRCCs as well as universal hypermethylation of the *CDKN2A* promoter. CIMP patients were younger (42 years old) and had a worse prognosis overall. Cluster analysis using profiles of mRNA expression, microRNA expression, and protein expression generated four clusters: C1 (composed of type 1 pRCC tumors), C2a and C2b (composed of type 2 pRCC tumors), and C2c (composed of CIMP-associated pRCC). The C1 cluster was associated with type 1 pRCC, gain of chromosome 7, *MET* mutation and early stage tumors (Stage 1 and 2). Cluster C2a was also associated with early stage of tumor development. Cluster C2b consisted of type 2 and unclassified pRCCs. This cluster was associated with DNA methylation cluster 1, advanced stage tumor development (Stage 3 or 4), and *SETD 2* mutations. Cluster C2c had predominantly the CIMP-associated tumor subtype. Patients with cluster C1 or cluster C2a tumors had the best survival probability, whereas patients with cluster C2c had the worst survival probability (Haake, Weyandt, & Kimryn Rathmell, 2016; The Cancer Genome Atlas Research Network, 2016).

Chromophobe renal cell carcinoma

Single nucleotide polymorphism array analysis detected loss of one copy of the entire chromosome for most or all of chromosomes 1, 2, 6, 10, 13, and 17 in 86% of cases. Less frequently, the loss of chromosomes 3, 5, 8, 9, 11, 18, and 21 was also noted. WES showed 0.4 exonic somatic mutations per megabase. The two most frequently mutated genes were *TP53* and *PTEN* in 32% and 9% of patient samples, respectively. A major driving event in chRCC appears to be complete inactivation of p53 signaling. An interesting finding was genomic alterations in the *TERT* promoter region resulting in elevated *TERT* gene expression. Electron transport chain (etc) complex I genes were altered in 18% of patients, with the most frequently altered gene being MT-ND5 ($n = 6$). This suggested increased utilization of the Krebs cycle and etc for ATP generation in chRCC (Davis et al., 2014). The findings are summarized in Table 17.1.

Prognostic implications of molecular markers

The most commonly used prognostic scoring systems that rely on clinical parameters are the Memorial Sloan Kettering Cancer Center (MSKCC) Prognostic model and International Metastatic RCC Database Consortium criteria (IMDC) (Heng et al., 2009; Motzer, Bacik, Murphy, Russo, & Mazumdar, 2002). Only recently have molecular markers such as *BAP-1* mutation, *PBRM-1* mutation, and PD-L1 expression been considered to categorize patient outcomes.

Hakimi et al. evaluated ~ 600 patients combined from an MSKCC and TCGA cohort for *BAP-1*, *SETD2*, *VHL*, and *PBRM-1* mutations. Patients with *BAP-1* mutations were found to have a median OS of 31.2 months versus 78.2 months for patients lacking a *BAP-1* mutation. A similar finding of an inferior prognosis was seen with *SETD2* mutations in the TCGA cohort (median OS: 62.7 months vs 78.2 months) but not in the MSKCC cohort. Truncating mutations of *BAP-1* also had a worse prognosis as compared to missense mutations. The impact of *PBRM-1* mutations is unclear with some studies suggesting an inferior survival in patients with *PBRM-1* mutations, while other studies have reported no difference in cancer-specific survival (CSS) with this mutation (Ari Hakimi et al., 2013; da Costa et al., 2014). One study of 1330 patients showed no change in CSS but a higher risk of metastases in *PRBM-1* deficient patients (Joseph et al., 2016). Voss et al. reported *PRBM-1* mutated patients with metastatic ccRCC treated with a tyrosine kinase inhibitor (TKI) had an improved progression-free survival (PFS) and OS (Voss et al., 2018). *PBRM-1* gene loss of function mutations have also been found to be associated with increased response to nivolumab (Miao et al., 2018).

PD-L1 (programmed cell death ligand) expressed on tumor cells and PD-1 (programmed cell death) located on T-cell, B-cells, dendritic cells, and Natural Killer (NK) cells negatively regulate T-cell immune response (Butte, Keir, Phamduy, Sharpe, & Freeman, 2007). Immunotherapies using anti-CTLA 4 (cytotoxic T-lymphocyte-associated antigen-4), anti-PD-1, and anti-PD-L1 antibodies that inhibit this PD-1 and PD-L1 pathway are now standard of care in patients with RCC. The Phase III COMPARZ trial evaluated the prognostic significance of tumor PD-L1 expression in 357 metastatic RCC patients treated with sunitinib and pazopanib. Tumor cell PD-L1 expression was evaluated based on H-score. Higher H-score (higher PD-L1 expression) was associated with shorter OS (Choueiri et al., 2015). In the CHECKMATE-025 trial comparing nivolumab versus everolimus in mRCC,

TABLE 17.1 Genomic Alterations in Renal Cell Carcinoma Subtypes.

RCC subtype	Incidence	Somatic copy number alterations[a]	Genetic mutations[a]	Mutations per megabase	Notable TCGA findings
Clear cell	65%–70%	Chromosome 3p loss (91%)	VHL, chromatin modifier genes-PBRM-1, BAP-1, and SETD2, mTOR, and TP53	1.1 ± 0.5 nonsilent mutations per megabase	m1-m4 clusters m1-PBMR-1 mutations, Improved OS m3-PTEN, CDKN2A mutations m4-BAP-1, mTOR mutations
Papillary	15%–20%	Gain of chromosome 7 and 17	MET, SETD2, NF2, KDM6A, and SMARCB1 Type 1-MET gene Type 2-CDKN2A gene alteration	1.45 nonsilent mutations per megabase	C1, C2a, C2b, C2c clusters C1-Type 1, gain of chromosome 7, MET mutation C2a-Type 2 C2b-Tpe 2 and unclassifiable pRCC C2c-CIMP-associated tumor subtype, worst prognosis
Chromophobe	5%	Loss of one copy of chromosome 1, 2, 6, 10, 13, and 17 (86%)	TP53 and PTEN	0.4 exonic somatic mutations per megabase	Elevated TERT gene expression, Altered etc complex-1 gene expression

chr, Chromosome; etc, electron transport chain; pRCC, papillary renal cell carcinoma; RCC, renal cell carcinoma; TCGA, The Cancer Genomic Atlas.
[a]Most common.

tumor PD-L1 expression was also associated with worse prognosis. Patients with mRCC receiving nivolumab with tumor PD-L1 expression >1% had an OS of 21.8 months [95% confidence interval (CI): 16.5−28.1] compared to 27.4 months (95% CI: 21.4 to not estimable) in patients with tumor PD-L1 expression <1% (Motzer et al., 2015). Similar results were obtained in a multicenter analysis with higher tumor PD-L1 expression associated with worse recurrence-free survival (RFS) and OS (Chipollini et al., 2019).

Predictive biomarkers and targeted therapeutics

PD-L1 has also been explored as a predictive biomarker for patients with RCC receiving immunotherapy. In the CHECKMATE-025 trial, patients receiving nivolumab had improved survival compared to everolimus, independent of PD-L1 expression (Motzer et al., 2015). In the CHECKMATE-214 trial, patients with intermediate/poor risk features who received ipilimumab and nivolumab had a higher objective response rate (ORR) compared to sunitinib, regardless of PD-L1 expression, but a significant PFS benefit for ipilimumab and nivolumab was only observed in patients with PD-L1 expression $\geq 1\%$ (Escudier et al., 2017). Similarly, in the IMmotion151 study evaluating the combination of atezolizumab and bevacizumab versus sunitinib, PFS benefit was observed for atezolizumab and bevacizumab in both the PD-L1 positive population (> 1% expression on tumor infiltrating cells) and intention-to-treat population (Rini et al., 2019). Thus the utility of PD-L1 expression as a predictive biomarker for immunotherapy has not been proven. This could be due to the use of different companion assays for PD-L1 testing, or PD-L1 staining on archival specimens that may not reflect the true status of expression and testing in metastatic sites of disease.

Molecular subtyping using gene expression analysis is under investigation to develop predictive biomarkers that suggest responsiveness to immunotherapy, VEGF inhibitors and mammalian target of rapamycin (mTOR) inhibitors. The IMmotion150 trial evaluated biological subgroups based on angiogenesis (Angio^{high} vs Angio^{Low}), Immune (T-eff^{high} vs T-eff^{Low}), and a Myeloid inflammation signature. Within the sunitinib treatment group, Angio^{high} was associated with a higher PFS [hazard ratio (HR): 0.31; 95% CI: 0.18−0.55) and ORR (46% in Angio^{High} vs 9% in Angio^{Low})]. Similarly, within the atezolizumab and bevacizumab cohort, the T-eff^{high} group had a higher PFS (HR: 0.50; 95% CI: 0.30−0.86) and ORR (49% in T-eff^{High} vs 16% in T-eff^{Low}) as compared with the T-eff^{Low} group. Myeloid inflammation is associated with suppression of the antitumor T-cell response (Gabrilovich & Nagaraj, 2009). Myeloid^{high} showed reduced PFS in atezolizumab alone (HR: 2.98; 95% CI: 1.68−5.29) and atezolizumab−bevacizumab arm (HR: 1.71; 95% CI: 1.01−2.88) but not in the sunitinib arm (McDermott et al., 2018). Meylan and colleagues reported an analysis of a cohort of 324 patients from NIVOREN GETUG-AFU 26 study which treated patients with nivolumab. Based on

IMmotion150 study analysis, investigators looked at the outcomes of 184 patients examining Teff and Angio signatures. Teff-high/Angio-low signature had better ORR and PFS (47%, $P = .01$; 10.1 months, $P = .0005$) as compared to Teff-low/Angio-low signature (ORR = 5%, $P = .01$) and median PFS = 2.6 months, $P = .0005$) (Albiges et al., 2019; Meylan et al., 2020). Kwiatkowski et al. reported a higher prevalence of activating mutations in *mTOR* and inactivating mutations in *TSC1* or *TSC2* in responders to mTOR inhibitors as compared to nonresponders (Kwiatkowski et al., 2016). Similarly, Beuselinck and colleagues evaluated the gene expression profiles of 121 ccRCC patients treated with sunitinib. Four ccrcc subtypes were identified (ccrcc 1, 2, 3, and 4). Patients in groups ccrcc2 and ccrcc3 responded to sunitinib therapy more than patients in the ccrcc1 and ccrcc4 groups. Patients in ccrcc1 and ccrcc4 groups had more poorly differentiated and higher grade tumors (76%) as compared to ccrcc2 and ccrcc3 groups (56%). Patients in the ccrcc4 group had tumors with strong CD8 + T-cell infiltrate, high lymphocytic PD-1 and tumor PD-L1 expression, and an "immune-high" signature. This group may respond to immune checkpoint inhibitors. Patients in the ccrcc1 group had an "immune low" signature, whereas patients in ccrcc2 and ccrcc3 had "angio-high" and "normal-like" signature, respectively (Becht et al., 2015; Beuselinck et al., 2015). The BIONIKK trial is a Phase II biomarker-driven trial (NCT02960906) exemplifying utilization of personalized medicine. Previously untreated metastatic ccRCC patients were classified in ccrcc groups 1−4 utilizing a 35-gene signature. Therapy was based on molecular group assessment. Patients in the ccrcc1 and ccrcc4 groups were randomized to treatment with nivolumab alone versus nivolumab−ipilimumab combination. Patients in ccrcc2 and ccrcc3 groups were allocated to treatment with TKI (sunitinib or pazopanib) or nivolumab−ipilimumab combination. At a median follow-up of 16 months, patients in the ccrcc4 ("immune high") group had similar response rate to both nivolumab alone and nivolumab−ipilimumab combination therapy. Patients in the ccrcc2 group ("pro-angiogenic") responded similarly to both the TKI and the nivolumab−ipilimumab combination. Patients in the ccrcc1 ("immune low") group responded better to the nivolumab−ipilimumab combination compared to nivolumab alone. Results from the BIONIKK trial appear promising and need to be further investigated in larger prospective clinical trials utilizing a similar precision medicine approach (Vano et al., 2020).

Information elucidated from genomic analysis is also being applied to develop targeted therapeutic agents against specific RCC tumor types. For example, type 1 pRCC is associated with mutations in the *MET* gene. The multikinase inhibitor foretinib that targets the MET, VEGF, RON, AXL, and TIE-2 receptors demonstrated antitumor activity in pRCC patients with *MET* mutations. The median PFS was 9.4 months (95% CI: 6.9−12.9 months), ORR was 13.5% (95% CI: 6.7%−23.5%), with one-year survival of 70% (Choueiri et al., 2013). The SAVIOR Phase III trial evaluated savolitinib, another potent and selective MET inhibitor, in patients with MET-driven unresectable, locally advanced/metastatic pRCC versus sunitinib. The median PFS was 7.0 months [95% CI: 2.8 to not calculated (NC)] in the savolitinib arm and 5.6 months (95% CI: 4.1−6.9) in the sunitinib arm (log rank two-sided $P = .31$). In patients receiving savolitinib, the median OS was not reached (95% CI: 11.9−NC) compared with a median OS of 13.2 months (95% CI: 7.6−NC) in patients treated with sunitinib (HR: 0.51; 95% CI: 0.2−1.2; $P = .11$). The ORR was 27% (95% CI: 13.3−45.5) and 7% (95% CI: 0.9−24.3) in the savolitinib and sunitinib cohorts, respectively (Choueiri et al., 2020). The oral HIF-2α inhibitor, MK-6482, has recently been evaluated in a Phase I/II study of ccRCC with a pathogenic *VHL* mutation. Patients were heavily pretreated (∼3 prior therapies) and 67% had been treated with anti-PD-1 and anti-VEGF therapies. Thirty-one patients (56%) had stable disease (SD), with a disease control rate (complete response + partial response + SD) of 80%. The drug was tolerated well with anemia, the most common adverse event (Choueiri et al., 2020). A Phase III trial is planned (NCT02974738). Srinivasan et al. also recently reported results from an open-label Phase II study of MK-6482 for previously untreated VHL-associated ccRCC (NCT03401788). The ORR was 36.1% (24.2−49.4), with 62.3% experiencing SD (Srinivasan et al., 2020). Loss of *BAP-1* may confer sensitivity to PARP (poly ADP ribose polymerase) inhibitors and niraparib is currently being evaluated in advanced tumors with *BAP-1* mutations along with the DNA damage response−deficient neoplasms (NCT03207347).

Conclusion

In summary, genomic studies have resulted in a greater understanding of the genomic characteristics of various types of kidney cancers. To date, there are no predictive biomarkers commonly used in clinical practice to guide therapy, although some early studies are promising. PD-L1 expression alone has not proven to be a reliable predictive immunotherapy biomarker. Widespread use of genomic expression analysis may soon prove to be a more useful tool to differentiate patients who will benefit from immunotherapy from those who are more likely to respond to anti-VEGF therapy. Several prognostic biomarkers such as *PBRM-1* and *BAP-1* are also currently being studied. In the future, combining several clinical, pathological, molecular, and genomic variables may be necessary to make the best treatment decisions and to deliver personalized cancer care for patients with RCC.

References

Albiges, L., Negrier, S., Dalban, C., et al. (2019). Safety and efficacy of nivolumab in metastatic renal cell carcinoma (mRCC): Final analysis from the NIVOREN GETUG AFU 26 study. *Journal of Clinical Oncology, 37*(7), abstract page 542.

Ari Hakimi, A., Ostrovnaya, I., Reva, B., Schultz, N., Chen, Y.-B., Gonen, M., et al. (2013). Adverse Outcomes in clear cell renal cell carcinoma with mutations of 3p21 epigenetic regulators BAP1 and SETD2: A report by MSKCC and the KIRC TCGA research network. *Clinical Cancer Research, 19*(12).

Becht, E., Giraldo, N., Beuselink., et al. (2015). Prognostic and theranostic impact of molecular subtypes and immune classifications in renal cell cancer (RCC) and colorectal cancer (CRC). *Oncoimmunology, 12.*

Beuselinck, B., Job, S., Becht, E., et al. (2015). Molecular subtypes of clear cell renal cell carcinoma are associated with sunitinib response in the metastatic setting. *Clinical Cancer Research: An Official Journal of the American Association for Cancer Research, 21,* 1329−1339.

Brannon, A. R., Reddy, A., Seiler, M., Arreola, A., Moore, D. T., Pruthi, R. S., et al. (2010). Molecular stratification of clear cell renal cell carcinoma by consensus clustering reveals distinct subtypes and survival patterns. *Genes & Cancer, 1*(2), 152−163, [PubMed: 20871783].

Bray, F., Ferlay, J., Soerjomataram, I., Siegel, R. L., Torre, L. A., & Jemal, A. (2018). Global cancer statistics 2018: GLOBOCAN estimates of incidence and mortality worldwide for 36 cancers in 185 countries. *CA: A Cancer Journal for Clinicians, 68*(6), 394−424, [PubMed: 30207593].

Butte, M. J., Keir, M. E., Phamduy, T. B., Sharpe, A. H., & Freeman, G. J. (2007). Programmed death-1 ligand 1 interacts specifically with the B7-1 costimulatory molecule to inhibit T cell responses. *Immunity, 27,* 111−122.

Cancer Genome Atlas Research Network. (2013). Comprehensive molecular characterization of clear cell renal cell carcinoma. *Nature, 499*(7456), 43−49.

Chipollini, J., Henriques da Costa, W., Werneck da Cunha, I., de Almeida e Paula, F., Salles, P. G. O., Azizi, M., et al. (2019). Prognostic value of PD-L1 expression for surgically treated localized renal cell carcinoma: Implications for risk stratification and adjuvant therapies. *Therapeutic Advances in Urology, 11,* 1−7.

Choueiri, T. K., Figueroa, D. J., Fay, A. P., Signoretti, S., Liu, Y., Gagnon, R., et al. (2015). Correlation of PD-L1 tumor expression and treatment outcomes in patients with renal cell carcinoma receiving sunitinib or pazopanib: Results from COMPARZ, a randomized controlled trial. *Clinical Cancer Research: An Official Journal of the American Association for Cancer Research, 21*(5), 1072−1077.

Choueiri, T. K., Heng, D. Y. C., Lee, J. L., Cancel, M., Verheijen, R. B., Mellemgaard, A., et al. (2020). Efficacy of savolitinib vs sunitinib in patients with MET-driven papillary renal cell carcinoma. *JAMA Oncology, 6*(8), 1247−1255.

Choueiri T.K., Plimack E.R., Bauer T.M., et al. (2020). Phase I/II study of the oral HIF-2a inhibitor MK-6482 in patients with advanced clear cell renal cell carcinoma. In *Proceedings of the 2020 genitourinary cancers symposium.* February 15, 2020. Abstract 611.

Choueiri, T. K., Vaishampayan, U., Rosenberg, J. E., et al. (2013). Phase II and biomarker study of the dual MET/VEGFR2 inhibitor foretinib in patients with papillary renal cell carcinoma. *Journal of Clinical Oncology: An Official Journal of the American Society of Clinical Oncology, 31,* 181−186.

da Costa, W. H., Rezende, M., Carneiro, F. C., Rocha, R. M., da Cunha, I. W., Carraro, D. M., . . . de Cassio Zequi, S. (2014). Polybromo-1 (PBRM1), a SWI/SNF complex subunit is a prognostic marker in clear cell renal cell carcinoma. *BJU International, 113,* E157−E163.

Davis, C. F., Ricketts, C., Wang, M., Yang, L., Cherniack, A. D., Shen, H., et al. (2014). The somatic genomic landscape of chromophobe renal cell carcinoma. *Cancer Cell, 26*(3), 319−330.

Eble, J. N., Sauter, G., & Epstein, J. (2004). *Tumours of the genitourinary and male genital organs.* Lyon: IARC Press, WHO Classification of Tumours.

Escudier, B., Tannir, N. M., McDermott, D. F., et al. (2017). CheckMate 214: Efficacy and safety of nivolumab + ipilimumab (N + I) v sunitinib (S) for treatment-naïve advanced or metastatic renal cell carcinoma (mRCC), including IMDC risk and PD-L1 expression subgroups. *Annals of Oncology: Official Journal of the European Society for Medical Oncology / ESMO, 28*(Suppl. 5). (abstr LBA5).

Gabrilovich, D. I., & Nagaraj, S. (2009). Myeloid-derived suppressor cells as regulators of the immune system. *Nature Reviews. Immunology, 9,* 162–174.

Haake, S. M., Weyandt, J. D., & Kimryn Rathmell, W. (2016). Insights into the genetic basis of the renal cell carcinomas from The Cancer Genome Atlas (TCGA). *Molecular Cancer Research, 14*(7), 589–598.

Heng, D. Y. C., Xie, W., Regan, M. M., Warren, M. A., Golshayan, A. R., et al. (2009). Prognostic factors for overall survival in patients with metastatic renal cell carcinoma treated with vascular endothelial growth factor-targeted agents: Results from a large, multicenter study. *Journal of Clinical Oncology: Official Journal of the American Society of Clinical Oncology, 27,* 5794–5799.

Hsieh, J. J., Le, V., Cao, D., Cheng, E. H., & Creighton, C. J. (2018). Genomic classifications of renal cell carcinoma: A critical step towards the future application of personalized kidney cancer care with pan-omics precision. *The Journal of Pathology, 244,* 525–537.

Hu, C. J., Wang, L. Y., Chodosh, L. A., Keith, B., & Simon, M. C. (2003). Differential roles of hypoxia-inducible factor 1alpha (HIF-1alpha) and HIF-2alpha in hypoxic gene regulation. *Molecular and Cellular Biology, 23*(24), 9361.

Joseph, R. W., Kapur, P., Serie, D. J., Parasramka, M., Ho, T. H., Cheville, J. C., ... Brugarolas, J. (2016). Clear cell renal cell carcinoma subtypes identified by BAP1 and PBRM1 expression. *The Journal of Urology, 195,* 180–187.

Kwiatkowski, D. J., Choueiri, T. K., Fay, A. P., et al. (2016). Mutations in TSC1, TSC2, and MTOR are associated with response to rapalogs in patients with metastatic renal cell carcinoma. *Clinical Cancer Research: An Official Journal of the American Association for Cancer Research, 22,* 2445–2452.

McDermott, D. F., Huseni, M. A., Atkins, M. B., Motzer, R. J., Rini, B. I., Escudier, B., et al. (2018). Clinical activity and molecular correlates of response to atezolizumab alone or in combination with bevacizumab vs sunitinib in renal cell carcinoma. *Nature Medicine, 24*(6), 749–757.

M. Meylan, Beuselinck B., Dalban C., et al. (2020). Kidney ccRCC immune classification (KIC) enhances the predictive value of T effector (Teff) and angiogenesis (Angio) signatures in response to nivolumab (N) (Vol. 31, Supplement 4, S553). In Proceedings of the *ESMO virtual congress,* September 19–21, 2020.

Miao, D., Margolis, C. A., Gao, W., et al. (2018). Genomic correlates of response to immune checkpoint therapies in clear cell renal cell carcinoma. *Science (New York, N.Y.), 59,* 801–806.

Moch, H., Humphrey, P. A., Ulbright, T. M., & Reuter, V. E. (2016). *WHO classification of tumours of the urinary system and male genital organs* (4th ed.). Lyon: IARC Press.

Motzer, R. J., Bacik, J., Murphy, B. A., Russo, P., & Mazumdar, M. (2002). *Journal of Clinical Oncology: Official Journal of the American Society of Clinical Oncology, 20*(1), 289–296.

Motzer, R. J., Escudier, B., McDermott, D. F., et al. (2015). CheckMate 025 Investigators. Nivolumab vs everolimus in advanced renal-cell carcinoma. *New England Journal of Medicine, 373,* 1803–1813.

Rini, B. I., Powles, T., Atkins, M. B., Escudier, B., McDermott, D. F., Suarez, C., et al. (2019). Atezolizumab plus bevacizumab vs sunitinib in patients with previously untreated metastatic renal cell carcinoma (IMmotion151): A multicentre, open-label, phase 3, randomised controlled trial. *Lancet, 393*, 2404–2415.

Solomon, J. P., & Hansel, D. E. (2015). Morphologic and molecular characteristics of bladder cancer. *Surgical Pathology Clinics, 8*(4), 663–676, [PubMed: 26612220].

Srinivasan, R., Donskov, F., Iliopoulos, O., et al. (2020). Phase II study of the oral HIF-2α inhibitor MK-6482 for Von Hippel-Lindau (VHL) disease-associated clear cell renal cell carcinoma (ccRCC): Update on RCC and non-RCC disease. *Annals of Oncology, 31*(Suppl. 4), S1142–S1215. Available from https://doi.org/10.1016/annonc/annonc325.

Sufan, R. I., Jewett, M. A. S., & Ohh, M. (2004). The role of von Hippel-Lindau tumor suppressor protein and hypoxia in renal clear cell carcinoma. *American Journal of Physiology. Renal Physiology, 287*, F1–F6.

The Cancer Genome Atlas Research Network. (2016). Comprehensive molecular characterization of papillary renal-cell carcinoma. *New England Journal of Medicine, 374*, 135–145.

Vano Y., Elaidi R., Bennamoun M., et al. (2020). Results from the phase 2 BIOmarker driven trial with nivolumab (N) and ipilimumab or VEGFR tyrosine kinase inhibitor (TKI) in naïve metastatic kidney cancer (m-ccRCC) patients (pts): The BIONIKK trial (NCT02960906). In *Proceedings of the ESMO virtual congress*, September 19–21, 2020. Abstract: LBA25.

Voss, M. H., Reising, A., Cheng, Y., Patel, P., Marker, M., Kuo, F., ... Hakimi, A. A. (2018). Genomically annotated risk model for advanced renal-cell carcinoma: A retrospective cohort study. *The Lancet Oncology, 19*, 1688–1698.

Chapter 18

Head and neck squamous cell carcinoma

Giovana R. Thomas[1], Jennifer H. Gross[2] and Vanessa C. Stubbs[3]

[1]*Division of Head & Neck Oncologic and Robotic Surgery, Department of Otolaryngology-Head & Neck Surgery, University of Miami Miller School of Medicine, Miami, FL, United States,* [2]*Department of Otolaryngology-Head & Neck Surgery, Emory University School of Medicine, Atlanta, GA, United States,* [3]*Department of Otolaryngology Head & Neck Surgery, Rutgers-Robert Wood Johnson School of Medicine, New Brunswick, NJ, United States*

Predisposition

The predominant environmental risk factors for developing head and neck squamous cell carcinoma (HNSCC) are the use of alcohol and tobacco, immunosuppression, and exposure to high-risk human papilloma virus (HPV) or Epstein–Barr virus (EBV). However, not all smokers and drinkers, and not all those exposed to HPV or EBV, develop cancers. In fact, only 10%–15% of smokers develop lung cancer, and an even smaller proportion is diagnosed with HNSCC (Ho, Wei, & Sturgis, 2007). Therefore HNSCC appears to be multifactorial, and genetic predisposition and environmental factors may play an equally important role in tumorigenesis (Lacko et al., 2014).

Processed tobacco contains at least 30 known carcinogens and cigarette smoke contains over 70 known carcinogens and procarcinogens (Jethwa & Khariwala, 2017). Electronic cigarettes have gained increasing popularity since their U.S. Food and Drug Administration (FDA) approval in 2004. The inhaled vapor contains a mixture of nicotine and other chemicals whose impact on health and carcinogenic properties are not yet known. Tobacco and alcohol are clearly implicated in the development of multiple types of malignancies, including HNSCC. Additionally, the synergistic effect of tobacco and alcohol on development of HNSCC is well established (Hashibe et al., 2009). However, the interplay of genetic predisposition, carcinogen metabolism and excretion, and immune competency modifies an individual's response to carcinogen exposure, and consequently the potential of the exposure to incite the development of HNSCC (Ho et al., 2007; Jethwa & Khariwala, 2017).

Genomic and Precision Medicine. DOI: https://doi.org/10.1016/B978-0-12-800684-9.00018-6

Molecular defenses against these (pro) carcinogens and their downstream effects include biotransformation, detoxification, DNA repair, and apoptosis. Genetic polymorphisms in the above listed defenses alter an individual's risk for carcinogenesis and development of HNSCC (Lacko et al., 2014).

The most significant, and the most heavily studied, of the (pro) carcinogens contained in combustible and smokeless forms of tobacco are polycyclic aromatic hydrocarbons (PAHs) and tobacco-specific nitrosamines (TSNAs). Both exert their mutagenic effect by either directly or indirectly (through downstream effects) binding to DNA to form DNA adducts (Pratt et al., 2011). The disrupted DNA structure then leads to miscoding followed by permanent mutations that may activate oncogenes or inactivate tumor suppressor genes (Jethwa & Khariwala, 2017). The repair of TSNA−DNA adducts can also lead to the production of reactive oxygen species, which in turn can lead to more DNA damage (Yalcin & de la Monte, 2016). Oral administration and inhalation of PAH have led to tumor development in every subsite of the upper aerodigestive tract (UADT) [International Agency for Research on Cancer (IARC Working Group on the Evaluation of Carcinogenic Risks to Humans, 2010)]. Smokers with HNSCC have elevated urinary levels of PAH and TSNA in a matched case−control study (Khariwala et al., 2013). In two separate Chinese case−control studies, urinary levels of TSNA were directly correlated in a dose-dependent manner with risk of lung and esophageal cancer (Yuan, Knezevich, et al., 2011; Yuan, Koh, et al., 2009). Multiple TSNAs have been shown to reproducibly induce the growth of head and neck tumors in rats (Balbo et al., 2013; Hecht, 1998).

An individual's capacity for DNA repair and carcinogen metabolism is intimately associated with susceptibility to the development of HNSCC (Hecht, 2003). The DNA repair capacity to remove and repair PAH−DNA adducts is shown to be an independent biomarker for the risk of developing smoking-related HNSCC (Cheng et al., 1998; Wang et al., 2010). Mutagen sensitivity, or an individual's susceptibility to free radical-induced chromosomal damage, is also relevant to the development of HNSCC (Schantz et al., 2000; Schantz, Zhang, Spitz, Sun, & Hsu, 1997). Studies by Schantz et al. have found that mutagen hypersensitivity was strongly associated with increased risk of both development of HNSCC and severity of disease.

Dietary intake of antioxidants, specifically vitamin C, may have a protective effect against free radical-induced chromosomal damage. Pooled data from 10 case−control studies (Edefonti et al., 2015) and a large cohort study with >100,000 participants (de Munter, Maasland, van den Brandt, Kremer, & Schouten, 2015) both show an inverse relationship between head and neck cancer and vitamin C intake.

The role of genomic instability on the development of HNSCC has been further elucidated by studying patients with Fanconi anemia, a rare autosomal recessive disorder characterized by the accumulation of DNA damage and defective DNA repair (Moldovan & D'Andrea, 2009). A study evaluating patients from the International Fanconi Anemia Registry revealed a 14% cumulative incidence of developing HNSCC by age 40, with a median age of onset of 31 years old (Kutler et al., 2003). Compared to the general population, the overall risk of developing HNSCC is 500−700 times greater in these patients. Their tumors are also more likely to develop within the oral cavity and in patients without any history of alcohol or tobacco use (Velleuer & Dietrich, 2014).

Although the incidence of laryngeal, oral cavity, and hypopharyngeal cancers has continued to decline since the 1980s, coinciding with the decline of active smokers, the incidence of oropharyngeal carcinomas continues to rise, particularly among individuals <45 years old (Pai & Westra, 2009). The increasing incidence of oropharyngeal cancer is directly attributed to HPV, with 70% of oropharyngeal carcinomas now showing HPV positivity (Saraiya et al., 2015). A landmark study by Ang et al. demonstrated that these tumors, when compared to HPV-negative oropharynx tumors, tend to occur in nondrinkers, nonsmokers, and those with a higher number of sexual partners (Ang et al., 2010). Patients with HPV-positive disease also have a significant survival advantage over those with HPV-negative disease, with a 3-year overall survival of 82% versus 57%. The increased understanding of HPV-related oropharyngeal cancer led to an amendment of the tumor-node-metastasis (TNM) staging system for HNSCC in the 8th American Joint Committee on Cancer (AJCC) system with HPV status becoming one of the key criteria to determine stage. For HPV-related oropharyngeal tumors with favorable prognosis, multiple de-escalation treatment studies are the current topic of interest and studies are underway.

EBV was the first human virus to be linked to carcinogenesis and is a causative agent in certain types of nasopharyngeal cancers. EBV has infected almost 90% of adults, generally as children who present with mild pharyngitis, which then becomes latent. EBV is expressed in 100% of undifferentiated nasopharyngeal cancer, 35% of nasopharyngeal squamous cell carcinomas (SCCs), and >50% of oral SCC (Goldenberg et al., 2001). Its role in the pathogenesis and behavior of HNSCC is still poorly understood.

Screening

The current standard of care for screening for mucosal HNSCC relies on physical exam, including evaluation of the oral cavity, oropharynx, and larynx, followed by tissue biopsy of any concerning lesions. Except perhaps for HPV-related oropharynx cancer, a multistep carcinogenesis process results in epigenetic and metabolic changes that give rise to histologically distinct precursor phenotypes that harbor-specific genetic alterations. This process is illustrated through the transition of normal healthy mucosa of the UADT first to the reversible premalignant stage of dysplasia, then to carcinoma in situ and finally to invasive carcinoma. Lesions of the UADT, however, can have different clinical presentations including leukoplakia (white patch) and erythroplakia (red patch). Leukoplakia may progress to invasive SCC in up to 18% of cases, whereas erythroplakia has a much greater potential for malignancy. Approximately 90% of lesions presenting as erythroplakia may demonstrate severe dysplasia, carcinoma in situ, or invasive SCC. The transformation rate of dysplasia to cancer has been reported as high as 36% in erythroplakia.

Improvements in overall survival in patients with HNSCC rest on early identification of premalignant lesions and intervention in patients at risk before the development of advanced-stage disease. Because the standard of screening by oral examination and subsequent tissue biopsy has only 64% sensitivity for oral cancer and 31% specificity for oral dysplasia or cancer, several clinical trials have been conducted proposing alternate screening methods utilizing biomarkers for early cancer detection (Brocklehurst et al., 2010; Epstein, Güneri, Boyacioglu, & Abt, 2012).

Serum molecular tests based on hypermethylation, RNA, and protein-based panels have been proposed. One study looked at patterns of CpG island hypermethylation within promoter regions of tumor suppressor genes. DNA methylation is thought to play a role in the epigenetic pathway of transcriptional inactivation of these tumor suppressor genes. Quantitative methylation-specific PCR (Q-MSP) was used to assess an expanded panel of CpG-rich promoters known to be differentially hypermethylated in HNSCC. After evaluating 21 separate genes in both patients with HNSCC and normal controls, they found that 100% of HNSCC cases showed methylation in tumor DNA for at least one of the study genes, indicating a significant role in promoter gene methylation as an epigenetic alteration in HNSCC. However, they also found some promoter methylation in normal controls that was associated with age, race, and tobacco and alcohol exposure. Though one panel of genes (CDH1, CCND2, TIMP3, HIC1, and PGP9.5) showed a sensitivity of 87.2%, the specificity was only 42.3% (Carvalho et al., 2008). While the area shows some promise for a serum-screening test, further study is needed for refinement before use.

Simultaneous testing of various cytokines, growth factors, and tumor antigens, or a serum multiplex panel of biomarkers, was also investigated for correlation with HNSCC disease status. A multimarker panel demonstrated 84.5% sensitivity with 98% specificity, classifying 92% of patients correctly. This panel comprises 25 biomarkers, including epidermal growth factor (EGF), epidermal growth factor receptor (EGFR), interleukin 8 (IL-8), granulocyte colony-stimulating factor, alpha-fetoprotein, matrix metalloproteinase-2, interferon gamma, and soluble vascular cell adhesion molecule among others (Linkov et al., 2007). Thus testing of serum for multiplex biomarkers appears to be another avenue for further study.

Salivary biomarkers have also been proposed as a simple, inexpensive, and painless screening mechanism. Soluble CD44 and total protein levels in saliva have been identified as one such biomarker. CD44 is a cell surface trans-membrane glycoprotein involved in cell proliferation, cell migration, and tumor initiation and is overexpressed in premalignant lesions (Ioachim et al., 1999). Soluble CD44 was shown to correlate with tissue CD44 staining and was also found in higher levels in saliva of HNSCC patients (Cohen et al., 2020; Franzmann et al., 2005). Furthermore, in a study of soluble CD44 and total protein levels in oral rinses of 150 HNSCC patients and 150 frequency-matched controls, CD44 level ≥ 5.33 ng/mL was highly associated in HNSCC patients versus reference group CD44 level <2.22 ng/mL. Total protein levels, thought elevated due to leakage from highly vascular tumors into the saliva, in combination with soluble CD44 levels provided a sensitivity of 80% and specificity of 48.7% (Pereira et al., 2016).

Circulating tumor DNA (ctDNA) in saliva has also been studied. When HNSCC patients were enrolled at the early stages (I and II) and late stages (III and IV), ctDNA, including both somatic tumor mutations and human papilloma virus genes, was detected in 100% of HNSCC patients enrolled at the early stages and in 95% enrolled at the late stages. Furthermore, ctDNA in saliva was found postsurgically in three patients before clinical diagnosis of recurrence and was not detected in patients without recurrence (Wang et al., 2015). No testing has yet been published with detection in normal healthy controls.

Salivary transcriptome profiling and microRNA have also recently shown promise in detecting HNSCC biomarkers on the RNA level. In microarray analysis, 1679 genes exhibited significantly different expression level in saliva of oral SCC patients compared to controls. Subsequent quantitative PCR analysis of mRNA identified seven cancer-related mRNA biomarkers [transcripts of IL8, interleukin-1,β (IL1B), dual specificity protein phosphatase 1, H3 histone, family 3A, ornithine decarboxylase antizyme 1 (OAZ1), S100 calcium binding protein P, and spermidine/spermine N1-acetyltransferase (SAT)] that were at least 3.5 times more elevated in cancer saliva. A panel including IL1B, OAZ1, SAT, and IL8 detected oral SCC with 91% sensitivity and 91% specificity (Li et al., 2004). Another study evaluated 50 microRNAs found in both whole and supernatant saliva samples of healthy and oral SCC patients and found miR-125a and miR-200a to be present at significantly lower levels in oral SCC than in control patients (Park et al., 2009). While many salivary biomarkers have been identified, further large studies are needed to enable widespread clinical use for screening of HNSCC and current standard screening remains physical examination.

Unlike HNSCC of the oral cavity, oropharynx, larynx, and hypopharynx, nasopharyngeal carcinoma is a separate entity and has been associated with the specific tumor marker of circulating cancer-derived EBV DNA in plasma. The pathogenesis of many nasopharyngeal carcinomas is highly associated with EBV and as such, the tumor marker of plasma EBV DNA has been established with a sensitivity of 96% and a specificity of 93% for identification of the disease (Lo et al., 1999). Furthermore, the utility of the marker as a screening agent for early asymptomatic nasopharygngeal carcinoma was assessed in a prospective trial of 20,174 asymptomatic participants. Those with persistently positive plasma EBV DNA levels (two positive tests) were subsequently evaluated by nasal endoscopic examination and magnetic resonance imaging (MRI). Those patients with nasopharyngeal carcinoma identified by screening were much more likely to be diagnosed with early stage disease (71%) compared to a historical cohort (20%) and had superior 3-year progression-free survival (PFS) (Chan et al., 2017). This illustrates the importance of establishing a reliable tumor marker for screening of head and neck cancers.

Diagnosis

Survival of HNSCC depends largely on stage at initiation of treatment, and a comprehensive head and neck physical examination is crucial for early detection of malignancy. However, the diagnosis of HNSCC is frequently delayed because noticeable symptoms, such as pain, dysphagia, and shortness of breath, occur in more advanced stages of disease. Laryngeal SCC is more likely to be detected at earlier stages due to the presence of persistent hoarseness, which can occur with mild alterations of the true vocal cord vibratory surfaces.

Radiologic imaging modalities such as ultrasound, computed tomography (CT), MRI, positron-emission tomography (PET), and combined PET/CT are critical tools that provide information on extent of tissue invasion, involvement of regional lymph nodes, and presence of distant metastatic disease. This information is critical for staging and subsequent treatment planning.

Exposure to carcinogens such as tobacco and alcohol may result in premalignant epithelial changes over a wide surface area of epithelium within the UADT. This clinical phenomenon is referred to as "field cancerization," and it may lead to a local recurrence or second primary tumor within the contiguous field of premalignant epithelial cells (Braakhuis, Tabor, Kummer, Leemans, & Brakenhoff, 2003). For this reason, it is imperative to evaluate mucosa of the UADT, including the esophagus and trachea, with directed biopsies during endoscopic examination.

However, there remains significant variability in outcomes for patients within the same TNM stage classification. Therefore attempting to identify high-risk patients through molecular markers is an active area of investigation. The federally funded The Cancer Genome Atlas (TCGA) began in 2006 and has generated the full genomic sequencing of >500 HNSCC tumors, along with reports of the patients' outcomes. These data are available to the public and provide unprecedented access to the genetic mutations found in these tumors. A paper published by the TCGA in 2015 detailed the profiles of tumor and healthy tissue from 279 patients with HNSCC, 80% of tumors were associated with tobacco use and 13% were HPV-positive. All 279 tumor samples showed 15 significantly mutated genes that included *CDKN2A*, *TP53*, *PIK3CA*, *NOTCH1*, *HRAS*, and *NEE2L2*. Among these, *PIK3CA* was mutated in ~21% of all samples. The studies also revealed that mutational burden and chromosomal alterations are higher in HPV-negative HNSCC compared to HPV-positive or related tumors. Additionally, each tumor genome had an average of 140 mutated genes (The Cancer Genome Atlas Network, 2015).

Molecular profiling of primary tumors from HNSCC remains an ongoing investigation for developing a "genetic fingerprint signature" with the potential for predicting the presence of lymph node metastasis at the time of diagnosis. By identifying patients at risk for cervical lymph node metastasis and extracapsular spread without the need for surgical node dissection, tissue or serum biomarkers could play a vital role in clinical decision-making. Resistance to anoikis, or programmed cell death after detachment from the extracellular matrix is a critical step in malignant cells' ability to metastasize (Paoli, Giannoni, & Chiarugi, 2013). Resistance to programmed cell death can be achieved by malignant cell integrin switch, epithelial-mesenchymal transition (losing cell−cell adhesion and gaining migratory and invasive properties), activating antiapoptotic signaling, and posttranslational gene regulation. DNA microarray gene expression of primary tumors of the oral cavity and oropharynx found that signature or predictor gene sets can detect local lymph node metastases using material from primary HNSCC with better performance than current clinical diagnosis (Belbin et al., 2005; O'Donnell et al., 2005; Roepman, Wessels, & Kettelarij, 2005). Other investigators have found that the gene expression profile of 53 genes with roles in cell differentiation, adhesion, signal transduction, and transcription regulation are associated with depth of invasion in patients with oral SCC (Toruner et al., 2004). Multiple genes aberrantly expressed to create metalloproteinases that are involved in the regulation of cell adhesion and cell-cycle-related proteins or act as tumor suppressors in OCSCC (Kornberg et al., 2005).

There are multiple studies in the literature utilizing microarray analysis to show genetic changes in oral SCC. However, each study uses different gene expression arrays and platforms, rendering direct comparison of data impossible. Also, no study tests the ability of the utilized genetic array(s) to predict the progression of a premalignant lesion to oral SCC against an independent validation dataset. Incorporation of "omics" technologies remains an active field of investigation (Nagaraj, 2009). Biomarker discovery, validation, and integration into clinical practice may eventuate early-stage disease detection and lead to improved outcomes.

Targeted biologic therapy has gained significant ground over the past decade, especially in the field of immunotherapy and immune checkpoint inhibitors. Programmed death receptor 1 (PD1) is one of the most widely studied immune checkpoint proteins. It is expressed by all active T-cells, and when bound with one of its ligands (PD-L1 and PD-L2), it leads to T-cell deactivation. This PD1 complex is overexpressed in many head and neck cancers (HNC) and has been directly linked with head and neck tumor immune invasion (Lepikhova et al., 2018). Testing recurrent or metastatic HNSCC patient's tumors for PD-L1 expression help inform more targeted therapy and determine which individuals may benefit from the addition of a PD1 inhibitor.

Prognosis

TNM staging currently remains the most important factor correlating with prognosis of HNSCC. Large primary tumor size, positive margins after surgical excision, perineural invasion, and the presence of lymph node metastasis have been reliable indicators of poor clinical outcome in patients with HNSCC. However, the single most important factor that determines survival is the metastatic status of the cervical lymph nodes at the time of diagnosis. Particularly, the presence of extracapsular spread in cervical lymph node metastasis remains the most significant clinical prognostic indicator of survival, local-regional recurrence, and distant metastasis in patients with HNSCC. Although these clinical prognostic parameters provide the best possible criteria for deciding adjuvant treatment modalities, they are limited in discerning future behavior of aggressive HNSCC.

The search for novel molecular prognostic markers with potentially significant predictive value for biological aggressiveness of HNSCC has exploded in the recent years. Better prediction of the risk of developing distant metastases would help introduce a more selective treatment approach, according to the biological aggressiveness of the tumor. In the era of biomarker-driven personalized cancer therapies, several molecular mediators of tumor progression, invasion, and metastasis that function in growth factor signaling, metastasis, and suppressor genes have been well investigated in HNSCC, and are described later in this chapter.

TP53 is a tumor suppressor gene located on 17p13. It consists of 11 exons that encode a protein, p53, and functions in carcinogenesis by initiating G1 arrest in response to certain DNA damage and apoptosis. The prevalence of *TP53* mutations is reportedly >50% in HNSCC. A number of studies have shown that *TP53* gene mutations are associated with increased risk for locoregional recurrence and poor outcome (van Ginkel, de Leng, de Bree, van Es, & Willems, 2016; Zhou, Liu, & Myers, 2016). Mutated p53 protein overexpression is also associated with reduced therapeutic responsiveness, tumor recurrence, poor overall survival rate, increased rates of locoregional failure, and decreased disease-free survival in HNSCC. These mutations confer varying degrees of dysregulation based upon where the mutation occurs at the chromosomal level, resulting in a variegation of clinical impact on the p53 protein structure, stability and DNA binding properties (Pai & Westra, 2009).

The EGFR is a transmembrane tyrosine kinase capable of promoting neo-plastic transformation. The downstream signaling events upon ligand binding include activation of tyrosine kinase and activation of intracellular Ras, Raf, and mitogen-activated protein kinase cascades. EGFR expression has been extensively studied in HNSCC, and its overexpression has been reported in more than 90% of HNSCCs (Pai & Westra, 2009). This marker is significantly associated with short disease-free survival and overall survival and poor prognosis in patients with HNSCC. However, only a small subset of these HNSCCs overexpressing EGFR actually demonstrates amplified copy numbers or mutational activation of the EGFR gene. Autocrine and paracrine loops take effect via high expression of the EGFR with various ligands, and binding of these ligands results in autophosphorylation of the intracellular kinase domain and subsequent activation of multiple oncogenic pathways. This scenario explains the only modest success of EGFR blockade as monotherapy in the treatment of patients with HNSCC (Boeckx et al., 2013; Bossi et al., 2016).

HPV, particularly type 16, is found in more than 70% of oropharyngeal HNSCCs in the United States, and this form of HNSCC behaves distinctly different than its smoking- and alcohol-related counterpart (D'Souza & Dempsey, 2011). HPV-positive HNSCCs express the viral oncoproteins E6 and E7, overexpress the p16 gene, and infrequently harbor p53 gene mutations. These HPV-positive tumors are associated with increased radiosensitivity and improved prognosis. In fact, Ang et al. found that those with HPV-positive oropharyngeal cancers had increased 3-year overall survival of 82% versus. 57% in HPV-negative oropharyngeal patients (Ang et al., 2010). This behavior is postulated to occur as a result of immune surveillance for viral-specific tumor antigens, an intact apoptotic response to radiation, and the absence of the widespread genetic alterations associated with smoking (Pai & Westra, 2009). In 2006 the HPV vaccine Gardasil was approved by the FDA as an effective means of preventing cervical cancer and precancerous lesions due to HPV types 6, 11, 16, and 18 (Pai & Westra, 2009). On March 3, 2019, the Gardasil vaccine was FDA approved for both men and women from the ages of 9−47 years old, expanding the use of the vaccine for HPV-related cancers. Though not specifically indicated for prevention of head and neck cancer, the increased knowledge of HPV and its relation to head and neck cancer has led to studies seeking to elucidate the effect of the vaccine on HNSCC prevention (Timbang et al., 2019).

Chromosomal changes have also been linked to differential outcomes in HNSCC. Rearrangements of chromosome 11q13 have been associated with poorer survival in HNSCC overall (Akervall et al., 1995). Furthermore, chromosomal gain of region 11q22.1−q22.2 and losses of 17p13.3 and 11q23−q25 were found to be associated with loco-regional recurrence in oral SCC specifically (Ambatipudi et al., 2011). Comparative genomic hybridization was also used in oral SCC to identify the correlation of 7p gain and 8p loss with node-positive oral SCC as well as 11q13 gain with high-grade oral SCC. 11q13 gain and 18q loss together were also found to be a strong bivariate predictor of poor prognosis (Noorlag et al., 2015; Pathare et al., 2011).

With the rise of systems analysis and -omics technologies, there has also been an increase in tumor data availability. For example, The Cancer Genome Atlas (TCGA) was initiated by the National Cancer Institute and has made the clinical and biological characteristics of over 500 HNSCC tumors available for data access. Through transcriptomic analysis, Saidak et al. were able to use this system to determine that three genes (*CCDC66*, *ZRANB2*, and *VCPKMT*) were all differentially expressed with significantly higher mRNA levels in tongue tumors with positive surgical margins compared to tumors with negative surgical margins (Saidak et al., 2019). As positive surgical margins have a strong negative prognostic impact, future systemic studies like these may offer insight into prognosis in HNSCC.

In larger studies of patients with HNSCC, DNA microarray analysis has been used to identify distinct gene expression signatures associated with clinical outcome in HNSCC. Distinct subtypes of HNSCC based on gene expression patterns obtained from tumor samples from patients with HNSCC were described. These subtypes had significant differences in clinical outcomes, including recurrence-free survival and overall survival, and patterns of expression were identified that could predict the presence of lymph node metastases in HNSCC tumors (Chung et al., 2004). Biomarker investigation remains an active area of ongoing research in the realm of screening, diagnosis, and prognosis.

Pharmacogenomics

While local HNSCCs are often treated with surgery, radiotherapy, and/or chemoradiotherapy, cancer recurrence is still frequent, and emergence of distant metastases portends significantly decreased survival. Systemic treatment for advanced HNSCC primarily consists of chemotherapeutic drugs, using mainly platinum salts and taxanes, but these drugs have only moderate efficacy (Lepikhova et al., 2018). Currently, the only approved targeted drugs for advanced HNSCC are cetuximab, a monoclonal antibody that binds to the EGFR, and, in 2016, the anti-PD1 antibodies nivolumab and pembrolizumab.

Immunotherapy options for treatment of HNSCC have recently been studied and made available for use. These have focused on inhibitors of the programmed death pathway including the PD1 and its ligand PD-L1. It has been found that overexpression of PD-L1 inhibits tumor-directed T-cell cytotoxicity and permits immune evasion and tumor growth (Moy, Moskovitz, & Ferris, 2017). Furthermore, blockade with anti-PD-(L/)1 antibodies promotes immune-mediated tumor rejection and destruction (Iwai et al., 2002). In

2016 FDA approved the use of anti-PD1 immune checkpoint inhibitors of nivolomab and pembrolizumab for recurrent HNSCC refractory to platinum-based therapy and in 2019, as first-line therapy for metastatic or unresectable, recurrent HNSCC. In a 2-year update to the nivolumab trial for recurrent and refractory HNSCC, it was found that nivolumab nearly tripled the estimated 24-month overall survival at 16.9% versus the investigators choice therapy with overall survival rate of 6.0% (Ferris et al., 2018). Interestingly, this improved survival was maintained regardless of PD-L1 expression level.

Despite remarkable improvements in radiation techniques and novel chemotherapeutic agents, however, overall survival for locally advanced HNSCC is still poor. Since there is significant inter- and intraindividual variability in clinical outcome, improvements in survival rates will likely require the identification of patients, prior to treatment, who are most likely to have chemoradiotherapeutic benefit and patients with the highest risk of suffering genotoxic side effects. Interindividual variability in treatment response frequently leads to treatment failure or treatment-related death.

The response to chemotherapy and radiation and their side effects in patients with HNSCC is dependent on several factors, such as site of primary tumor, disease stage, patient characteristics, and comorbidities. The interplay among these factors is still poorly understood. Genetic covariables may influence toxicity and tumor responses to chemotherapy and/or radiotherapy. Knowledge of interindividual pharmacokinetic variability and genetic profiling provides a novel scientific basis for an improved and individualized therapeutic approach.

Multidrug resistance (MDR) to chemotherapeutic agents and radiotherapy occurs in many types of tumors as well as in HNSCC, presents a major obstacle to the effectiveness of chemoradiotherapy, and subsequently leads to treatment failure. MDR is a process in which cells acquire simultaneous resistance to a group of drugs that appear to be unrelated structurally and functionally. The main mechanism that gives rise to the MDR phenotype in cancer is the overexpression of drug efflux transporters in the plasma membrane. ATP-binding cassette (ABC; ATP = adenosine triphosphate) transporters contribute to drug resistance via ATP-dependent drug efflux, extruding anticancer agents or their metabolites from cells. P-glycoprotein (Pgp), which is encoded by the MDR1 gene, confers resistance to certain anticancer agents. Very little information is known about the importance of MDR and Pgp expression in HNSCC. Although Pgp levels and mRNA have been noted in recurrent oral SCC and in oral mucosa with increasing severity of dysplasia, insignificant Pgp levels have been found in oral SCC cell lines (Pérez-Sayáns et al., 2010). Treatment of different cell lines with vincristine shows that Pgp can be induced by genetic induction of *MDR1* gene. Theile and colleagues studied HNSCC cell lines for drug transporter expression and susceptibility to cisplatin, paclitaxel and 5-FU, and found that cisplatin and paclitaxel resistances were inversely correlated. However, none of the cell lines expressed the well-established Pgp/ABCB1 drug transporter. Other ABC transporters not linked to MDR were induced. These findings questioned the significance of Pgp in MDR in HNSCC (Theile et al., 2011).

Multiple other mechanisms of MDR have also been described, which include increased detoxification of drugs by glutathione S-transferase (GST), downregulation of pharmacologic targets like DNA topoisomerase I and II, increased catabolism of drugs by MDR-associated proteins, inadequate anabolism of prodrugs, and altered regulation of acid pH in the tumor microenvironment. Decreasing GST activity has been shown to increase the cytotoxicity of doxorubicin in tongue cancer cells (Qin et al., 2013). Using the same cell line, Zhang et al. (2016) observed that cisplatin significantly increased the expression of xCT, a functional subnunit of the glutamate transporter, leading to an increase in glutathione levels and subsequent cisplatin resistance. Changes in cytosolic pH also play an important role in chemotherapy drug resistance. The more acidic extracellular pH in solid tumors interferes with the absorption of basic chemotherapy drugs, reducing their effect on tumor cells. Recent evidence suggests that vacuolar ATPases (V-ATPase) may play a role in acidification of the tumor microenvironment by secreting protons through the plasma membrane. Pretreatment with proton pump inhibitors has been found to sensitize tumor cell lines to the effects of different chemotherapy drugs (Pérez-Sayáns et al., 2010). Although pH regulation in oral SCC lines is mediated by vacuolar proton-pump ATPases, no V-ATPase inhibitors have been proven to be useful in oral SCC. Nevertheless, whether the above mechanisms play a significant role in MDR in HNSCC remains uncertain.

A clinical review from 2016 reported that EGFR targeting has synergistic effects with chemotherapy in HNSCC and reverses chemoresistance of epithelial tumors (Bossi et al., 2016). However, increasing evidence suggests that despite high rates of EGFR overexpression, most patients do not respond to cetuximab, a highly specific EGFR monoclonal antibody which binds the extracellular domain of EGFR and downregulates EGFR overexpression. In addition, those who show initial response ultimately become refractory to treatment, suggesting the development of acquired resistance (Alsahafi et al., 2019). Potential mechanisms of resistance to EGFR-targeted therapies involve *EGFR* and *RAS* mutations, mesenchymal-epithelial transition, and activation of alternative and downstream pathways.

Multiple studies have noted that resistance to cisplatin chemotherapy is significantly correlated with expression of mutant p53 (Bradford et al., 2003; Cabelguenne et al., 2000) and overexpression of antiapoptotic proteins Bcl-2 and Bcl-x_L (Bauer et al., 2005). Low expression of Bcl-x_L in tumor specimens from patients with HNSCC is correlated with response to induction chemotherapy (Andrews et al., 2004; Bradford et al., 2003) reported that induction of mutant p53 in HNSCC lines led to decreased expression of Bcl-2 and increased susceptibility to cisplatin-induced apoptosis, implicating

Bcl-2 in the deregulation of p53-induced apoptosis. Overexpression of mutant-type p53 in HNSCC is also associated with increased sensitivity to ionizing radiation (Hutchinson, Mierzwa, & D'Silva, 2020). Tumor cells with mutated p53 maintain a decreased ability to repair radiation-induced DNA damage, thereby accumulating genetic mutations that confer resistance. Complete loss-of-function p53 mutations disable cell-cycle arrest and apoptosis, leading to radiation failure and cell survival.

In the future, individualization of treatment for patients with advanced HNSCC by genetic profiling will require prospective, properly powered, randomized clinical trials. However, research for effective ways of overcoming MDR in HNSCC is still in its infancy.

Monitoring

Patients treated for HNSCC are followed clinically for evidence of recurrent disease, development of second primary lesions or distant metastasis. The chance of developing a second primary tumor has been estimated at 2%−3% per year in patients with HNSCC. In addition, 20%−30% of patients treated for HNSCC will develop recurrent disease at the primary site, and such recurrence is the most common cause of treatment failure. Because prognosis of late-stage recurrent disease is dismal, early detection is imperative. Distant metastatic disease occurs in 11%−15% of patients treated for HNSCC, and at this stage treatment is palliative. Identifying molecular markers in primary tumors that are associated with locoregional relapse may allow for early identification of patients needing additional surveillance and treatment, and may have the potential to decrease the probability of distant disease.

There was an initial anticipation that EGFR expression would serve as predictive biomarker for likelihood of response to cetuximab therapy, since 80%−100% of HNSCC show EGFR overexpression, and overexpression has been shown to correlate with decreased survival, resistance to radiation, local treatment failure, and increased distant metastasis (Byeon, Ku, & Yang, 2019). However, studies have shown that IHC-based assays measuring EGFR expression are not predictors for response to cetuximab therapy. In addition, no mutations in EGFR have been identified to date that are reliable predictors for antibody-based EGFR therapies.

Telomeres are specialized DNA structures located at the ends of chromosomes and are essential for stabilizing chromosomes by protecting them from end-to-end fusion and DNA degradation. Studies have shown that telomere aberrations are consistently found in HNSCC and in mucosa surrounding preoplastic areas and invasive upper aerodigestive tract carcinomas (Boscolo-Rizzo et al., 2016). Telomerase activation, which is detected in 90% of malignancies, give cancer cells the capability to replicate indefinitely and is associated with poor outcomes and disease aggressiveness. Therefore it could be potential marker of cancer risk and disease outcome. While telomerase activity is detectable in most tumors, it is usually absent in normal somatic cells. Multiple studies have concluded that SCC arises in telomere-shortened epithelial field characterized by genetic instability and a prone-to-transformation status. Telomere shortening, therefore, can be considered a biosensor for field cancerization that can identify patients at risk of local relapses or second primary tumors.

Gene expression signatures using DNA microarray technology have potential utility as biomarkers to predict patients at risk for locoregional recurrence. Several gene expression signatures from HNSCC tumors from various anatomical sites in the head and neck have been identified. However, it was not until a few years ago that new important studies on gene expression signatures in HNSCC emerged. These studies show an increase in the sample size and a more accurate selection of cases, the publication of a growing number of studies applying a computational integration (*meta*-analysis) of different microarray datasets addressing similar clinical/biological questions, the increased use of molecular subclassification of tumors according to their gene expression, and the release of the largest publicly available dataset in HNSCC by TCGA (Tonella, Giannoccaro, Alfieri, Canevari, & De Cecco, 2017).

Novel and emerging therapeutics

Surgery has historically been the primary treatment modality for malignant neoplasia of the head and neck until the Veterans Affairs larynx trial published in 1991. Since that study and many following, the combination of radiation therapy and chemotherapy administered concurrently has allowed organ preservation and treatment of locally advanced HNSCC with improved outcomes in most patients.

Surgery

The goal of surgery in HNSCC is complete resection with negative margins. A positive margin is an indisputable poor prognostic factor, and it significantly increases the likelihood of recurrence and decreased overall survival. Surgery combined with a regimen consisting of postoperative radiation or chemoradiotherapy is reserved for locally advanced disease. In the surgical realm, a few technological advances have dominated surgical treatment of patients with HNSCC including the emphasis on function-preserving surgical techniques, anatomical and functional reconstruction after ablative surgeries, the use of sentinel lymph node biopsy and minimally invasive transoral surgeries including robotic-assisted transoral surgery (TORS). TORS has made an impact on potentially improving quality of life in HPV-related HNSCC patients expected to have a prolonged survival after treatment (Bekeny & Ozer, 2016; Byrd & Ferris, 2016).

The ECOG-ACRIN Cancer Research Group designed and conducted the ECOG3311 Phase II clinical trial with funding from the National Cancer Institute, National Institutes of Health. The trial, conducted in patients undergoing TORS, tested reduced postoperative radiation therapy in patients with HPV-related oropharynx SCC at intermediate risk for recurrence. Specifically, patients at low risk were observed after transoral resection. Patients at intermediate risk were randomized to two arms of radiation alone, both at postoperative doses lower (50 or 60 Gy) than usual (60−66 Gy). At the time the trial opened in 2013, the optimal dose of radiation therapy for HPV-related oropharynx cancer patients was not defined. Patients at high risk were assigned to usual radiation therapy plus chemotherapy. The study reported that transoral resection of p16 + HPV-related oropharynx SCC is safe and results in good oncologic outcome, presenting a promising deintensification approach. For patients with low-risk disease, 2-year PFS is favorable without postoperative therapy. For those with uninvolved surgical margins, <5 involved nodes, and minimal (<1 mm) extranodal extension, reduced dose postoperative radiation therapy without chemotherapy appears sufficient. Transoral surgery plus 50 Gy should be compared to optimal nonsurgical therapy in a Phase III trial (Ferris et al., 2020).

Radiation therapy

Advances in computer-assisted radiological techniques over the past two decades such as intensity-modulated radiotherapy, functional MRI for radiotherapy planning, and volumetric modulated have revolutionized radiotherapy planning by allowing for better dose conformation to rumor target and sparing of surrounding normal tissues. However, despite these advances, radiation resistance is frequently encountered in HNSCC and is a complex process depending on many biological factors and cellular mechanisms regulated by intrinsic cell signaling network. Much effort is being placed on elucidation of radioresistance mechanisms as well as discovery of new prognostic and predictive biomarkers in response to ionizing radiation. Ionizing radiation generates highly reactive oxygen species, which results of DNA double-strand breaks, genomic instability, and cell death. Resistance to radiation, therefore, may results from any cellular mechanisms or tumor microenvironmental factors that interfere with these radiation-induced cell death pathways. The following have been associated with radioresistance in HNSCC: hypoxia-induced neovascularization, deregulation of TP53-associated intrinsic apoptosis, cancer stem cells, microRNAs, and alterations in the EGFR, PIK/AKT, and RAS pathways (Ahmad et al., 2017). MicroRNA are endogenous, evolutionary conserved, small noncoding RNAs 18−25 nucleotide in length, which regulate gene expression by binding to targeted messenger RNA. There is growing evidence supporting the role of microRNA in radioresistance of HNSCC either by their upregulation or downregulation (Ahmad et al., 2017).

Targeted therapies

There are several FDA-approved molecular targeting agents available to treat HNSCC. They include cetuximab (EGFR inhibitor), nivolumab, and pembrolizumab (mAbs that inhibit the interaction between PD1 and its ligand PD-L1 and PD-L2). However, the development of drug resistance—whether intrinsic or acquired—remains a significant challenge. Intrinsic resistance occurs when cancer cells are inherently insensitive to a treatment, while acquired resistance occurs when treated cells become insensitive after an initial period of treatment benefit. Better identification of patients who may benefit from specific treatments is likely to improve outcomes and decrease cost. Genetic heterogeneity in HNSCC provides a significant obstacle to identification of biomarkers and selection of appropriate treatment for this disease. Cetuximab in combination with chemotherapy or radiotherapy improves survival. However, the response rate as a single agent has been disappointingly low (Byeon et al., 2019). There is a need, therefore, for novel therapeutic agents that overcome drug resistance. The current EGFR-targeted drugs can be either monoclonal antibodies or tyrosine kinase inhibitors. In ongoing clinical trials, the monoclonal antibodies panitumumab, nimotuzumab, zalutumumab, and duligotuzumab are being investigated for their ability to provide a lower immunogenicity profile and therefore, decrease life-threatening hypersensitivity reactions (Lee, Johnson, & Grandis, 2018). So far, the benefits of these agents in improving overall survival have been disappointing in Phase II and III clinical trials (Ausoni et al., 2016). Tyrosine kinase inhibitors target the downstream signaling of EGFR tyrosine kinase domain, eventually blocking the proliferation of tumors cells. Agents such as lapatinib and afatinib are under development in Phase III clinical trials as second-line treatment in recurrent and or metastatic HNSCC. Multiple other targeted agents are in various stages of investigation with the purpose of overcoming EGFR resistance. These include an IGF-1 receptor targeted agent, VEGF/VEGFR angiogenesis inhibitors, SRC kinase inhibitors, PI3K/Akt/mTOR pathway inhibitors, gene therapy such as gendicine (SBN-1), and H-101 (Wen & Grandis, 2015).

Conclusion

Despite the recent advances in clinical cancer diagnosis and treatment, survival rates of patients with advanced HNSCC remain low. Genomic complexity, intra- and intertumoral genetic heterogeneity, and epigenetic alterations make the goal of precision cancer medicine much more onerous. However, present and future research contributions from TCGA—with identification of positive and negative predictive biomarkers—and the emergence of immunotherapy have the potential to mitigate treatment resistance, improve treatment response rates, and improve overall survival. To this end research on molecular profiling of tumors and immune profiling of tumor–host microenvironments must converge and integrate seamlessly to arrive at personalized treatment decisions.

References

Ahmad, P., Sana, J., Slavik, M., Slampa, P., Smilek, P., & Slaby, O. (2017). MicroRNAs involvement in radioresistance of head and neck cancer. *Disease Markers*, *2017*, 8245345.

Akervall, J. A., Jin, Y., Wennerberg, J. P., Zätterström, U. K., Kjellén, E., Mertens, F., & Mitelman, F. (1995). Chromosomal abnormalities involving 11q13 are associated with poor prognosis in patients with squamous cell carcinoma of the head and neck. *Cancer*, *76*(5), 853–859.

Alsahafi, E., Begg, K., Amelio, I., Rauff, N., Lucarelli, P., Sauter, T., . . . Tavassoli, M. (2019). Clinical update on head and neck cancer: Molecular biology and ongoing challenges. *Cell Death & Disease*, *10*(8), 540.

Ambatipudi, S., Gerstung, M., Gowda, R., Pai, P., Borges, A. M., Schäffer, A. A., . . . Mahimkar, M. B. (2011). Genomic profiling of advanced-stage oral cancers reveals chromosome 11q alterations as markers of poor clinical outcome. *PLoS One*, *6*(2), e17250.

Andrews, G. A., Xi, S., Pomerantz, R. G., Lin, C. J., Gooding, W. E., Wentzel, A. L., . . . Grandis, J. R. (2004). Mutation of p53 in head and neck squamous cell carcinoma correlates with Bcl-2 expression and increased susceptibility to cisplatin-induced apoptosis. *Head & Neck*, *26*(10), 870–877.

Ang, K. K., Harris, J., Wheeler, R., Weber, R., Rosenthal, D. I., Nguyen-Tân, P. F., . . . Gillison, M. L. (2010). Human papillomavirus and survival of patients with oropharyngeal cancer. *The New England Journal of Medicine*, *363*(1), 24–35.

Ausoni, S., Boscolo-Rizzo, P., Singh, B., da Mosto, M. C., Spinato, G., Tirelli, G., . . . Azzarello, G. (2016). Targeting cellular and molecular drivers of head and neck squamous cell carcinoma: Current options and emerging perspectives. *Cancer Metastasis Reviews*, *35*(3), 413–426.

Balbo, S., James-Yi, S., Johnson, C. S., O'Sullivan, M. G., Stepanov, I., Wang, M., . . . Hecht, S. S. (2013). S)-N'-nitrosonornicotine, a constituent of smokeless tobacco, is a powerful oral cavity carcinogen in rats. *Carcinogenesis*, *34*(9), 2178–2183.

Bauer, J. A., Trask, D. K., Kumar, B., Los, G., Castro, J., Lee, J. S.-J., . . . Carey, T. E. (2005). Reversal of cisplatin resistance with a BH3 mimetic, (-)-gossypol, in head and neck cancer cells: Role of wild-type p53 and Bcl-xL. *Molecular Cancer Therapeutics*, *4*(7), 1096–1104.

Bekeny, J. R., & Ozer, E. (2016). Transoral robotic surgery frontiers. *World Journal of Otorhinolaryngology Head & Neck Surgery*, *2*(2), 130–135.

Belbin, T. J., Singh, B., Smith, R. V., Socci, N. D., Wreesmann, V. B., Sanchez-Carbayo, M., & Childs, G. (2005). Molecular profiling of tumor progression in head and neck cancer. *Archives of Otolaryngology–Head & Neck Surgery*, *131*(1), 10–18.

Boeckx, C., Baay, M., Wouters, A., Specenier, P., Vermorken, J. B., Peeters, M., & Lardon, F. (2013). Anti-epidermal growth factor receptor therapy in head and neck squamous cell carcinoma: Focus on potential molecular mechanisms of drug resistance. *The Oncologist*, *18*(7), 850–864.

Boscolo-Rizzo, P., Da Mosto, M., Rampazzo, E., Giunco, S., Del Mistro, A., Menegaldo, A., . . . De Rossi, A. (2016). Telomeres and telomerase in head and neck squamous cell carcinoma: From pathogenesis to clinical implications. *Cancer Metastasis Reviews*, *35*(3), 457–474.

Bossi, P., Resteghini, C., Paielli, N., Licitra, L., Pilotti, S., & Perrone, F. (2016). Prognostic and predictive value of EGFR in head and neck squamous cell carcinoma. *Oncotarget, 7*(45), 74362−74379.

Braakhuis, B. J. M., Tabor, M. P., Kummer, J. A., Leemans, C. R., & Brakenhoff, R. H. (2003). A genetic explanation of slaughter's concept of field cancerization: Evidence and clinical implications. *Cancer Research, 63*(8), 1727−1730.

Bradford, C. R., Zhu, S., Ogawa, H., Ogawa, T., Ubell, M., Narayan, A., . . . Carey, T. E. (2003). P53 mutation correlates with cisplatin sensitivity in head and neck squamous cell carcinoma lines. *Head & Neck, 25*(8), 654−661.

Brocklehurst, P., Kujan, O., O'Malley, L. A., Ogden, G., Shepherd, S., & Glenny, A.-M. (2010). Screening programmes for the early detection and prevention of oral cancer. *Cochrane Database of Systematic Reviews*, CD004150.

Byeon, H. K., Ku, M., & Yang, J. (2019). Beyond EGFR inhibition: Multilateral combat strategies to stop the progression of head and neck cancer. *Experimental & Molecular Medicine, 51*(1), 8.

Byrd, J., & Ferris, R. L. (2016). Is there a role for robotic surgery in the treatment of head and neck cancer? *Current Treatment Options in Oncology, 17*(6), 29.

Cabelguenne, A., Blons, H., de Waziers, I., Carnot, F., Houllier, A. M., Soussi, T., . . . Laurent-Puig, P. (2000). p53 alterations predict tumor response to neoadjuvant chemotherapy in head and neck squamous cell carcinoma: A prospective series. *Journal of Clinical Oncology, 18*(7), 1465−1473.

Carvalho, A. L., Jeronimo, C., Kim, M. M., Henrique, R., Zhang, Z., Hoque, M. O., . . . Califano, J. A. (2008). Evaluation of promoter hypermethylation detection in body fluids as a screening/diagnosis tool for head and neck squamous cell carcinoma. *Clinical Cancer Research, 14*(1), 97.

Chan, K. C. A., Woo, J. K. S., King, A., Zee, B. C. Y., Lam, W. K. J., Chan, S. L., . . . Lo, D. (2017). Analysis of plasma Epstein-Barr virus DNA to screen for nasopharyngeal cancer. *The New England Journal of Medicine, 377*(6), 513−522.

Cheng, L., Eicher, S. A., Guo, Z., Hong, W. K., Spitz, M. R., & Wei, Q. (1998). Reduced DNA repair capacity in head and neck cancer patients. *Cancer Epidemiology, Biomarkers & Prevention, 7*(6), 465−468.

Chung, C. H., Parker, J. S., Karaca, G., Wu, J., Funkhouser, W. K., Moore, D., . . . Perou, C. M. (2004). Molecular classification of head and neck squamous cell carcinomas using patterns of gene expression. *Cancer Cell, 5*(5), 489−500.

Cohen, E. R., Reis, I. M., Gomez-Fernandez, C., Smith, D., Pereira, L., Freiser, M. E., . . . Franzmann, E. J. (2020). CD44 and associated markers in oral rinses and tissues from oral and oropharyngeal cancer patients. *Oral Oncology, 106*, 104720.

de Munter, L., Maasland, D. H. E., van den Brandt, P. A., Kremer, B., & Schouten, L. J. (2015). Vitamin and carotenoid intake and risk of head-neck cancer subtypes in the netherlands cohort study. *The American Journal of Clinical Nutrition, 102*(2), 420−432.

D'Souza, G., & Dempsey, A. (2011). The role of HPV in head and neck cancer and review of the HPV vaccine. *Preventive Medicine, 53*(Suppl. 1), S5−S11.

Edefonti, V., Hashibe, M., Parpinel, M., Turati, F., Serraino, D., Matsuo, K., . . . Decarli, A. (2015). Natural vitamin C intake and the risk of head and neck cancer: A pooled analysis in the international head and neck cancer epidemiology consortium. *International Journal of Cancer, 137*(2), 448−462.

Epstein, J. B., Güneri, P., Boyacioglu, H., & Abt, E. (2012). The limitations of the clinical oral examination in detecting dysplastic oral lesions and oral squamous cell carcinoma. *Journal of the American Dental Association (1939), 143*, 1332−1342.

Ferris, R. L., Blumenschein, G., Jr., Fayette, J., Guigay, J., Colevas, D., Licitra, L., ... Gillison, M. L. (2018). Nivolumab vs investigator's choice in recurrent or metastatic squamous cell carcinoma of the head and neck: 2-year long-term survival update of CheckMate 141 with analyses by tumor PD-L1 expression. *Oral Oncology, 81*, 45−51.

Ferris, R. L., Flamand, Y., Weinstein, G., Li, S., Quon, H., Mehra, R., ... Burtness, B. (2020). Transoral robotic surgical resection followed by randomization to low- or standard-dose IMRT in resectable p16 + locally advanced oropharynx cancer: A trial of the ECOG-ACRIN Cancer Research Group (E3311). *Journal of Clinical Oncology, 38*. Available from https://doi.org/10.1200/JCO.2020.38.15_suppl.6500.

Franzmann, E. J., Reategui, E. P., Carraway, K. L., Hamilton, K. L., Weed, D. T., & Goodwin, W. J. (2005). Salivary soluble CD44: A potential molecular marker for head and neck cancer. *Cancer Epidemiology, Biomarkers & Prevention, 14*(3), 735−739.

Goldenberg, D., Golz, A., Netzer, A., Rosenblatt, E., Rachmiel, A., Goldenberg, R. F., & Joachims, H. Z. (2001). Epstein-Barr virus and cancers of the head and neck. *American Journal of Otolaryngology, 22*(3), 197−205.

Hashibe, M., Brennan, P., Chuang, S.-C., Boccia, S., Castellsague, X., Chen, C., ... Boffetta, P. (2009). Interaction between tobacco and alcohol use and the risk of head and neck cancer: Pooled analysis in the international head and neck cancer epidemiology consortium. *Cancer Epidemiology, Biomarkers & Prevention, 18*(2), 541−550.

Hecht, S. S. (1998). Biochemistry, biology, and carcinogenicity of tobacco-specific N-nitrosamines. *Chemical Research in Toxicology, 11*(6), 559−603.

Hecht, S. S. (2003). Tobacco carcinogens, their biomarkers and tobacco-induced cancer. *Nature Reviews. Cancer, 3*(10), 733−744.

Ho, T., Wei, Q., & Sturgis, E. M. (2007). Epidemiology of carcinogen metabolism genes and risk of squamous cell carcinoma of the head and neck. *Head & Neck, 29*, 682−699.

Hutchinson, M.-K. N. D., Mierzwa, M., & D'Silva, N. J. (2020). Radiation resistance in head and neck squamous cell carcinoma: Dire need for an appropriate sensitizer. *Oncogene, 39* (18), 3638−3649.

IARC Working Group on the Evaluation of Carcinogenic Risks to Humans. (2010). Some non-heterocyclic polycyclic aromatic hydrocarbons and some related exposures. *IARC Monographs on the Evaluation of Carcinogenic Risks to Humans, 92*, 1−853.

Ioachim, E., Assimakopoulos, D., Goussia, A. C., Peschosl, D., Skevas, A., & Agnantis, N. J. (1999). Glycoprotein CD44 expression in benign, premalignant and malignant epithelial lesions of the larynx: An immunohistochemical study including correlation with Rb, p53, Ki-67 and PCNA. *Histology and Histopathology, 14*, 1113−1118.

Iwai, Y., Ishida, M., Tanaka, Y., Okazaki, T., Honjo, T., & Minato, N. (2002). Involvement of PD-L1 on tumor cells in the escape from host immune system and tumor immunotherapy by PD-L1 blockade. *Proceedinhs of the National Academy of Sciences of the United States of America, 99* (19), 12293−12297.

Jethwa, A. R., & Khariwala, S. S. (2017). Tobacco-related carcinogenesis in head and neck cancer. *Cancer and Metastasis Reviews, 36*(3), 411−423.

Khariwala, S. S., Carmella, S. G., Stepanov, I., Fernandes, P., Lassig, A. A., Yueh, B., ... Hecht, S. S. (2013). Elevated levels of 1-hydroxypyrene and N'-nitrosonornicotine in smokers with head and neck cancer: A matched control study. *Head & Neck, 35*(8), 1096−1100.

Kornberg, L. J., Villaret, D., Popp, M., Lui, L., McLaren, R., Brown, H., ... McFadden, M. (2005). Gene expression profiling in squamous cell carcinoma of the oral cavity shows abnormalities in several signaling pathways. *The Laryngoscope, 115*(4), 690−698.

Kutler, D. I., Auerbach, A. D., Satagopan, J., Giampietro, P. F., Batish, S. D., Huvos, A. G., ... Singh, B. (2003). High incidence of head and neck squamous cell carcinoma in patients with fanconi anemia. *Archives of Otolaryngology—Head & Neck Surgery, 129*(1), 106–112.

Lacko, M., Braakhuis, B. J. M., Sturgis, E. M., Boedeker, C. C., Suárez, C., Rinaldo, A., ... Takes, R. P. (2014). Genetic susceptibility to head and neck squamous cell carcinoma. *International Journal of Radiation Oncology, Biology, Physics, 89*(1), 38–48.

Lee, Y. S., Johnson, D. E., & Grandis, J. R. (2018). An update: Emerging drugs to treat squamous cell carcinomas of the head and neck. *Expert Opinion on Emerging Drugs, 23*(4), 283–299.

Lepikhova, T., Karhemo, P. R., Louhimo, R., Yadav, B., Murumägi, A., Kulesskiy, E., ... Monni, O. (2018). Drug-sensitivity screening and genomic characterization of 45 HPV-negative head and neck carcinoma cell lines for novel biomarkers of drug efficacy. *Molecular Cancer Therapeutics, 17*(9), 2060–2071.

Li, Y., St., John, M. A., Zhou, X., Kim, Y., Sinha, U., Jordan, R. C. K., ... Wong, D. T. (2004). Salivary transcriptome diagnostics for oral cancer detection. *Clinical Cancer Research, 10*(24), 8442–8450.

Linkov, F., Lisovich, A., Yurkovetsky, Z., Lisovich, A., Yurkovetsky, Z., Marrangoni, A., ... Ferris, R. L. (2007). Early detection of head and neck cancer: Development of a novel screening tool using multiplexed immunobead-based biomarker profiling. *Cancer Epidemiology, Biomarkers & Prevention, 16*(1), 102–107.

Lo, Y. M., Chan, L. Y., Lo, K. W., Leung, S. F., Zhang, J., Chan, A. T., ... Huang, D. P. (1999). Quantitative analysis of cell-free Epstein-Barr virus DNA in plasma of patients with nasopharyngeal carcinoma. *Cancer Research, 59*(6), 1188–1191.

Moldovan, G.-L., & D'Andrea, A. D. (2009). How the fanconi anemia pathway guards the genome. *Annual Review of Genetics, 43*, 223–249.

Moy, J. D., Moskovitz, J. M., & Ferris, R. L. (2017). Biological mechanisms of immune escape and implications for immunotherapy in head and neck squamous cell carcinoma. *European Journal of Cancer, 76*, 152–166.

Nagaraj, N. S. (2009). Evolving 'omics' technologies for diagnostics of head and neck cancer. *Briefings in Functional Genomics & Proteomics, 8*(1), 49–59.

Noorlag, R., van Kempen, P., Stegeman, I., Koole, R., van Es, R. J., & Willems, S. M. (2015). The diagnostic value of 11q13 amplification and protein expression in the detection of nodal metastasis from oral squamous cell carcinoma: A systematic review and *meta*-analysis. *Virchows Archiv: An International Journal of Pathology, 466*(4), 363–373.

O'Donnell, R. K., Kupferman, M., Wei, S. J., Singhal, S., Weber, R., O'Malley, B., ... Muschel, R. J. (2005). Gene expression signature predicts lymphatic metastasis in squamous cell carcinoma of the oral cavity. *Oncogene, 24*(7), 1244–1251.

Pai, S. I., & Westra, W. H. (2009). Molecular pathology of head and neck cancer: Implications for diagnosis, prognosis, and treatment. *Annual Review of Pathology, 4*, 49–70.

Paoli, P., Giannoni, E., & Chiarugi, P. (2013). Anoikis molecular pathways and its role in cancer progression. *Biochimica et Biophysica Acta, 1833*(12), 3481–3498.

Park, N. J., Zhou, H., Elashoff, D., Henson, B. S., Kastratovic, D. A., Abemayor, E., & Wong, D. T. (2009). Salivary microRNA: Discovery, characterization, and clinical utility for oral cancer detection. *Clinical Cancer Research, 15*(17), 5473–5477.

Pathare, S. M., Gerstung, M., Beerenwinkel, N., Schäffer, A. A., Kannan, S., Pai, P., ... Mahimkar, M. B. (2011). Clinicopathological and prognostic implications of genetic alterations in oral cancers. *Oncology Letters, 2*(3), 445–451.

Pereira, L. H. M., Reis, I. M., Reategui, E. P., Gordon, C., Saint-Victor, S., Duncan, R., ... Franzmann, E. J. (2016). Risk stratification system for oral cancer screening. *Cancer Prevention Research, 9*(6), 445–455.

Pérez-Sayáns, M., Somoza-Martín, J. M., Barros-Angueira, F., Diz, P. G., Gandara Rwy, J. M., & García-García, A. (2010). Multidrug resistance in oral squamous cell carcinoma: The role of vacuolar ATPases. *Cancer Letters, 295*(2), 135–143.

Pratt, M. M., John, K., MacLean, A. B., Afework, S., Phillips, D. H., & Poirier, M. C. (2011). Polycyclic aromatic hydrocarbon (PAH) exposure and DNA adduct semi-quantitation in archived human tissues. *International Journal of Environmental Research and Public Health 8, 7*, 2675–2691.

Qin, Q., Ma, P.-F., Kuang, X.-C., Gao, M.-X., Mo, D.-H., Shuang, X., ... Lin, C.-W. (2013). Novel function of N,N-bis(2-chloroethyl)docos-13-enamide for reversal of multidrug resistance in tongue cancer. *European Journal of Pharmacology, 721*(1–3), 208–214.

Roepman, P., Wessels, L. F., Kettelarij, N., Kemmeren, P., Miles, A. J., Lijnzaad, P., ... Holstege, F. C. P. (2005). An expression profile for diagnosis of lymph node metastases from primary head and neck squamous cell carcinomas. *Nature Genetics, 37*(2), 182–186.

Saidak, Z., Pascual, C., Bouaoud, J., Galmiche, L., Clatot, F., Dakpé, S., ... Galmiche, A. (2019). A three-gene expression signature associated with positive surgical margins in tongue squamous cell carcinomas: Predicting surgical resectability from tumour biology? *Oral Oncology, 94*, 115–120.

Saraiya, M., Unger, E. R., Thompson, R. D., Lynch, C. F., Hernandez, B. Y., Lyu, C. W., ... HPV Typing of Cancers Workgroup. (2015). US assessment of HPV types in cancers: Implications for current and 9-valent HPV vaccines. *JNCI Journal of the National Cancer Institute, 107*(6).

Schantz, S. P., Huang, Q., Shah, K., Murty, V. V., Hsu, T. C., Yu, G., ... Chaganti, R. S. (2000). Mutagen sensitivity and environmental exposures as contributing causes of chromosome 3p losses in head and neck cancers. *Carcinogenesis, 21*(6), 1239–1246.

Schantz, S. P., Zhang, Z. F., Spitz, M. S., Sun, M., & Hsu, T. C. (1997). Genetic susceptibility to head and neck cancer: Interaction between nutrition and mutagen sensitivity. *The Laryngoscope, 107*(6), 765–781.

The Cancer Genome Atlas Network. (2015). Comprehensive genomic characterization of head and neck squamous cell carcinomas. *Nature, 517*, 576–582.

Theile, D., Ketabi-Kiyanvash, N., Herold-Mende, C., Dyckhoff, G., Efferth, T., Bertholet, V., ... Weiss, J. (2011). Evaluation of drug transporters' significance for multidrug resistance in head and neck squamous cell carcinoma. *Head & Neck, 33*(7), 959–968.

Timbang, M. R., Sim, M. W., Bewley, A. F., Farwell, D., Mantravadi, A., & Moore, M. G. (2019). HPV-related oropharyngeal cancer: A review on burden of the disease and opportunities for prevention and early detection. *Human Vaccines & Immunotherapeutics, 15*(7–8), 1920–1928.

Tonella, L., Giannoccaro, M., Alfieri, S., Canevari, S., & De Cecco, L. (2017). Gene expression signatures for head and neck cancer patient stratification: Are results ready for clinical application? *Current Treatment Options in Oncology, 18*, 32.

Toruner, G. A., Ulger, C., Alkan, M., Galante, A. T., Rinaggio, J., Wilk, R., ... Dermody, J. J. (2004). Association between gene expression profile and tumor invasion in oral squamous cell carcinoma. *Cancer Genetics and Cytogenetics, 154*(1), 27–35.

van Ginkel, J. H., de Leng, W., de Bree, R., van Es, R., & Willems, S. M. (2016). Targeted sequencing reveals TP53 as a potential diagnostic biomarker in the post-treatment surveillance of head and neck cancer. *Oncotarget., 7*(38), 61575–61586.

Velleuer, E., & Dietrich, R. (2014). Fanconi anemia: Young patients at high risk for squamous cell carcinoma. *Molecular and Cellular Pediatrics, 1*.

Wang, L., Hu, Z., Sturgis, E. M., Spitz, M. R., Strom, S. S., Amos, C. I., ... Wei, Q. (2010). Reduced DNA repair capacity for removing tobacco carcinogen-induced DNA adducts contributes to risk of head and neck cancer but not tumor characteristics. *Clinical Cancer Research, 16*(2), 764–774.

Wang, Y., Springer, S., Mulvey, C. L., Silliman, N., Schaefer, J., Sausen, M., … Agrawal, N. (2015). Detection of somatic mutations and HPV in the saliva and plasma of patients with head and neck squamous cell carcinomas. *Science Translational Medicine, 7*(293), 293ra104.

Wen, Y., & Grandis, J. R. (2015). Emerging drugs for head and neck cancer. *Expert Opinion on Emerging Drugs, 20*(2), 313–329.

Yalcin, E., & de la Monte, S. (2016). Tobacco nitrosamines as culprits in disease: Mechanisms reviewed. *Journal of Physiology and Biochemistry, 72*(1), 107–120.

Yuan, J.-M., Knezevich, A. D., Wang, R., Gao, Y.-T., Hecht, S. S., & Stepanov, I. (2011). Urinary levels of the tobacco-specific carcinogen N'-nitrosonornicotine and its glucuronide are strongly associated with esophageal cancer risk in smokers. *Carcinogenesis, 32*(9), 1366–1371.

Yuan, J.-M., Koh, W.-P., Murphy, S. E., Fan, Y., Wang, R., Carmella, S. G., … Hecht, S. S. (2009). Urinary levels of tobacco-specific nitrosamine metabolites in relation to lung cancer development in two prospective cohorts of cigarette smokers. *Cancer Research, 69*(7), 2990–2995.

Zhang, P., Wang, W., Wei, Z., Xu, L., Yang, X., & Yuanhong, D. (2016). xCT expression modulates cisplatin resistance in Tca8113 tongue carcinoma cells. *Oncology Letters, 12*(1), 307–314.

Zhou, G., Liu, Z., & Myers, J. N. (2016). TP53 mutations in head and neck squamous cell carcinoma and their impact on disease progression and treatment response. *Journal of Cellular Biochemistry, 117*(12), 2682–2692.

Chapter 19

Melanoma

Norma E. Farrow[1], Aaron Therien[1], Douglas S. Tyler[2] and
Georgia M. Beasley[1]
*[1]Department of Surgery, Duke University, Durham, NC, United States, [2]Department of Surgery,
University of Texas Medical Branch, Galveston, TX, United States*

Introduction

The incidence of melanoma is increasing at a rate faster than any other cancer, with an estimated 96,480 new cases in 2019 (American Cancer Society, 2020). The treatment for advanced melanoma has shifted dramatically in recent years with eight therapeutic FDA-approved agents since 2011; seven of eight approved agents were associated with improvements in overall survival (OS) (Andtbacka et al., 2015; Robert et al., 2015; Schachter et al., 2017). In this chapter, we will summarize current knowledge regarding genetic and molecular basis of melanoma and the clinical applications of molecular tumor profiling.

Melanoma progression

Melanoma arises from mutated melanocytes that have escaped normal growth control. Melanocytes derive from neural crest cells and migrate predominantly to the skin and hair, where the melanin produced by these cells determines the pigmentation and acts to absorb ultraviolet radiation, providing protection to the skin from sun-induced damage (Ibrahim & Haluska, 2009). Melanocytes also migrate to other sites such as the uveal tract of the eye and ectodermal mucosa. Melanoma can be grouped into three families based on the primary site of the tumor: cutaneous, mucosal, and ocular. The dominant site of melanoma occurrence is the skin, with cutaneous melanoma accounting for 91.2% of all melanomas and the remainder occurring at ocular (5.8%), mucosal (1.3%), or unknown sites. Cutaneous melanoma has historically involved four histopathological descriptions—superficial spreading, lentigo maligna, nodular, and acral lentiginous (Duncan, 2009; Ibrahim & Haluska, 2009). Although sun exposure is considered the predominant environmental risk factor for melanoma, the relationship between sun exposure

Genomic and Precision Medicine. DOI: https://doi.org/10.1016/B978-0-12-800684-9.00005-8
319

and melanoma is in fact complex, with melanoma occurring not only on chronic and intermittent sun-exposed sites, but also at sites with little to no sun exposure, such as acral (palm and sole) and mucosal melanoma. Furthermore, sun-exposed areas (head, neck) have the highest rates of genetic alterations while acral, mucosal, and uveal have the lowest rates of mutation (Curtin et al., 2005). This genetic complexity of melanoma has therapeutic implications in that highly mutated melanoma may be more sensitive to immunotherapy while also having mutations that confer resistance to targeted therapy (BRAF and MEK inhibitors) (Luke, Flaherty, Ribas, & Long, 2017).

Progression from melanocyte to melanoma generally begins with a benign nevus—a clonal population of melanocytes that have proliferated into a hyperplastic lesion, but which are in a state of cellular senescence and hence do not progress. There are several factors, both inherited genetic risk factors and environmental risk factors such as sun exposure, which likely contribute to this aberrant proliferative state. In response to appropriate stimuli, these hyperplastic lesions exit senescence and begin to grow as dysplastic nevi and then progress to a radial growth phase (RGP), where the lesions spread superficially in an area confined to the epidermis, with little invasive potential. Eventually, these RGP lesions progress to a vertical growth phase, at which point they begin to invade the dermis and eventually metastasize. It is estimated that 1 in 10,000 benign nevi eventually transforms into melanoma (Tsao, Mihm, & Sheehan, 2003). Metastatic spread of cutaneous melanoma usually occurs as regional lymph node metastases, loco-regional satellite or in-transit metastases, or distant metastases (in 50%, 20%, and 30% of patients with recurrent melanoma, respectively) (Leiter, Meier, Schittek, & Garbe, 2004).

Genetics of melanoma

Inherited genetic lesions

Although familial, or hereditary, melanoma accounts for only a small percentage of all cases of melanoma (<7%) (Olsen, Carroll, & Whiteman, 2010), it can provide valuable insight into genetic factors contributing to the aberrant growth of a melanocyte. The cyclin-dependent kinase inhibitor 2A (*CDKN2A*) locus on chromosome 9 is the most clearly linked, high-risk gene associated with melanoma, with estimates of 25%–50% of familial melanoma patients harboring mutations at this locus (Nelson & Tsao, 2009). Two tumor suppressor proteins critical in the regulation of the cell cycle are encoded by the *CDKN2A* locus, p16INK4a and p14ARF, and genetic mutations at this locus can lead to deregulated signaling in both the retinoblastoma (Rb) and p53 pathways. Mutation of p16INK4a (9% of melanomas) or loss of p16INK4a (frequently due to hypermethylation; 50% of melanomas) can thus lead to increased cell cycle progression. p14ARF binds human homolog of murine Mdm2 (HDM2), sequestering it in the nucleolus and preventing HDM2 from interacting with and destabilizing the tumor suppressor protein p53. Decreased p14ARF activity due to loss or mutation leads to increased interaction of HDM2 with p53, the subsequent ubiquitination and degradation of p53 and ultimately to genomic instability, as the cell is unable to detect genetic damage, signal for DNA repair, or activate apoptotic pathways when DNA damage is too extensive for repair. Another gene associated with melanoma is the melanocortin-1 receptor (MC1R) gene, which is important in the regulation of skin color and when activated by α-melanocyte stimulating hormone triggers melanocytes to switch production from pheomelanin (red/yellow melanin) to eumelanin (brown/black melanin) (Ibrahim & Haluska, 2009). A high incidence of inactivating mutations in *TP53* occurs in many cancer types, and germline mutations in *TP53* occur in melanoma, but only rarely ($\sim 9\%$) (Sekulic et al., 2008).

Acquired genetic lesions

For most melanomas ($\sim 90\%$), the genetic lesions contributing to progression of the disease are acquired after birth, rather than inherited, and are termed sporadic melanomas. To date, over 1000 melanomas have been sequenced, illuminating key pathogenic alterations involved (Hayward et al., 2017). Mutations that activate the mitogen-activated protein kinase (MAPK) are ubiquitous. Furthermore, tumor suppressor genes *CDKN2A* or *PTEN* can cooperate with MAPK pathway genes to drive genesis of melanoma (Vogelstein, Papadopoulos, Velculescu, Zhou, & Diaz, 2013). Activating mutations in the oncogene *BRAF*, a key component of the MAPK pathway, are the most commonly identified mutations in melanoma, and over 50% of melanomas have been shown to harbor a single base-pair change leading to a glutamate for valine substitution at codon 600 in the kinase domain (*BRAF* V600E) (Davies et al., 2002). This mutation leads to constitutive activation of BRAF kinase, with the resultant unchecked stimulation of the MAPK/ MEK and extracellular signal-regulated kinase (ERK) pathway leading to increased activity in several critical pathways involved in proliferation and survival.

Phosphatase and tensin homolog (*PTEN*) was identified as a frequently deleted gene in melanoma more than 20 years ago (Parmiter, Balaban, Clark, & Nowell, 1988). *PTEN* is located on chromosome 10q23−24, a region that is frequently lost at an early stage in melanoma (Fountain, Bale, Housman, & Dracopoli, 1990). PTEN functions as a tumor suppressor protein regulating the levels of phosphatidylinositol phosphate. The high frequency of PTEN loss in melanoma, compared to the relatively low incidence of *PTEN* mutation, suggests that epigenetic silencing of PTEN is important in melanomagenesis, and the multiple methylation sites identified in the *PTEN* promoter region lend support to this theory (Palmieri et al., 2009). *PTEN* and *BRAF* mutations are frequent concurrent events in melanoma, with this dual mutation acting to deregulate both the MAPK and the Akt signaling pathways (Haluska, 2006).

Somatic loss of function mutations in *CDKN2A* is present in nearly 50% of melanomas (Zeng, Judson-Torres, & Shain, 2020). Melanoma in situ tends to have heterozygous *CDKN2A* mutations whereas invasive melanomas harbor bi-allelic alterations in the *CDKN2A* gene (Shain et al., 2018); this loss is associated with invasion and metastasis in in vivo models (Ackermann et al., 2005).

Although less common, activating mutations in *NRAS* also occur in ~15% to 30% of melanomas, leading to similar deregulation in pathways such as MAPK and ERK that are important in proliferation and survival (Haluska, 2006). NF1 is a negative regulator of RAS signaling and inactivating mutations in *NF1* appear to be another distinct melanoma subtype. *NF1* and *NRAS* mutations do not appear to be as essential to MAPK signaling as BRAF that has therapeutic implications for BRAF/MEK-targeted therapy.

c-Kit is a tyrosine kinase receptor that is important in melanocyte migration from the neural crest to the dermis during development (Alexeev & Yoon, 2006; Masson & Rönnstrand, 2009). Activating mutations are found in 1% of all melanomas, but in 10% of mucosal and acral melanomas (Curtin, Busam, Pinkel, & Bastian, 2006). See Table 19.1 for a summary of the genetics of melanoma.

The progression of melanoma is complex and likely a nonbinary event in which pathways are either turned "on" or "off." As we continue to understand the molecular basis for melanoma, the concept that multiple mutations in the same pathway cannot cooccur has largely been disproven (Zeng et al., 2020). While *BRAF* V600E and *NRAS* codon 61 mutations seem to be exclusive, mutations may be dynamic and melanoma progression may occur as result of incremental pathway dysregulation.

Clinical applications of genomics in melanoma

Targeted therapeutics

In current clinical melanoma practice, the only standard molecular analysis that guides treatment decision-making is *BRAF* V600 mutations. *BRAF*-mutant melanoma comprises over 50% of clinical cases (Cancer Genome Atlas Network, 2015; Davies et al., 2002) and treatments for advanced and metastatic cases have evolved significantly in recent years (Algazi et al., 2016; Luke et al., 2013). V600E is the most common (74%−86%) followed by V600K (10%−30%); both mutations were included in regulatory approvals of BRAF/MEK inhibitors (Long et al., 2011).

TABLE 19.1 Genetics of melanoma.

	Protein	Normal function	Effect of lesion	Affected cellular process	Oncogenic signaling pathway
Inherited genetic lesions					
CDKN2A	p16/INK4α	Tumor suppressor	Copy number loss (9p21.3), inactivating mutation	Cell cycle progression	E2F1
	p14/ARF	Tumor suppressor	Copy number loss (9p21.3), inactivating mutation	Cell cycle progression	p53
CDK4	Cdk4	Oncogene	Activating mutation	Cell cycle progression	E2F1
MC1R	MC1 receptor	Melanogenesis	Variants	Pigmentation	
TP53	p53	Tumor suppressor	Inactivating mutation	Cell cycle progression	p53
Acquired genetic lesions					
BRAF	BRaf	Oncogene	Activating mutation, copy number gain (7q34)	Proliferation; survival	Ras
NRAS	Nras	Oncogene	Activating mutation	Proliferation; survival	Ras, PI3K
PTEN	PTEN	Tumor suppressor	Copy number loss (10q23.31)	Cell cycle progression; survival	PI3K, Akt
CDKN2A	P16 INK4A	Tumor suppressor	Loss of expression	Loss of cell cycle checkpoint	E2F1

(Continued)

TABLE 19.1 (Continued)

	Protein	Normal function	Effect of lesion	Affected cellular process	Oncogenic signaling pathway
MITF	MITF	Transcription factor	Mutation; copy number gain (3p13)	Pigmentation	
TRPM1	TRPM1	Ion channel	Loss of expression	Pigmentation	
CKIT	c-Kit	Oncogene	Activating mutation	Proliferation; differentiation	Src, PI3K

The first BRAF-inhibitors developed for advanced-stage melanoma, vemurafenib and dabrafenib, both demonstrated similar clinical benefits in randomized controlled trials when compared to previously standard dacarbazine, with dabrafenib increasing the overall response rate (ORR) from 6% to 50%, and improving the median progression free survival (PFS) from 2.7 to 5.1 months (Eisenhauer et al., 2009; Hauschild et al., 2012; McArthur et al., 2014). In an effort to further improve clinical responses, ensuing BRAF-targeted melanoma therapies sought to combine BRAF inhibitors with MEK inhibitors, as BRAF signaling is dependent on MEK1/2 activation (Luke et al., 2017). MEK inhibitors, such as trametinib, cobimetinib, and binimetinib, are essential to disrupt a deleterious signaling circuit between tumor cells and macrophages, which prohibit the entry of effector T-cells into the tumor environment and consequently exacerbate melanoma growth (Wang et al., 2015). Results from an early phase I and II trial combining dabrafenib with trametinib revealed higher ORR, PFS, and median OS, 76%, 9.4 months (Flaherty et al., 2012), and 27.4 months, respectively, than in the aforementioned studies that relied on dabrafenib (Hauschild et al., 2012) or trametinib (Flaherty et al., 2012) monotherapy (Flaherty et al., 2014).

The international phase III COMBI-d and COMBI-v trials further explored clinical outcomes of BRAF/MEK combination therapy (Long et al., 2015). A recent pooled update from the COMBI-v and COMBI-d trials showed an OS of 37% at 4 years and 34% at 5 years for patients who received dabrafenib plus trametinib; and for the patients who achieved a complete response (19%), the 5-year OS was 71% (95% CI, 62−79) (Robert et al., 2019).

The coBRIM study was a phase III trial of vemurafenib plus cobimetinib or vemurafenib plus placebo, with ORR of 70% versus 50% and median PFS of 12.3 months versus 7.2 months (Ascierto et al., 2016). Similarly, the COLUMBUS trial randomized patients to encorafenib plus binimetinib, or encorafenib alone, or vemurafenib alone; the median PFS was 14.9 months in the encorafenib/binimetinib group, 9.6 months in the encorafenib group, and 7.3 months in the vemurafenib group (Dummer et al., 2018). These three BRAF/MEK inhibitor combinations (dabrafenib/trametinib, vemurafenib/cobimetinib, encorafenib/binimetinib) are now currently approved for BRAF-mutant advanced melanoma; dabrafenib/trametinib is also approved in the adjuvant setting after resection of high-risk melanoma (Long et al., 2017). The choice between the different targeted BRAF/MEK regimens or between targeted therapies or immune therapy and order of sequence of these therapies is based on a number of clinical and patient factors, including side-effect profiles and prior therapies.

Immunotherapy

The overarching goal of immune therapy is to engage the patient's immune system in order to mount a specific antitumor response. Historic immune therapies for melanoma such as the cytokine-based agents interferon and interleukin-2 (IL-2), as well as vaccine strategies, failed to show improvements in survival in randomized clinical trials. The modern era of immune therapy is largely focused on using immune checkpoint blockade to augment cell-mediated immunity and has revolutionized the landscape of care for patients with melanoma, and indeed many cancer types. Three checkpoint inhibitors have been approved since 2011, including cytotoxic T-lymphocyte-associated antigen-4 (CTLA-4) and programmed cell-death protein-1 (PD-1) inhibitory antibodies, which have been shown to improve both melanoma-specific and OS. These modern immune therapies are now standard of care for patients with advanced melanoma in both the metastatic and adjuvant settings, regardless of BRAF-mutation status. In addition to immune checkpoint blockade, immune therapy with intratumoral oncolytic viruses can promote an inflammatory response in the tumor microenvironment and causes tumor cell lysis; the genetically modified oncolytic herpes virus talimogene laherparepvec (T-VEC) was approved in 2015 for treatment of unresectable melanoma based on the OPTiM trial results showing an improvement in durable response rate over granulocyte-macrophage colony-stimulating factor (Andtbacka et al., 2015).

The modern era of immune therapeutics characterized by antibody-mediated checkpoint blockade began in 2010 with the development of ipilimumab, a CTLA-4 inhibitory antibody. After years of trials failing to show an improvement in survival for patients with metastatic melanoma, the phase III CA184−002 trial comparing ipilimumab, ipilimumab plus a glycoprotein 100 (gp100) peptide vaccine or gp100 alone, found that ipilimumab improved OS from 6.4 months with gp100 alone to 10.1 months with ipilimumab alone for patients with stage III or IV melanoma (Hodi et al., 2010). Ipilimumab when used alone has an objective response rate in metastatic melanoma of 12%−15%, with a subset of patients having durable long-term responses (Ascierto et al., 2017). Ipilimumab achieved FDA approval for treatment of metastatic melanoma in 2011.

With the success of anti-CTLA4 blockade, the development of anti-PD-1 therapies soon followed, leading to the approval of anti-PD-1 antibodies pembrolizumab and nivolumab in 2014. While CTLA-4 is a broad regulator of T-cell priming, PD-1 engages with its ligand PD-L1, which is expressed on tumor cells and upregulated in many somatic cells during inflammatory responses and elicits T-cell exhaustion and the regulation of effector function against tumor cells. Perhaps due to this increased specificity for regulation of antitumor T cells, both pembrolizumab and nivolumab have been shown to have superior clinical efficacy and more limited toxicity compared to ipilimumab, with ORR between 30% and 45% (Hodi et al., 2018; Robert et al., 2015; Schadendorf et al., 2015). The phase III Checkmate-067 trial comparing nivolumab plus ipilimumab versus nivolumab alone versus ipilimumab alone also showed that this clinical efficacy in terms of ORR, PFS, and OS can be further improved by combining anti-PD-1 therapy with anti-CTLA-4 therapies, though at a cost of increased toxicity (Puzanov et al., 2016).

Predicting response to immunotherapy

Despite the great progress made in therapies for melanoma in recent years, with a significant proportion of patients now achieving durable responses, the majority of patients will eventually go on to develop resistance to current therapies. There is great interest in identifying patients who are at greater risk for progression versus disease control, and identification of biomarkers for selection of appropriate treatments. High-quality biomarkers for response to immune therapy have been elusive, though several have been studied.

Melanoma is particularly suited to immune therapy for a variety of reasons, including its relatively high-mutational burden and tumor infiltration by immune cells (Luke et al., 2017). Somatic mutation burden, or the total number of mutations present in a tumor, is therefore emerging potential biomarker to predict response to immune therapies. There is evidence suggesting that patients with higher tumor mutational burdens, who are likely producing more neoantigens, may have higher response rates to checkpoint blockades (Snyder et al., 2014). Melanoma has a relatively high-mutational burden compared to many cancers, contributing to its role at the forefront of immune therapy development. Furthermore, desmoplastic melanoma, a rare subtype of melanoma with a particularly high-mutational burden, has exceptionally high response rates to PD-1 blockade, at around 70% (Eroglu et al., 2018).

Another promising area of ongoing research into immune therapy responses is the role of immune cell infiltration, and particularly CD8 + T-cell infiltration, of the tumor microenvironment. Immune therapy with checkpoint inhibition is largely thought to work via the presence of preexisting populations of tumor-specific T cells that are held at bay by these checkpoints. Once unleashed, these antitumor T cells recognize tumor neoantigens, leading to T cell—mediated killing of melanoma cells as well as an interferon-gamma predominate pro-inflammatory cascade. Gene expression profiles consistent with inflamed tumor microenvironments are associated with improved response rates to immune therapy and improved survival (Morrison et al., 2018). Along those lines, combination therapy with T-VEC or injectable therapies that increase intratumoral CD8$^+$ T-cell infiltration may have a clinical benefit and improve response rates to checkpoint blockade (Puzanov et al., 2016).

Current treatment landscape

While there are multiple potential biomarkers and prognostic genetic tests being investigated for the ability to provide prognostic value as well as predict response to therapy, none are currently used in regular clinical practice. Currently, the preferred regimens for systemic therapy include the anti-PD1 agents nivolumab and pembrolizumab, or combination BRAF/MEK inhibitor therapy for BRAF V600-mutant melanoma. The presence of a *BRAF* V600 mutation allows for identification of patients eligible for BRAF/MEK-targeted therapy, though does not in itself predict likelihood of response to these therapies. There is also investigation into BRAF/MEK therapy combined with immunotherapy. In addition, there are numerous studies combined checkpoint therapy with other innate immune agonists to augment response.

Conclusion

The American Cancer Society recently reported the 1-year survival rate for patients with metastatic melanoma improved from 42% from 2008−10 to 55% from 2013−15. While targeted and checkpoint therapies have improved survival for some patients, primary and acquired resistance to these therapies are still common in the majority of patients. Continued investigation into therapy directed at the molecular basis for oncogenesis holds promise for further declines in melanoma mortality.

References

Ackermann, J., Frutschi, M., Kaloulis, K., Mckee, T., Trumpp, A., & Beermann, F. (2005). Metastasizing melanoma formation caused by expression of activated N-RasQ61Kon an INK4a-deficient background. *Cancer Research*, *65*(10), 4005−4011, May 15.

Alexeev, V., & Yoon, K. (2006). Distinctive role of the cKit receptor tyrosine kinase signaling in mammalian melanocytes. *Journal of Investigative Dermatology*, *126*(5), 1102−1110, May.

Algazi, A. P., Tsai, K. K., Shoushtari, A. N., Munhoz, R. R., Eroglu, Z., Piulats, J. M., ... Sullivan, R. J. (2016). Clinical outcomes in metastatic uveal melanoma treated with PD-1 and PD-L1 antibodies. *Cancer*, *122*(21), 3344−3353, Nov 15.

American Cancer Society. (2020). Melanoma skin cancer statistics [Internet]. American Cancer Society. [cited Mar 27]. <https://www.cancer.org/cancer/melanoma-skin-cancer/about/key-statistics.html>.

Andtbacka, R. H., Kaufman, H. L., Collichio, F., Amatruda, T., Senzer, N., Chesney, J., ... Coffin, R. S. (2015). Talimogene laherparepvec improves durable response rate in patients with advanced melanoma. *Journal of Clinical Oncology*, *33*(25), 2780−2788, Sep 1.

Ascierto, P. A., Del Vecchio, M., Robert, C., Mackiewicz, A., Chiarion-Sileni, V., Arance, A., ... Maio, M. (2017). Ipilimumab 10 mg/kg vs ipilimumab 3 mg/kg in patients with unresectable or metastatic melanoma: A randomised, double-blind, multicentre, phase 3 trial. *The Lancet Oncology*, *18*(5), 611−622, May 1.

Ascierto, P. A., Mcarthur, G. A., Dréno, B., Atkinson, V., Liszkay, G., Giacomo, A. M. D., ... Larkin, J. (2016). Cobimetinib combined with vemurafenib in advanced BRAFV600-mutant melanoma (coBRIM): Updated efficacy results from a randomised, double-blind, phase 3 trial. *The Lancet Oncology*, *17*(9), 1248−1260, Sep 17.

Cancer Genome Atlas Network. (2015). Genomic classification of cutaneous melanoma. *Cell, 16*(7), 1681−1696, Jun 18.

Curtin, J. A., Busam, K., Pinkel, D., & Bastian, B. C. (2006). Somatic activation of KIT in distinct subtypes of melanoma. *Journal of Clinical Oncology, 24*(26), 4340−4346, Sep 21.

Curtin, J. A., Fridlyand, J., Kageshita, T., Patel, H. N., Busam, K. J., Kutzner, H., ... Bastian, B. C. (2005). Distinct sets of genetic alterations in melanoma. *New England Journal of Medicine, 353*(20), 2135−2147, Nov 17.

Davies, H., Bignell, G. R., Cox, C., Stephens, P., Edkins, S., Clegg, S., ... Futreal, P. A. (2002). Mutations of the BRAF gene in human cancer. *Nature, 417*(6892), 949−954, Jun 9.

Dummer, R., Ascierto, P. A., Gogas, H. J., Arance, A., Mandala, M., Liszkay, G., ... Ghosh, A. K. (2018). Encorafenib plus binimetinib vs vemurafenib or encorafenib in patients with BRAF -mutant melanoma (COLUMBUS): A multicentre, open-label, randomised phase 3 trial. *The Lancet Oncology, 19*(5), 603−615, May 19.

Duncan, L. M. D. (2009). The classification of cutaneous melanoma. *Hematology/Oncology Clinics of North America, 23*(3), 501−513, Jun.

Eisenhauer, E., Therasse, P., Bogaerts, J., Schwartz, L., Sargent, D., Ford, R., ... Verweij, J. (2009). New response evaluation criteria in solid tumours: Revised RECIST guideline (version 1.1). *European Journal of Cancer, 45*(2), 228−247, Jan.

Eroglu, Z., Zaretsky, J. M., Hu-Lieskovan, S., Kim, D. W., Algazi, A., Johnson, D. B., ... Gherardini, P. F. (2018). High response rate to PD-1 blockade in desmoplastic melanomas. *Nature, 553*(7688), 347−350, Jan.

Flaherty, K., Daud, A., Weber, J. S., Sosman, J. A., Kim, K., Gonzalez, R., ... Flaherty, F. (2014). Updated overall survival (OS) for BRF113220, a phase 1−2 study of dabrafenib (D) alone vs combined dabrafenib and trametinib (D T) in pts with BRAF V600 mutation-positive metastatic melanoma (MM). *Journal of Clinical Oncology, 32*(15−suppl.), 9010, May 20.

Flaherty, K. T., Infante, J. R., Daud, A., Gonzalez, R., Kefford, R. F., Sosman, J., ... Weber, J. (2012). Combined BRAF and MEK inhibition in melanoma with BRAF V600 mutations. *New England Journal of Medicine, 367*(18), 1694−1703, Nov 1.

Flaherty, K. T., Robert, C., Hersey, P., Nathan, P., Garbe, C., Milhem, M., ... Schadendorf, D. METRIC Study Group. (2012). Improved survival with MEK inhibition in BRAF-mutated melanoma. *New England Journal of Medicine, 367*(2), 107−114, Jul 12.

Fountain, J. W., Bale, S. J., Housman, D. E., & Dracopoli, N. C. (1990). Genetics of melanoma. *Cancer Survival, 9*(4), 645−671.

Haluska, F. G. (2006). Genetic alterations in signaling pathways in melanoma. *Clinical Cancer Research, 12*(7), Apr 11.

Hauschild, A., Grob, J.-J., Demidov, L. V., Jouary, T., Gutzmer, R., Millward, M., ... Chapman, P. B. (2012). Dabrafenib in BRAF-mutated metastatic melanoma: A multicentre, open-label, phase 3 randomised controlled trial. *The Lancet, 380*(9839), 358−365, Jul 28.

Hayward, N. K., Wilmott, J. L., Waddell, N., Johansson, P. A., Field, M. A., Nones, K., ... Mann, G. J. (2017). Whole-genome landscapes of major melanoma subtypes. *Nature, 545,* 175−180, May 3.

Hodi, F. S., Chiarion-Sileni, V., Gonzalez, R., Grob, J. J., Rutkowski, P., Cowey, C. L., ... Ferrucci, P. F. (2018). Nivolumab plus ipilimumab or nivolumab alone vs ipilimumab alone in advanced melanoma (CheckMate 067): 4-year outcomes of a multicentre, randomised, phase 3 trial. *The Lancet Oncology, 19*(11), 1480−1492, Nov 1.

Hodi, F. S., O'Day, S. J., McDermott, D. F., Weber, R. W., Sosman, J. A., Haanen, J. B., ... Urba, W. J. (2010). Improved survival with ipilimumab in patients with metastatic melanoma. *New England Journal of Medicine, 363*(8), 711−723, Aug 19.

Ibrahim, N., & Haluska, F. G. (2009). Molecular pathogenesis of cutaneous melanocytic neoplasms. *Annual Review of Pathology: Mechanisms of Disease, 4*(1), 551–579, Oct 29.

Leiter, U., Meier, F., Schittek, B., & Garbe, C. (2004). The natural course of cutaneous melanoma. *Journal of Surgical Oncology, 86*(4), 172–178, Jun 21.

Long, G. V., Hauschild, A., Santinami, M., Atkinson, V., Mandalà, M., Chiarion-Sileni, V., ... Schadendorf, D. (2017). Adjuvant dabrafenib plus trametinib in stage IIIBRAF-mutated melanoma. *New England Journal of Medicine, 377*(19), 1813–1823, Nov9.

Long, G. V., Menzies, A. M., Nagrial, A. M., Haydu, L. E., Hamilton, A. L., Mann, G. J., ... Kefford, R. F. (2011). Prognostic and clinicopathologic associations of oncogenic BRAF in metastatic melanoma. *Journal of Clinical Oncology, 29*(10), 1239–1246, Apr 1.

Long, G. V., Stroyakovskiy, D., Gogas, H., Levchenko, E., Braud, F. D., Larkin, J., ... Flaherty, K. (2015). Dabrafenib and trametinib vs dabrafenib and placebo for Val600 BRAF-mutant melanoma: A multicentre, double-blind, phase 3 randomised controlled trial. *The Lancet, 386*(9992), 444–451, Aug 1.

Luke, J. J., Callahan, M. K., Postow, M. A., Romano, E., Ramaiya, N., Bluth, M., ... Carvajal, R. D. (2013). Clinical activity of ipilimumab for metastatic uveal melanoma. *Cancer, 119* (20), 3687–3695, Oct 15.

Luke, J. J., Flaherty, K. T., Ribas, A., & Long, G. V. (2017). Targeted agents and immunotherapies: Optimizing outcomes in melanoma. *Nature Reviews Clinical Oncology, 14*, 463–482, Apr 4.

Masson, K., & Rönnstrand, L. (2009). Oncogenic signaling from the hematopoietic growth factor receptors c-Kit and Flt3. *Cellular Signalling, 21*(12), 1717–1726, Dec.

McArthur, G. A., Chapman, P. B., Robert, C., Larkin, J., Haanen, J. B., Dummer, R., ... Hauschild, A. (2014). Safety and efficacy of vemurafenib in BRAFV600E and BRAFV600K mutation-positive melanoma (BRIM-3): Extended follow-up of a phase 3, randomised, open-label study. *The Lancet Oncology, 15*(3), 323–332, Mar.

Morrison, C., Pabla, S., Conroy, J. M., Nesline, M. K., Glenn, S. T., Dressman, D., ... Qin, M. (2018). Predicting response to checkpoint inhibitors in melanoma beyond PD-L1 and mutational burden. *Journal for Immunotherapy of Cancer, 6*(1), 32, Dec.

Nelson, A. A., & Tsao, H. (2009). Melanoma and genetics. *Clinics in Dermatology, 27*(1), 46–52.

Olsen, C. M., Carroll, H. J., & Whiteman, D. C. (2010). Familial melanoma: A *meta*-analysis and estimates of attributable fraction. *Cancer Epidemiology Biomarkers & Prevention, 19*(1), 65–73, Jan 6.

Palmieri, G., Capone, M., Ascierto, M., Gentilcore, G., Stroncek, D. F., Casula, M., ... Ascierto, P. A. (2009). Main roads to melanoma. *Journal of Translational Medicine, 7*(1), 86, Oct 14.

Parmiter, A. H., Balaban, G., Clark, W. H., & Nowell, P. C. (1988). Possible involvement of the chromosome region 10q24→q26 in early stages of melanocytic neoplasia. *Cancer Genetics and Cytogenetics, 30*(2), 313–317, Feb 1.

Puzanov, I., Milhem, M. M., Minor, D., Hamid, O., Li, A., Chen, L., ... Kaufman, H. L. (2016). Talimogene laherparepvec in combination with ipilimumab in previously untreated, unresectable stage IIIB-IV melanoma. *Journal of Clinical Oncology, 34*(22), 2619–2626, Aug 1.

Robert, C., Grob, J. J., Stroyakovskiy, D., Karaszewska, B., Hauschild, A., Levchenko, E., ... Long, G. V. (2019). Five-year outcomes with dabrafenib plus trametinib in metastatic melanoma. *New England Journal of Medicine, 381*(7), 626–636, Aug 15.

Robert, C., Karaszewska, B., Schachter, J., Rutkowski, P., Mackiewicz, A., Stroiakovski, D., ... Schadendorf, D. (2015). Improved overall survival in melanoma with combined dabrafenib and trametinib. *New England Journal of Medicine, 372*(1), 30–39, Jan.

Robert, C., Schachter, J., Long, G. V., Arance, A., Grob, J. J., Mortier, L., ... Ribas, A. (2015). Pembrolizumab vs ipilimumab in advanced melanoma. *New England Journal of Medicine*, *372*(26), 2521−2532, Jun 25.

Schachter, J., Ribas, A., Long, G. V., Arance, A., Grob, J.-J., Mortier, L., ... Robert, C. (2017). Pembrolizumab vs ipilimumab for advanced melanoma: Final overall survival results of a multicentre, randomised, open-label phase 3 study (KEYNOTE-006). *The Lancet*, *390* (10105), 1853−1862, Apr 16.

Schadendorf, D., Hodi, F. S., Robert, C., Weber, J. S., Margolin, K., Hamid, O., ... Wolchok, J. D. (2015). Pooled analysis of long-term survival data from phase II and phase III trials of ipilimumab in unresectable or metastatic melanoma. *Journal of Clinical Oncology*, *33*(17), 1889, Jun 10.

Sekulic, A., Haluska, P., Miller, A. J., De Lamo, J. G., Ejadi, S., Puldio, J. S., ... Markovic, S. N. (2008). Malignant melanoma in the 21st century: The emerging molecular landscape. *Mayo Clinic Proceedings*, *83*(7), 825−846, Jul.

Shain, A. H., Joseph, N. M., Yu, R., Benhamida, J., Liu, S., Prow, T., ... Bastian, B. C. (2018). Genomic and transcriptomic analysis reveals incremental disruption of key signaling pathways during melanoma evolution. *Cancer Cell*, *34*(1), 45−55, Jul 9.

Snyder, A., Makarov, V., Merghoub, T., Yuan, J., Zaretsky, J. M., Desrichard, A., ... Hollmann, T. J. (2014). Genetic basis for clinical response to CTLA-4 blockade in melanoma. *New England Journal of Medicine*, *371*(23), 2189−2199, Dec 4.

Tsao, H., Mihm, M. C., & Sheehan, C. (2003). PTEN expression in normal skin, acquired melanocytic nevi, and cutaneous melanoma. *Journal of the American Academy of Dermatology*, *49*(5), 865−872, Nov.

Vogelstein, B., Papadopoulos, N., Velculescu, V. E., Zhou, S., & Diaz, L. A. (2013). Cancer genome landscapes. *Science (New York, N.Y.)*, *339*(6127), 1546−1558, Mar 29.

Wang, T., Xiao, M., Ge, Y., Krepler, C., Belser, E., Lopez-Coral, A., ... Kaufman, R. E. (2015). BRAF inhibition stimulates melanoma-associated macrophages to drive tumor growth. *Clinical Cancer Research*, *21*(7), 1652−1664, Apr 1.

Zeng, H., Judson-Torres, R. L., & Shain, A. H. (2020). The evolution of melanoma − Moving beyond binary models of genetic progression. *Journal of Investigative Dermatology*, *140*(2), 291−297, Oct 14.

Chapter 20

Precision cancer immunotherapy

Ahmed Galal

Division of Hematologic Malignancies and Cellular Therapy, Department of Medicine, Duke University Medical Center, Durham, NC, United States

Introduction

Since the time of William Coley's initial description of tumor regression in the setting of contemporaneous infection (Coley, 1991), cancer physicians and researchers have sought to eradicate tumors via immunotherapy (Yang, 2015). Immunotherapy now plays an important role as standard of care treatments across the wide array of solid tumors and hematologic malignancies and genomic tools now play an integral role in prediction of responses to immunotherapies. In fact, the approval of immune checkpoint inhibitor (ICIs)−based therapy for tumor mutational burden high (TMB-high) solid tumors represents one of the first instances whereby a therapeutic is used based on the underlying genomic features of a given tumor as opposed to assignment of treatment based on histologic diagnosis (Yu, 2018). Across a wide array of malignant entities, next-generation sequencing has been leveraged to understand targets for precision cancer immunotherapy and this effort as benefited additionally from the same technology used to understand the host tumor environment and mechanisms underlying immunotherapy success or failure. These efforts have led to the adoption of multiple therapeutic strategies in cancer including allogeneic bone marrow transplant and chimeric antigen receptor−modified T cell (CAR T) therapies in hematologic malignancies as well as CTLA-4 antagonist and ICIs in solid tumors. Multiple lines of investigation into these and other modalities hold promise for improving cancer outcomes through precision immunotherapeutics (Havel, Chowell, & Chan, 2019; Mukherjee, 2019; Sahin & Türeci, 2018).

Immune effector cells, recognition of cancer cells as "foreign" and the cancer microenvironment

We have come to understood over the past couple of decades that cancer cells exist in a complex microenvironment of immune effector cells. Immune

Genomic and Precision Medicine. DOI: https://doi.org/10.1016/B978-0-12-800684-9.00019-8

cells operate via innate as well as adaptive mechanisms and these immune compartments play key roles in immune surveillance of malignant cells. The goal of cancer immunotherapy is to manipulate these underlying immune compartments to effective control tumors; conversely tumor cells suppress effective immune responses via a multitude of molecular/genomic altera-tions. Fig. 20.1 gives an overview of key interactions between tumor cells, the T cell synapsed, and tumor microenvironment as well as the potential for genomic and precision medicine to interface with computational techniques to direct targeted immunotherapy for improved cancer outcomes. Innate immune effectors lack clonal gene rearrangements that direct B- or T-cells to specific antigens (Demaria et al., 2019; Düwell, Heidegger, & Kobold, 2019; Gajewski, Schreiber, & Fu, 2013). Cytotoxic NK cells are innate immune cells capable of recognizing and killing abnormal cells with signs of stress or class I human leukocyte antigen (HLA) molecular downregulation without otherwise encountering activation molecules or other signaling effectors (Minetto, Guolo, & Pesce, 2019; Shimasaki, Jain, & Campana, 2020;

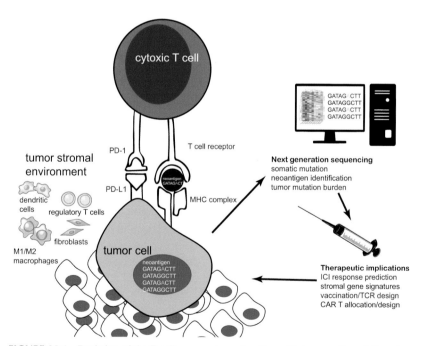

FIGURE 20.1 Depiction of the T-cell synapse in conjunction with tumor cells and the micro-environment. Somatic mutations result in production neoantigens that can be assayed with geno-mic assays (NGS DNA sequencing and RNA-seq) and precision immunotherapy approaches including allocation of checkpoint inhibition (with or without CTLA-4 antibodies) or design of novel tumor vaccination, adoptive T-cell therapies including chimeric antigen receptor−modified T cells may be directed based on genomic methods.

Valipour et al., 2019). Similarly, macrophages are immune cells that play a variety of roles in regulating the immune response within tissues and macrophages appear to develop into at least two subtypes, with M1 macrophages being capable of producing an antitumor response (Shapouri-Moghaddam, Mohammadian, & Vazini, 2018; Yunna, Mengru, Lei, & Weidong, 2020). Currently available targeted therapies such as rituximab, a monoclonal antibody recognizing CD20 (Grillo-López, White, & Dallaire, 2000; McLaughlin, Grillo-López, & Link, 1998), rely on macrophage-mediated antibody-dependent cellular cytotoxicity (Flieger, Renoth, Beier, Sauerbruch, & Schmidt-Wolf, 2000) as well as other mechanisms for effective tumor responses. In addition, host features expected to abrogate the response to rituximab and other monoclonal antibodies such as Fc-γ receptor (FCGR3A gene polymorphisms) or inhibitory molecules (Clynes, Towers, Presta, & Ravetch, 2000) do indeed appear to impact clinical response to these agents and assaying this a priori may be possible with multiplexed tumor or germline assays. Macrophage-mediated antitumor responses may be resisted by expression of molecules such as CD47 on tumor cells, which delivers a "don't eat me" signal upon SIRPa ligation on macrophages (Armant, Avice, & Hermann, 1999; Vernon-Wilson et al., 2000) and this is one way that tumors avoid macrophage recognition and surveillance. Strategies such as anti-CD47 antibodies such as Hu5F9-G4 or SIRPα-Fc fusion proteins have been developed to abrogate these inhibitory signals on macrophages and appear to reinvigorate phagocytic response to tumors (Bibeau, Lopez-Crapez, & Di Fiore, 2009; Cartron, 2009; Paiva et al., 2008; Quartuccio, Fabris, & Pontarini, 2014; Ruyssen-Witrand, Rouanet, & Combe, 2012; Ziakas, Poulou, & Zintzaras, 2016). Tumor responses clinically to Hu5F9-G4 have been documented and many clinical trials are underway to evaluate this strategy across various neoplasm subtypes (Advani, Flinn, & Popplewell, 2018; Sikic, Lakhani, & Patnaik, 2019). Other myeloid-derived cells play important roles as immune suppressor cells in the tumor microenvironment and manipulating these cellular compartments may prove essential in overcoming resistance to current immunotherapies (Bronte, Serafini, Apolloni, & Zanovello, 2001; Law, Valdes-Mora, & Gallego-Ortega, 2020).

Much of the focus of cancer immunotherapy development has centered on T cells. T cells are an important component of the adaptive immune system and recognize antigen presented by major histocompatibility complex based on the specificity of T cell receptors (TCR) that rearrange in order to foster antigen recognition diversity. T cells become activated against antigen (including tumor-associated antigens) in a two-step process on antigen-presenting cells that activates costimulatory signaling pathways upon antigen recognition by the TCR. TCRs then recognize antigen on target cells in order to effect a cytotoxic immune "attack." An important concept in T cell–based immunotherapy for cancer is the fact that somatic alterations in cancer cells produce "neoantigens" (Cohen, Sibal, & Fink, 1974; Whittingham & Mackay 1977) that are recognized as foreign by host immune cells and these can be thus targeted in an adaptive immune approach to produce successful cancer immunotherapy. Cancer immunotherapy holds great promise in the era of next-generation sequencing as neoantigens can be detected at the whole genome or whole transcriptome level (and aided in some instances by RNA-seq performed at the single-cell level) and this may be capable of directing a precision immunotherapy approach targeting specific neoantigens. In addition, novel neoantigens can be detected using such approaches, thus facilitating the design of immunotherapies that home in on these targets.

T-cell immunotherapy approaches in cancer

Unfortunately, T-cell immunity may fail to eradicate cancer cells via various mechanisms including expression of T-cell inhibitory molecules or inhibitory cytokines, expansion, or stimulation of suppressive myeloid or regulatory T-cell subsets as well as downregulation of antigen presentation; current T-cell immunotherapy approaches have attempted to overcome these obstacles. Allogeneic bone marrow transplantation is one example of an adoptive therapy utilized in hematologic malignancies based on the fact that a HLA-mismatched bone marrow allograft can exhibit a potent antitumor effect and is often utilized in high risk or relapsed/refractory cases of leukemia, lymphoma, and myeloma. The field of transplantation has benefited significantly from our ability to better match patients to allografts and next-generation sequencing approaches are now used for donor selection as well as prediction of graft versus host disease (GvHD) risk (Mayor, Robinson, & McWhinnie, 2015). GvHD may be further mitigated by the development of wider screens of host and donor features via NGS in the future.

Other important targeted approaches to T-cell immunotherapy include inhibition of immune checkpoints with monoclonal antibodies, and CAR T bi-specific T-cell engagers (BiTEs). BiTEs are molecules capable of ligation to cells with two different cell surface proteins, including T cells to malignant cells (Offner, Hofmeister, Romaniuk, Kufer, & Baeuerle, 2006; Schlereth, Fichtner, & Lorenczewski, 2005). When host T cells are brought into proximity with a cancer cell via ligation to a BiTE (e.g., CD3 + T cells and CD19 + leukemia or lymphoma cells), a T-cell cytotoxic response may occur. Blinatumomab is a CD3/CD19 BiTE that is now approved in acute lymphoblastic leukemia and multiple other B-cell lineage targeted BiTEs are in clinical trials across B lymphoid malignancies (Bargou, Leo, & Zugmaier, 2008; Topp, Kufer, & Gökbuget, 2011). B-cell cancers have also been effective targeted utilizing CAR T cells. CAR T cells utilize a novel construct whereby a Fc fragment specific for a cell surface protein is fused in a chimeric fashion to activation (usually CD3-ζ) and a costimulatory molecule (such as CD28 or 4−1BB) and this construct is expressed on autologous or allogeneic T cells in order to produce a powerful cytotoxic attack on cells bearing the antigen recognized by the engineered Fc fragment (Srivastava & Riddell, 2015). Cell products transduced with early iterations of CAR constructs consisting of Fc fragments fused with activation motifs and lacking costimulatory molecules lacked significant clinical efficacy, but there are currently at least four autologous CAR T products targeting CD19 approved for acute lymphoblastic leukemia and non-Hodgkin lymphoma subtypes in the United States. Similar approaches in solid tumors have been limited by increased on target toxicity relative to the targeting of B-cell antigens, but clinical trials are ongoing using multiple products target nonlymphoid lineage cells. Theoretically, CAR T cells can be engineered to target specific neoantigens and even utilized engineered TCR for precise targeting of tumor-associated antigens, but the development of this technology is limited by underlying genomic heterogeneity in tumors as well as the diversity of the T-cell repertoire target specific antigens as well as by the complexity and cumbersome nature of producing and administering CAR T products in the clinic.

Cytotoxic T lymphocyte—associated antigen 4 (CTLA-4) is one of several receptors expressed on the surface of T cells that suppress T-cell activation in the presence of their cognate ligand. CTLA-4 negatively regulates T-cell activation via competition with CD28 for B7 ligands on APCs. CTLA-4 competition for B7 can be blocked with monoclonal antibodies, thus reversing inhibition of T-cell activation in response to tumor antigens on APCs (Wolchok & Saenger, 2008). Ipilimumab is an FDA-approved anti-CTLA-4 antibody first approved after a randomized trial of ipilimumab with or without a gp100-peptide vaccine versus a gp100-peptide vaccine alone showed an overall survival of 10 months in ipilimumab-treated patients as compared to 6.4 months in the gp100 arm ($P = 0.03$) (Hodi, O'Day, & McDermott, 2010). Ipilimumab has been approved in other indications including in combination with ICIs targeting the programmed cell death axis (PD-1:PD-L1/2). The immune inhibitory inhibitors the programed cell death 1 (PD-1) receptor (CD279), and its ligands PD-L1 (B7-H1, CD274) and PD-L2 (B7-DC, CD273) are important regulators of the T-cell immune response (Hodi et al., 2010). Immune attack on normal cells is inhibited by the expression of PD-L1 and cancer cells may similarly utilized this pathway to avoid immune surveillance via PD-L1 expression. PD-L1 and PD-L2 may be overexpressed in tumors via a variety of mechanisms, including gene amplification. Genetic amplification of PD-L1 (CD279) on an amplicon also containing *JAK2* and *CD273* occurs frequently in the malignant cells in Hodgkin lymphoma and high levels of amplification appear to be associated with worse survival with standard chemotherapy approaches (Joos, Granzow, & Holtgreve-Grez, 2003; Roemer, Advani, & Ligon, 2016). PD-L1 ligation to PD-1 receptors on cytotoxic T cells results in abrogation of cytotoxic immune responses in the tumor microenvironment. This process can be reversed with monoclonal antibodies similar to strategies above for blocking CD47 or CTLA-4. Approved PD-1 agents include nivolumab and pembrolizumab whereas atezolizumab is an approved anti-PD-L1 agent and other agents are likely to be approved for clinical use in the future. In addition, there are at least several indications where PD-1 inhibitors are combined with CTLA-4 antagonistic agents such as ipilimumab. Ipilimumab and nivolumab combination therapy in *BRAF* wild-type melanoma was shown to have a response rate of 61% with 22% of patients achieved complete remission as compared to a response rate of 11% (no complete responding patients) with single-agent ipilimumab in a randomized trial (Roemer et al., 2016). Trials in multiple other cancers have shown promise of combined CTLA-4 and PD-1 or PD-L1 inhibitors and it is likely there will be many indications for this combination therapy in the future. However, this may not be a "one size fits all" approach as there is the potential for increased immune toxicity with combination treatment and entities such as diffuse large B-cell lymphoma do not appear to benefit from a combined approach.

Prediction of response to immune checkpoint inhibitors and other immunotherapy agents

Whole exome/genome/transcriptome techniques have been used identify host and tumor-specific features predictive of immunotherapy, even in a dynamic fashion. Much of this work has focused on features associated with treatment success and resistance in common clinical situations where ICIs have been used the longest such as melanoma, lung cancer and cancers exhibiting microsatellite instability (MSI-high) such colorectal tumors occurring in Lynch syndrome. The fact that the first and fourth approved cancer drugs for histology agnostic indications [pembrolizumab in MSI-high tumors per the KEYNOTE-158 and KEYNOTE-164 studies (Le, Kim, & Van Cutsem, 2020; Marabelle, Le, & Ascierto, 2020) and pembrolizumab for advanced solid tumors with TMB ≥ 10 mutations per megabase of tumor DNA (based on KEYNOTE-158 and KEYNOTE-028 that screened additional biomarkers for enrollment (Ott, Bang, & Piha-Paul, 2019))] depend on genomic assays to guide therapy highlight the progress that has been made in leveraging genomics tools for precision immunotherapy in cancer.

Host and tumor features associated with ICI response have been probed in clinical studies utilizing genomic correlatives in many tumor subtypes. Tumor features such as PD-1/PD-L1 expression, identification of microenvironmental features including acquired somatic mutations, and TMB have been associated with response to immunotherapy, particularly ICI therapy in many clinical studies; TMB appears also to be associated with relatively improved survival in several different cancer entities (Brown, Warren, & Gibb, 2014; Le, Uram, & Wang, 2015; Rizvi, Hellmann, & Snyder, 2015; Snyder, Makarov, & Merghoub, 2014; Van Allen, Miao, & Schilling, 2015). For the most part, features associated with immunotherapy appear intuitive considering our understanding of the immune tumor microenvironment and concepts of neoantigen acquisition in tumors and the immune response to neoantigens.

Surrogates for neoantigen burden in tumors include detection of microsatellite instability as well as TMB. TMB represents the number of mutations per sequenced megabase (mut/Mb) and requires next-generation sequencing for quantitation. Numerous studies now have examined the relationship between TMB and neoantigen load as well as the clinical response to cancer immunotherapy. ICI responses tend to be higher in TMB-high tumors (as they do in MSI-high tumors) ostensibly due to the presence of more neoantigens as supported by studies across tumor types. Cancer types such as melanoma and lung cancer tend to have more TMB and were the first cancer subtypes to show relatively high responses to ICIs and this is reflective in the fact that FDA approvals for ICIs used alone or in combination occurred first in these diseases without criteria for testing for high TMB. TMB as a predictive biomarker to direct ICI therapy has been studied in multiple clinical studies and there appears to be a "plateau" effect where response rates peak above a certain TMB threshold and this threshold may be tumor-type dependent. Pembrolizumab gained FDA approval in solid tumors based on the KEYNOTE-158 study, which was a multicohort study that enrolled 10 cancer types and treatment was employed with single-agent pembrolizumab with response rates then analyzed in subgroups defined by tumor TMB (tTMB) less than or greater than 10 mut/Mb. Of note, 105/790 (13%) of patients had tTMB-high status and most of the tTMB-high cases were one of five subtypes. In tTMB-high cases, the ORR was 29% with a median DOR not reached at a median 37.1 m of f/u. While these data drove approval for pembrolizumab across all tumor subtypes, response rates vary widely between tTMB-high tumors based on histology; in addition, this study lacked breast and prostate cancer patients and these cancer types generally exhibit low response rate to ICIs. ICIs appear to benefit glioma patients with low tTMB more so than tTMB-high patients, and so it is likely that there are multiple features that dictate the response to ICIs and other cancer immunotherapies beyond the tumor neoantigen content.

Beyond neoantigen burden, there appears to be other features associated with the response to available immunotherapies in cancer. In melanoma, specific neoantigen profiles appear to be associated with higher response to ICI therapy (Van Allen et al., 2015), so the overall mutational pattern occurring in a tumor may be important in dictating the response to ICIs. ICI treatment may also be thwarted by acquired tumor somatic mutations. In a study of paired samples from melanoma patients before and after development of ICI resistance, mutations in *JAK1*, *JAK2*, and *B2M* were documented (Zaretsky, Garcia-Diaz, & Shin, 2016) and other studies have found similar findings regarding primary ICI resistance (Horn et al., 2018; Shin, Zaretsky, & Escuin-Ordinas, 2017). It was postulated that loss of JAK-STAT coupling to upstream interferon-receptor signaling resulted in loss of the cytostatic activity mediated by microenvironmental interferons while β-2-microglobulin loss could abrogate TCR-mediated recognition of neoantigen at the immune synapse. PD-L1 expression may be another feature that identifies tumors more likely to respond to ICI. ICI combination therapy with nivolumab and ipilimumab was studied in the CheckMate012 study documented an ORR of 41% in tumors with PD-L1 expression on >1% of cells via IHC, compared to 15% in tumor with PD-L1 <1% (Horn et al., 2018; Shin et al., 2017). Separate from PD-L1, TMB testing showed no correlation between TMB and PD-L1 expression and there were increased response rates in patients with higher TMB with a plateau in response rates at a cut-off TMB ≥ 10 mut/Mb similar to other studies. PD-L1 expression may be an important marker for predicting response to therapy in lymphomas as well. In Hodgkin lymphoma, PD-L1 is frequently amplified and both nivolumab and pembrolizumab have FDA approval for treatment in relapsed/refractory patients based on relatively high response rates to ICIs (Ansell, Lesokhin, & Borrello, 2015; Armand, Engert, & Younes, 2018; Chen, Zinzani, & Fanale, 2017). ICI therapy is prescribed in Hodgkin lymphoma without testing of PD-L1 expression, although level of amplification may be associated with prognosis with standard therapies. In diffuse large B-cell lymphoma, a small proportion of cases have been described that have similar gains of chromosome 9q24 in an amplicon containing PD-L1 and PD-L2 as noted frequently in Hodgkin lymphoma and subtypes such as testicular and CNS lymphoma frequently having these alterations appear to respond to ICI at higher rates than other DLBLC subtypes. High tTMB may additionally be a predictor of response but high TMB is rare in diffuse large B-cell lymphoma and there are less data surrounding the association with TMB and ICI response in lymphoma.

Summary

Cancer immunotherapy has been a focus of laboratory and clinical investigation for many years. Fortunately, we now have effective agents that are standard of care therapies for a wide range of malignancies. Technologic advances including next-generation sequencing and single-cell analysis of tumor cells and the immune microenvironment hold significant promise in advancing our understanding of the mechanisms, which tumors avoid immune surveillance as well as the development of predictive tools for which patients will best respond to available immunotherapies. Finally, there is hope that these studies will additionally illuminate targets for the design of more effective and safer immunotherapies that will increase the lifespan of patients afflicted with cancer.

References

Advani, R., Flinn, I., Popplewell, L., et al. (2018). CD47 blockade by Hu5F9-G4 and rituximab in non-Hodgkin's lymphoma. *The New England Journal of Medicine, 379*(18), 1711−1721. Available from https://doi.org/10.1056/NEJMoa1807315.

Ansell, S. M., Lesokhin, A. M., Borrello, I., Halwani, A., Scott, E. C., Gutierrez, M. D., . . . Armand, P. (2015). PD-1 blockade with nivolumab in relapsed or refractory Hodgkin's lymphoma. *The New England Journal of Medicine*, *372*(4), 311–319. Available from https://doi.org/10.1056/NEJMoa1411087.

Armand, P., Engert, A., Younes, A., Fanale, M., Santoro, A., Zinzani, P. L., . . . Ansell, S. M. (2018). Nivolumab for relapsed/refractory classic Hodgkin lymphoma after failure of autologous hematopoietic cell transplantation: Extended follow-up of the multicohort single-arm phase II CheckMate 205 trial. *Journal of Clinical Oncology: Official Journal of the American Society of Clinical Oncology*, *36*(14), 1428–1439. Available from https://doi.org/10.1200/jco.2017.76.0793.

Armant, M., Avice, M. N., Hermann, P., Rubio, M., Kiniwa, M., Delespesse, G., & Sarfati, M. (1999). CD47 ligation selectively downregulates human interleukin 12 production. *The Journal of Experimental Medicine*, *190*(8), 1175–1182. Available from https://doi.org/10.1084/jem.190.8.1175.

Bargou, R., Leo, E., Zugmaier, G., Klinger, M., Goebeler, M., Knop, S., . . . Kufer, P. (2008). Tumor regression in cancer patients by very low doses of a T cell-engaging antibody. *Science (New York, NY)*, *321*(5891), 974–977. Available from https://doi.org/10.1126/science.1158545.

Bibeau, F., Lopez-Crapez, E., Di Fiore, F., Thezenas, S., Ychou, M., Blanchard, F., . . . Boissière-Michot, F. (2009). Impact of Fc{gamma}RIIa-Fc{gamma}RIIIa polymorphisms and KRAS mutations on the clinical outcome of patients with metastatic colorectal cancer treated with cetuximab plus irinotecan. *Journal of Clinical Oncology: Official Journal of the American Society of Clinical Oncology*, *27*(7), 1122–1129. Available from https://doi.org/10.1200/jco.2008.18.0463.

Bronte, V., Serafini, P., Apolloni, E., & Zanovello, P. (2001). Tumor-induced immune dysfunctions caused by myeloid suppressor cells. *Journal of Immunotherapy (Hagerstown, Md: 1997)*, *24*(6), 431–446. Available from https://doi.org/10.1097/00002371-200111000-00001.

Brown, S. D., Warren, R. L., Gibb, E. A., Martin, S. D., Spinelli, J. J., Nelson, B. H., & Holt, R. A. (2014). Neo-antigens predicted by tumor genome *meta*-analysis correlate with increased patient survival. *Genome Research*, *24*(5), 743–750. Available from https://doi.org/10.1101/gr.165985.113.

Cartron, G. (2009). FCGR3A polymorphism story: A new piece of the puzzle. *Leukemia & Lymphoma*, *50*(9), 1401–1402. Available from https://doi.org/10.1080/10428190903161109.

Chen, R., Zinzani, P. L., Fanale, M. A., Armand, P., Johnson, N. A., Brice, P., . . . KEYNOTE-087. (2017). Phase II study of the efficacy and safety of pembrolizumab for relapsed/refractory classic Hodgkin lymphoma. *Journal of Clinical Oncology: Official Journal of the American Society of Clinical Oncology*, *35*(19), 2125–2132. Available from https://doi.org/10.1200/jco.2016.72.1316.

Clynes, R. A., Towers, T. L., Presta, L. G., & Ravetch, J. V. (2000). Inhibitory Fc receptors modulate in vivo cytotoxicity against tumor targets. *Nature Medicine*, *6*(4), 443–446. Available from https://doi.org/10.1038/74704.

Cohen, M. H., Sibal, L. R., & Fink, M. A. (1974). Relative importance of viral and neoantigens in cytotoxic reaction against murine leukaemia cells. *Immunology*, *26*(1), 37–48.

Coley, W. B. (1991). The treatment of malignant tumors by repeated inoculations of erysipelas. With a report of ten original cases. 1893. *Clinical Orthopaedics and Related Research*, *262*, 3–11, In eng.

Demaria, O., Cornen, S., Daëron, M., Morel, Y., Medzhitov, R., & Vivier, E. (2019). Harnessing innate immunity in cancer therapy. *Nature*, *574*(7776), 45–56. Available from https://doi.org/10.1038/s41586-019-1593-5.

Düwell, P., Heidegger, S., & Kobold, S. (2019). Innate immune stimulation in cancer therapy. *Hematology/Oncology Clinics of North America*, *33*(2), 215−231. Available from https://doi.org/10.1016/j.hoc.2018.12.002.

Flieger, D., Renoth, S., Beier, I., Sauerbruch, T., & Schmidt-Wolf, I. (2000). Mechanism of cytotoxicity induced by chimeric mouse human monoclonal antibody IDEC-C2B8 in CD20-expressing lymphoma cell lines. *Cellular Immunology*, *204*(1), 55−63. Available from https://doi.org/10.1006/cimm.2000.1693.

Gajewski, T. F., Schreiber, H., & Fu, Y. X. (2013). Innate and adaptive immune cells in the tumor microenvironment. *Nature Immunology*, *14*(10), 1014−1022. Available from https://doi.org/10.1038/ni.2703.

Grillo-López, A. J., White, C. A., Dallaire, B. K., Varns, C. L., Shen, C. D., Wei, A., ... Rosenberg, J. (2000). Rituximab: The first monoclonal antibody approved for the treatment of lymphoma. *Current Pharmaceutical Biotechnology*, *1*(1), 1−9. Available from https://doi.org/10.2174/1389201003379059.

Havel, J. J., Chowell, D., & Chan, T. A. (2019). The evolving landscape of biomarkers for checkpoint inhibitor immunotherapy. *Nature Reviews Cancer*, *19*(3), 133−150. Available from https://doi.org/10.1038/s41568-019-0116-x.

Hodi, F. S., O'Day, S. J., McDermott, D. F., Weber, R. W., Sosman, J. A., Haaner, J. B., ... Urba, W. J. (2010). Improved survival with ipilimumab in patients with metastatic melanoma. *The New England Journal of Medicine*, *363*(8), 711−723. Available from https://doi.org/10.1056/NEJMoa1003466.

Horn, S., Leonardelli, S., Sucker, A., Schadendorf, D., Griewank, K. G., & Paschen, A. (2018). Tumor CDKN2A-associated JAK2 loss and susceptibility to immunotherapy resistance. *Journal of the National Cancer Institute*, *110*(6), 677−681. Available from https://doi.org/10.1093/jnci/djx271.

Joos, S., Granzow, M., Holtgreve-Grez, H., Siebert, R., Harder, L., Martín-Subero, J. I., ... Jauch, A. (2003). Hodgkin's lymphoma cell lines are characterized by frequent aberrations on chromosomes 2p and 9p including REL and JAK2. *International Journal of Cancer*, *103*(4), 489−495. Available from https://doi.org/10.1002/ijc.10845.

Law, A. M. K., Valdes-Mora, F., & Gallego-Ortega, D. (2020). Myeloid-derived suppressor cells as a therapeutic target for cancer. *Cells*, *9*(3). Available from https://doi.org/10.3390/cells9030561.

Le, D. T., Kim, T. W., Van Cutsem, E., Geva, R., Jäger, D., Hara, H., ... André, T. (2020). Phase II open-label study of pembrolizumab in treatment-refractory, microsatellite instability-high/mismatch repair-deficient metastatic colorectal cancer: KEYNOTE-164. *Journal of Clinical Oncology: Official Journal of the American Society of Clinical Oncology*, *38*(1), 11−19. Available from https://doi.org/10.1200/jco.19.02107.

Le, D. T., Uram, J. N., Wang, H., Bartlett, B. R., Kemberling, H., Eyring, A. D., ... Diaz, L. A. (2015). PD-1 blockade in tumors with mismatch-repair deficiency. *The New England Journal of Medicine*, *372*(26), 2509−2520. Available from https://doi.org/10.1056/NEJMoa1500596.

Marabelle, A., Le, D. T., Ascierto, P. A., Di Giacomo, A. M., De Jesus-Acosta, A., Delord, J.-P., ... Diaz, L. A. (2020). Efficacy of pembrolizumab in patients with noncolorectal high microsatellite instability/mismatch repair-deficient cancer: Results from the phase II KEYNOTE-158 study. *Journal of Clinical Oncology: Official Journal of the American Society of Clinical Oncology*, *38*(1), 1−10. Available from https://doi.org/10.1200/jco.19.02105.

Mayor, N. P., Robinson, J., McWhinnie, A. J., Ranade, S., Eng, K., Midwinter, W., ... Marsh, S. G. (2015). HLA typing for the next generation. *PLoS One*, *10*(5), e0127153. Available from https://doi.org/10.1371/journal.pone.0127153.

McLaughlin, P., Grillo-López, A. J., Link, B. K., Levy, R., Czuczman, M. S., Williams, M. E., . . . Dallaire, B. K. (1998). Rituximab chimeric anti-CD20 monoclonal antibody therapy for relapsed indolent lymphoma: Half of patients respond to a four-dose treatment program. *Journal of Clinical Oncology: Official Journal of the American Society of Clinical Oncology, 16*(8), 2825−2833. Available from https://doi.org/10.1200/jco.1998.16.8.2825.

Minetto, P., Guolo, F., Pesce, S., Greppi, M., Obino, V., Ferrett, E., . . . Marcenaro, E. (2019). Harnessing NK cells for cancer treatment. *Frontiers in Immunology, 10*, 2836. Available from https://doi.org/10.3389/fimmu.2019.02836.

Mukherjee, S. (2019). Genomics-guided immunotherapy for precision medicine in cancer. *Cancer Biotherapy & Radiopharmaceuticals, 34*(8), 487−497. Available from https://doi.org/10.1089/cbr.2018.2758.

Offner, S., Hofmeister, R., Romaniuk, A., Kufer, P., & Baeuerle, P. A. (2006). Induction of regular cytolytic T cell synapses by bispecific single-chain antibody constructs on MHC class I-negative tumor cells. *Molecular Immunology, 43*(6), 763−771. Available from https://doi.org/10.1016/j.molimm.2005.03.007.

Ott, P. A., Bang, Y. J., Piha-Paul, S. A., Abdul Razak, A. R., Bennouna, J., Soria, J.-C., . . . Lunceford, J. K. (2019). T-cell-inflamed gene-expression profile, programmed death ligand 1 expression, and tumor mutational burden predict efficacy in patients treated with pembrolizumab across 20 cancers: KEYNOTE-028. *Journal of Clinical Oncology: Official Journal of the American Society of Clinical Oncology, 37*(4), 318−327. Available from https://doi.org/10.1200/jco.2018.78.2276.

Paiva, M., Marques, H., Martins, A., Ferreira, P., Catarino, R., & Medeiros, R. (2008). FcgammaRIIa polymorphism and clinical response to rituximab in non-Hodgkin lymphoma patients. *Cancer Genetics and Cytogenetics, 183*(1), 35−40. Available from https://doi.org/10.1016/j.cancergencyto.2008.02.001.

Quartuccio, L., Fabris, M., Pontarini, E., Salvin, S., Zabotti, A., Benucci, M., . . . de Vita, S. (2014). The 158VV Fcgamma receptor 3A genotype is associated with response to rituximab in rheumatoid arthritis: results of an Italian multicentre study. *Annals of the Rheumatic Diseases, 73*(4), 716−721. Available from https://doi.org/10.1136/annrheumdis-2012-202435.

Rizvi, N. A., Hellmann, M. D., Snyder, A., Kvistborg, P., Makarov, V., Havel, J. J., . . . Chan, T. A. (2015). Cancer immunology. Mutational landscape determines sensitivity to PD-1 blockade in non-small cell lung cancer. *Science (New York, NY), 348*(6230), 124−128. Available from https://doi.org/10.1126/science.aaa1348.

Roemer, M. G., Advani, R. H., Ligon, A. H., Natkunam, Y., Redd, R. A., Homer, H., . . . hipp, M. A. (2016). PD-L1 and PD-L2 genetic alterations define classical Hodgkin lymphoma and predict outcome. *Journal of Clinical Oncology: Official Journal of the American Society of Clinical Oncology, 34*(23), 2690−2697. Available from https://doi.org/10.1200/jco.2016.66.4482.

Ruyssen-Witrand, A., Rouanet, S., Combe, B., Dougados, M., Loët, X. L., Sibilia, J., . . . Constantin, A. (2012). Fcγ receptor type IIIA polymorphism influences treatment outcomes in patients with rheumatoid arthritis treated with rituximab. *Annals of the Rheumatic Diseases, 71*(6), 875−877. Available from https://doi.org/10.1136/annrheumdis-2011-200337.

Sahin, U., & Türeci, Ö. (2018). Personalized vaccines for cancer immunotherapy. *Science (New York, NY), 359*(6382), 1355−1360. Available from https://doi.org/10.1126/science.aar7112.

Schlereth, B., Fichtner, I., Lorenczewski, G., Kleindienst, P., Brischwein, K., da Silva, A., . . . Baeuerle, P. A. (2005). Eradication of tumors from a human colon cancer cell line and from ovarian cancer metastases in immunodeficient mice by a single-chain Ep-CAM-/CD3-bispecific antibody construct. *Cancer Research, 65*(7), 2882−2889. Available from https://doi.org/10.1158/0008-5472.Can-04-2637.

Shapouri-Moghaddam, A., Mohammadian, S., Vazini, H., Taghadosi, M., Esmaeili, S.-A., Mardani, F., ... Sahebkar, A. (2018). Macrophage plasticity, polarization, and function in health and disease. *Journal of Cellular Physiology, 233*(9), 6425−6440. Available from https://doi.org/10.1002/jcp.26429.

Shimasaki, N., Jain, A., & Campana, D. (2020). NK cells for cancer immunotherapy. *Nature Reviews. Drug Discovery, 19*(3), 200−218. Available from https://doi.org/10.1038/s41573-019-0052-1.

Shin, D. S., Zaretsky, J. M., Escuin-Ordinas, H., Garcia-Diaz, A., Hu-Lieskovan, S., Kalbasi, A., ... Ribas, A. (2017). Primary resistance to PD-1 blockade mediated by JAK1/2 mutations. *Cancer Discovery, 7*(2), 188−201. Available from https://doi.org/10.1158/2159-8290.Cd-16-1223.

Sikic, B. I., Lakhani, N., Patnaik, A., Shah, S. A., Chandana, S. R., Rasco, D., ... Padda, S. K. (2019). First-in-human, first-in-class phase I trial of the anti-CD47 antibody Hu5F9-G4 in patients with advanced cancers. *Journal of Clinical Oncology: Official Journal of the American Society of Clinical Oncology, 37*(12), 946−953. Available from https://doi.org/10.1200/jco.18.02018.

Snyder, A., Makarov, V., Merghoub, T., Yuan, J., Zaretsky, J. M., Desrichard, A., ... Chan, T. A. (2014). Genetic basis for clinical response to CTLA-4 blockade in melanoma. *The New England Journal of Medicine, 371*(23), 2189−2199. Available from https://doi.org/10.1056/NEJMoa1406498.

Srivastava, S., & Riddell, S. R. (2015). Engineering CAR-T cells: Design concepts. *Trends in Immunology, 36*(8), 494−502. Available from https://doi.org/10.1016/j.it.2015.06.004.

Topp, M. S., Kufer, P., Gökbuget, N., Goebeler, M., Klinger, M., Neumann, S., ... Bargou, R. C. (2011). Targeted therapy with the T-cell-engaging antibody blinatumomab of chemotherapy-refractory minimal residual disease in B-lineage acute lymphoblastic leukemia patients results in high response rate and prolonged leukemia-free survival. *Journal of Clinical Oncology: Official Journal of the American Society of Clinical Oncology, 29*(18), 2493−2498. Available from https://doi.org/10.1200/jco.2010.32.7270.

Valipour, B., Velaei, K., Abedelahi, A., Karimipour, M., Darabi, M., & Charoudeh, H. N. (2019). NK cells: An attractive candidate for cancer therapy. *Journal of Cellular Physiology, 234*(11), 19352−19365. Available from https://doi.org/10.1002/jcp.28657.

Van Allen, E. M., Miao, D., Schilling, B., Shukla, S. A., Blank, C., Zimmer, L., ... Garraway, L. A. (2015). Genomic correlates of response to CTLA-4 blockade in metastatic melanoma. *Science (New York, NY), 350*(6257), 207−211. Available from https://doi.org/10.1126/science.aad0095.

Vernon-Wilson, E. F., Kee, W. J., Willis, A. C., Barclay, A. N., Simmons, D. L., & Brown, M. H. (2000). CD47 is a ligand for rat macrophage membrane signal regulatory protein SIRP (OX41) and human SIRPalpha 1. *European Journal of Immunology, 30*(8), 2130−2137. Available from https://doi.org/10.1002/1521-4141(2000)30:8/2130::Aid-immu2130/3.0.Co;2-8.

Whittingham, S., & Mackay, I. R. (1977). Tissue antigens: Autoantigens, alloantigens, xenoantigens and neoantigens. *Australian and New Zealand Journal of Medicine, 7*(2), 172−194. Available from https://doi.org/10.1111/j.1445-5994.1977.tb04689.x.

Wolchok, J. D., & Saenger, Y. (2008). The mechanism of anti-CTLA-4 activity and the negative regulation of T-cell activation. *The Oncologist, 13*(4), 2−9. Available from https://doi.org/10.1634/theoncologist.13-S4-2, Suppl.

Yang, Y. (2015). Cancer immunotherapy: Harnessing the immune system to battle cancer. *The Journal of Clinical Investigation*, *125*(9), 3335–3337. Available from https://doi.org/10.1172/jci83871.

Yu, Y. (2018). Molecular classification and precision therapy of cancer: Immune checkpoint inhibitors. *Frontiers in Medicine*, *12*(2), 229–235. Available from https://doi.org/10.1007/s11684-017-0581-0.

Yunna, C., Mengru, H., Lei, W., & Weidong, C. (2020). Macrophage M1/M2 polarization. *European Journal of Pharmacology*, *877*, 173090. Available from https://doi.org/10.1016/j.ejphar.2020.173090.

Zaretsky, J. M., Garcia-Diaz, A., Shin, D. S., Escuin-Ordinas, H., Hugo, W., Hu-Lieskovan, S., ... Ribas, A. (2016). Mutations associated with acquired resistance to PD-1 blockade in melanoma. *The New England Journal of Medicine*, *375*(9), 819–829. Available from https://doi.org/10.1056/NEJMoa1604958.

Ziakas, P. D., Poulou, L. S., & Zintzaras, E. (2016). FcγRIIa-H131R variant is associated with inferior response in diffuse large B cell lymphoma: A *meta*-analysis of genetic risk. *Journal of BUON: Official Journal of the Balkan Union of Oncology*, *21*(6), 1454–1458.

Index

Note: Page numbers followed by "*f*" and "*t*" refer to figures and tables, respectively.